Edited by
Jacques Ganoulis, Alice Aureli,
and Jean Fried

Transboundary Water
Resources Management

Related Titles

Edited by
Jacques Ganoulis, Alice Aureli, and Jean Fried

Transboundary Water Resources Management

A Multidisciplinary Approach

WILEY-VCH

WILEY-VCH Verlag GmbH & Co. KGaA

The Editors

Prof. Dr. Jacques Ganoulis
UNESCO Chair and Network INWEB
Aristotle University of Thessaloniki
Department of Civil Engineering
Division of Hydraulics and Environmental
Engineering
54124 Thessaloniki
Greece

Dr. Alice Aureli
UNESCO
International Hydrological Programme
1, rue Miollis
75732 Paris
France

Prof. Dr. Jean Fried
University of California Irvine
School of Social Ecology
Irvine, CA 92697-7050
USA

Library of Congress Card No.: applied for

British Library Cataloguing-in-Publication Data
A catalogue record for this book is available from the British Library.

Bibliographic information published by the Deutsche Nationalbibliothek
The Deutsche Nationalbibliothek lists this publication in the Deutsche Nationalbibliografie; detailed bibliographic data are available on the Internet at http://dnb.d-nb.de.

Cover Design   Adam Design, Weinheim
Typesetting   Thomson Digital, Noida, India
Printing and Binding   Fabulous Printers Pte Ltd

Printed in Singapore
Printed on acid-free paper

Print ISBN: 978-3-527-33014-0
ePDF ISBN: 978-3-527-63667-9
ePub ISBN: 978-3-527-63666-2
mobi ISBN: 978-3-527-63668-6
oBook ISBN: 978-3-527-63665-5

Foreword: Transboundary Water Management
A Multidisciplinary Approach

For centuries, political and strategic considerations have been *the* major drivers behind the delineation of boundaries across the globe. Mountains, rivers, lakes and entire ecosystems (not to mention human settlements) have been assigned to the jurisdiction of different states, provinces and other administrative entities with little regard to their environmental cycles or effective management. Yet natural resources, and freshwater in particular, know no man-made boundaries, and indeed require internationally coordinated actions to be sustainably and effectively managed. It is only in recent years that transboundary waters, both surface and groundwater, have taken centre stage in international dialogue, as issues of water and food security force policy makers to take a more holistic view. Climate and global change are rapidly placing added pressures on the world's water reserves and the time has come to strengthen cooperation and build peace amongst states.

UNESCO's mission to 'contribute to the building of peace, eradication of poverty, sustainable development and intercultural dialogue through education, the sciences, culture, communication and information' is achieved fully through its international water initiatives coordinated by the UNESCO International Hydrological Programme (IHP). UNESCO-IHP, established in 1975, is the only global scientific intergovernmental programme of the UN system devoted entirely to water resources, emphasizing the formulation of policy-relevant strategies for their sustainable management. Through its ISARM (Internationally Shared Aquifer Resources Management) and PCCP (From Potential Conflict to Cooperation Potential) programmes, UNESCO provides Member States with high level expertise and knowledge and assists them in the elaboration of policies for the sustainable management of transboundary waters.

Transboundary Water Management, edited by J. Ganoulis, A. Aureli and J. Fried, is the result of several years' of research in the field of international water resources. The UNESCO Chair that coordinates the International Network of Water-Environment Centres for the Balkans played an important role in organizing both the compilation of existing knowledge and the elaboration of sound policy recommendations. It is with great pleasure, therefore, that I welcome the publication of this title and commend it to Member States. A multidisciplinary approach to the management of shared natural resources is indeed paramount to finding solutions to multi-faceted challenges and I trust that future water managers, policy-makers and academics will find pleasure in reading this publication as well as benefit from its findings.

Gretchen Kalonji
UNESCO Assistant Director-General
for Natural Sciences

Contents

Preface

This book uses the term 'transboundary waters', as in Transboundary Waters Resources Management (TWRM), to mean waters crossing the borders of different riparian countries, which therefore are by definition countries sharing common surface and/or groundwater resources. The term is synonymous with 'internationally shared waters' and is in accordance with the terminology used by UNESCO in its international hydrological initiatives, such as the UNESCO/ISARM (Internationally Shared-Transboundary-Aquifer Resources Management) and the UNESCO/PC-CP (Potential Conflict-Cooperation Potential) programmes. It is considered to be a better choice than other similar expressions such as 'international waters', 'multinational waters' or 'regional waters', and avoids misunderstandings due to political sensitivities over national sovereignty in regions located near the borders.

'Boundaries' may also exist with different connotations between administrative regions or between cultural or ethnic entities located within the same country. In these cases both surface waters (rivers and lakes) and groundwaters (aquifers) may involve different administrations or various communities and their shared management should aim to resolve issues of potential regional or local conflicts in terms of water needs, water quality, environmental preservation or differences in legislation and economic issues. When the boundary is international and waters cross the borders of different riparian countries, then TWRM faces the major challenge of potential political conflicts and even war. The main issue in this case is how to convert these potential conflicts into collaborative actions. Such global TWRM challenges and general tools with which they may be addressed are explained in *Part I: A Global View.*

The book aims to serve as a practical guide for enhancing models of collaborative activities between riparian countries. In this context 'collaboration' means the active involvement of partners and institutions from both sides of the border, which includes exchange of information, interaction and dialogue between partners, in order to reach common decisions and find unified solutions to TWRM problems. In this sense, 'collaboration' is considered to be a more advanced stage of 'cooperation' or 'coordination'. The first step in cooperation can be achieved by a simple exchange of information with no further interaction between partners; this may be called 'passive cooperation'. A more advanced second step is engaging in dialogue and developing a consultation process; this may be called 'coordinated cooperation' and is a prerequisite condition for the third step, which is 'active collaboration'. Only with active and effective collaboration can sustainable governance of transboundary water resources be achieved.

Since there is no single universal model for a collaborative approach to TWRM, this book presents an analysis of various effective models illustrated by case studies from around the world. Even though case studies are particular and not easily transferred to different situations, they are very helpful in showing relationships between different more or less independent variables, such as physical, hydrologic, hydrogeological, ecological, socio-economic conditions, institutional structures, stakeholders participation, legal agreements and political willingness. The main dependent variable that emerges from this process is the need for active collaboration and effective governance in TWRM.

Models of collaborative actions in TWRM depend on the approach used, for example, whether the model is developed by a particular scientific discipline, by a professional community or by different kinds of scientists.

For engineers, hydrologists, hydrogeologists or environmental professionals emphasis is placed on modelling the physical and ecological transboundary hydro-systems in terms of (i) delineating their natural borders (hydrologic basins for transboundary rivers and lakes or hydrogeological boundaries for groundwater aquifers), (ii) analysing relationships between physical and ecological variables such as precipitation, river flow, pollutant inputs, lake water quality, biodiversity or groundwater recharge and (iii) suggesting structural or non-structural measures in order to obtain solutions and improve TWRM. These models, conceptual or mathematical, are more or less accurate subject to data availability and precision and various assumptions and simplifications in modelling. They are useful for understanding how the physical and ecological transboundary systems behave under natural and anthropogenic inputs in terms of water quantity and environmental impacts. These kinds of models for transboundary aquifers, lakes and rivers are presented in *Part II: Physical, Environmental and Technical Approaches.*

For lawyers and social scientists (geographers, economists, sociologists) emphasis is placed on human factors, which can be very complex and difficult to analyse or predict, such as institutional cooperation, stakeholder participation and negotiation strategies. For lawyers the emphasis is on regulating provisions and duties of riparian countries in terms of access, utilization, protection, preservation and management of transboundary waters. The codification of such legal rules is very useful to the international community, even though this process may be somewhat general and unable to cover all specific cases. The main challenge is whether different national administrations will agree to implement international rules at the national level and at the same time coordinate their activities with riparian countries through bilateral or regional collaborative agreements. This challenge may be faced by raising public and stakeholders' awareness in participatory processes involving national institutions, academic partners and international organizations. All these approaches are presented in *Part III: Legal, Socio-Economic and Institutional Approaches.*

In the real world all the above issues and approaches coexist and are interrelated. To achieve effective TWRM these models, whether descriptive or prescriptive, should merge. In Chapter 8 of *Part IV: Bridging the Gaps*, two main strategies for achieving such integration are presented: (i) through effective capacity building and training in TWRM and (ii) by analysing a general framework of conflict resolution, based on how riparian countries may share benefits and risks. Both these strategies are supported by UNESCO's ISARM and PC-CP programmes.

The main contents of the book are based on updated papers first presented at the 'IV International Symposium on Transboundary Water Management', Thessaloniki, Greece, October 2008. Recommendations of this Conference on how to bridge the gaps are summarized in the 'Thessaloniki Statement', which is reported in Chapter 9 of *Part IV*.

I am very grateful to all authors and contributors to this book for their excellent collaboration during and after the conference. Personally and on behalf of my co-editors, Alice Aureli and Jean Fried, I would like to thank Dr. Frank Weinreich, manager of Wiley-VCH Water & Environmental books programme, for giving us the opportunity to publish this book, and to Lesley Belfit, Project Editor at Wiley-VCH, for her help with the publication process. My appreciation and special thanks go to Katie Quartano at the UNESCO Chair, Aristotle University of Thessaloniki, for her professional contribution to the reviewing and proofreading processes.

Thessaloniki, Greece   *Jacques Ganoulis*
January 2011

List of Contributors

Thomas K. Alexandridis
Aristotle University of Thessaloniki
School of Agriculture
Laboratory of Applied Soil Science
Thessaloniki
Greece

Manolia Andredaki
Democritus University of Thrace
Department of Civil Engineering
Vas. Sofias 12
67100 Xanthi
Greece

Francesca Antonelli
World Wildlife Fund
European Policy Office
Via PO, 25/C
00198 Rome
Italy

Bo Appelgren
UNESCO International Hydrological
Programme
N. Colesanti 13
01023 Bolsena
Italy

and

UNESCO International Hydrological
Programme
1 rue Miollis
75732 Paris
France

Majed Atwi Saab
University of Zaragoza
Faculty of Economics and Business
Administration
Department of Economic Analysis
Gran Vía 2
50005 Zaragoza
Spain

Marina Babić Mladenović
'Jaroslav Cerni' Institute for the
Development of Water Resources
Jaroslava Cernog 80
11226 Belgrade
Serbia

Alexey V. Babkin
State Hydrological Institute
Laboratory of Water Resources
and Water Balance
Second Line, 23 V.O.
199053 St. Petersburg
Russia

Evangelos A. Baltas
Aristotle University of Thessaloniki
School of Agriculture
Department of Hydraulics, Soil Science
and Agricultural Engineering
Laboratory of General and Agricultural
Hydraulics and Land Reclamation
54124 Thessaloniki
Greece

Djana Bejko
University 'Luigj Gurakuqi'
Faculty of Natural Sciences
Sheshi '2 Prilli'
L. Qemal Stafa, Rr. Vasil Shanto, Nr 21
4001 Shkoder
Albania

Georg Berthold
Hessian Agency for Environment
and Geology (HLUG)
Rheingaustraße 186
65203 Wiesbaden
Germany

Roberto Bertoni
C.N.R. Institute of Ecosystem Study
Largo Tonolli 50
28922 Verbania Pallanza
Italy

Adriane Blum
Bureau de Recherches Géologiques et
Minières (BRGM)
3 avenue Claude-Guillemin
45060 Orléans
France

Ognjen Bonacci
University of Split
Faculty of Civil Engineering
and Architecture
Matice hrvatske 15
21000 Split
Croatia

Sabine Brels
University of Laval
Faculty of Law
2325 rue de l'Université Québec
Québec City, Québec
Canada G1V 0A6

Mitja Brilly
University of Ljubljana
Faculty of Civil and Geodetic
Engineering
Jamova 2
1000 Ljubljana
Slovenia

Serge Brouyere
University of Liège
HG-GeomaC
4000 Sart Tilman, Liège
Belgium

Anne Browning-Aiken
University of Arizona
Udall Centre for Studies in Public Policy
Tucson, AZ
USA

Brilanda Bushati
University 'Luigj Gurakuqi'
Faculty of Economic Sciences
Sheshi '2 Prilli'
L. Qemal Stafa, Rr. Zog i Pare, Nr. 37
4001 Shkoder
Albania

Zsuzsanna Buzás
Ministry for Environment and Water
Főutca 44-50
1011 Budapest
Hungary

Devinder Kumar Chadha
Global Hydrogeological Solutions
G-66 (Ground Floor)
Vikaspuri
New Delhi - 110 018
India

Eleni Charou
National Centre for Scientific Research
"Demokritos"
Institute of Informatics &
Telecommunications
153 10 Aghia Paraskevi
Greece

Ioannis Chronis
Aristotle University of Thessaloniki
School of Agriculture
Laboratory of Applied Soil Science
Thessaloniki
Greece

David Coates
Secretariat of the Convention
on Biological Diversity
413 Saint Jacques Street
Montreal, Québec
Canada QC H2Y 1N9

Ana Carolina Coelho
Colorado State University
Department of Civil Engineering
Engineering Building - Campus
Delivery 1372
Fort Collins, CO 80523-1372
USA

Alain Dassargues
University of Liège
DepartmentArGEnCo
4000 Sart Tilman, Liège
Belgium

Hubert Machard de Gramont
BRGM
Water Division
3 avenue Claude Guillemin
BP 36009-45060 Orléans
France

Lilian Del Castillo-Laborde
University of Buenos Aires
School of Law
Av. Figueroa Alcorta 2263
1425 Buenos Aires
Argentina

Mónica D'Elia
National University of El Litoral
Faculty of Engineering and Water
Sciences
Ciudad Universitaria
Ruta Nacional 168-Km 472
S3000 Santa Fe
Argentina

Eglantina Demiraj
Polytechnic University of Tirana
Institute of Energy, Water and
Environment
Durresi Street 219
Tirana
Albania

Milan Dimkić
'Jaroslav Cerni' Institute for the
Development of Water Resources
Jaroslava Cernog 80
11226 Belgrade
Serbia

Dragan Dolinaj
University of Novi Sad
Faculty of Natural Sciences and
Mathematics
Climatology and Hydrology Research
Centre
Trg Dositeja Obradovica 3
21000 Novi Sad
Serbia

Jean-François Donzier
International Network of Basin
Organizations
c/o International Office for Water
21 rue de Madrid
75008 Paris
France

Radu Drobot
Technical University of Civil
Engineering
Bd. Lacul Tei 124, Sector 2
020396 Bucharest
Romania

Viktor A. Dukhovny
Scientific Information Centre
of Interstate Coordination Water
Commission of Aral Sea Basin
(SIC ICWC)
Massiv Karasu 4, Building 11
100187 Tashkent
Uzbekistan

Eleni Eleftheriadou
Aristotle University of Thessaloniki
Civil Engineering Department
Hydraulics Laboratory
54124 Thessaloniki
Greece

Zsuzsanna Engi
West-Transdanubian Environmental
and Water Directorate[JA20]
Department for Prevention and
Protection from Water Damages
Gyor
Hungary

Darrell Fontane
Colorado State University
Department of Civil Engineering
Fort Collins, CO 80523
USA

Jean Fried
University of California
School of Social Ecology
Department of Planning, Policy and
Design
Irvine, CA 92697
USA

and

UNESCO
Paris
France

Hans-Gerhard Fritsche
Hessian Agency for Environment and
Geology (HLUG)
Rheingaustraße 186
65203 Wiesbaden
Germany

Jacques Ganoulis
UNESCO Chair and Network INWEB
Aristotle University of Thessaloniki
Department of Civil Engineering
Division of Hydraulics and
Environmental Engineering
54124 Thessaloniki
Greece

Miltos Gletsos
Society for the Protection of Prespa
530 77 Aghios Germanos
Greece

Piero Guilizzoni
C.N.R. Institute of Ecosystem Study
Largo Tonolli 50
28922 Verbania Pallanza
Italy

Bojan Hajdin
University of Belgrade
Faculty of Mining & Geology
Department of Hydrogeology
Djusina 7
11000 Belgrade
Serbia

André Hernandes
Ministry of Transport
National Department of Transport
Infrastructure (DNIT)
Parana Waterway Administration
(AHRANA)
Av. Brigadeiro Faria Lima
SP-CEP 01451-000 Sao Paulo
Brazil

Vlassios Hrissanthou
Democritus University of Thrace
Department of Civil Engineering
Vas. Sofias 12
67100 Xanthi
Greece

Natacha Jacquin
L'Office International de l'Eau OIEAU
15 rue Edouard Chamberland
87065 Limoges Cedex
France

Andreas Kallioras
Technical University of Darmstadt
Institute of Applied Geosciences
Hydrogeology Group
Karolinenplatz 5
64289 Darmstadt
Germany

Kamal Karaa
Litani River Authority
Bechara el Khoury Street
Ghannageh Buld.
3732 Beirut
Lebanon

Katharina Kober
Mediterranean Network of Basin
Organizations
Avda. Blasco Ibañez 48
46010 Valencia
Spain

Elpida Kolokytha
Aristotle University of Thessaloniki
Department of Civil Engineering
Division of Hydraulics and
Environmental Engineering
54124 Thessaloniki
Greece

Stanka Koren
Environmental Agency of the Republic
of Slovenia
Vojkova 1b
1000 Ljubljana
Slovenia

Vladimir Kotov
EcoPolicy Research and Consulting
Moscow
Russia

Nikolaos Kotsovinos
Democritus University of Thrace
Department of Civil Engineering
Vas. Sofias 12
67100 Xanthi
Greece

Alexei V. Kouraev
Université de Toulouse
UPS (OMP-PCA)
LEGOS
14 Av. Edouard Belin
F-31400 Toulouse
France

and

State Oceanography Institute
St. Petersburg Branch
St. Petersburg
Russia

Balázs Kovács
University of Szeged
Department of Mineralogy,
Geochemistry and Petrology
Egyetem 2-6
6722 Szeged
Hungary

Péter Kozák
ATIKOVIZIG
Directorate for Environmental
Protection and Water Management of
Lower Tisza District
Stefania 4
6701 Szeged
Hungary

Neno Kukuric
IGRAC – International Groundwater
Resources Assessment Centre
3508 AL Utrecht
The Netherlands

Ralf Kunkel
Research Centre Jülich
Agrosphere Institute (ICG-4)
Leo-Brandt-Strasse
52425 Jülich
Germany

Richard Laster
Hebrew University
Faculty of Law and Faculty of
Environmental Studies
Jerusalem
Israel

and

Laster Gouldman Law Offices
48 Azza Street
92384 Jerusalem
Israel

Efthalia Lazaridou
Omikron LTD
Environmental Department
Agricultural Road Straitsa
57001 Thessaloniki
Greece

Maria Lazaridou
Aristotle University of Thessaloniki
Department of Biology
Laboratory of Zoology
Thessaloniki
Greece

Milojko Lazić
University of Belgrade
Faculty of Mining & Geology
Department of Hydrogeology
Djusina 7
11000 Belgrade
Serbia

Louis Lebel
Chiang Mai University
Unit for Social and Environmental
Research
239 Huay Kaew Road
50200 Chiang Mai
Thailand

Lászlò Lenart
University of Miskolc
3515 Miskolc-Egyetemvaros
Hungary

Flavia Rocha Loures
World Wildlife Fund (WWF)
International Law and Policy
Freshwater Program
1250 24th Street, NW
Washington, DC 20037-1193
USA

Rodrigo Maia
Universidade do Porto
Department of Civil Engineering
Rua Dr. Roberto Frias
4200-465 Porto
Portugal

Sotir Mali
University of Elbasan
Rruga Rinia
Elbasan
Albania

Daphne Mantziou
Society for the Protection of Prespa
530 77 Aghios Germanos
Greece

Daene C. McKinney
The University of Texas at Austin
Center for Research in Water Resources
10100 Burnet Rd., Bldg 119
Austin, TX 78703
USA

Petra Meglič
Geological Survey of Slovenia
Dimičeva ulica 14
1000 Ljubljana
Slovenia

Saša Milanović
University of Belgrade
Faculty of Mining & Geology
Department of Hydrogeology
Djusina 7
11000 Belgrade
Serbia

Dragana Milovanović
Ministry of Agriculture
Forestry and Water Management
Directorate for Water
Bulevar umetnosti 2a
11070 Belgrade
Serbia

Miodrag Milovanović
'Jaroslav Cerni' Institute for the
Development of Water Resources
Jaroslava Cernog 80
11226 Belgrade
Serbia

Marin-Nelu Minciuna
National Institute of Hydrology and
Water Management
Sos. Bucuresti-Ploiesti 97
013686 Bucharest
Romania

Jean-Marie Monget
Mines ParisTech
Earth & Environmental Sciences
60 Boulevard Saint-Michel
75272 Paris
France

Barbara J. Morehouse
University of Arizona
Institute of the Environment
Marshall Building
845 N. Park Avenue
Tucson, AZ 85721
USA

Rosario Mosello
C.N.R. Institute of Ecosystem Study
Largo Tonolli 50
28922 Verbania Pallanza
Italy

Jacques Mudry
University of Besançon
UMR Chrono-Environnement
F-25030 Besançon
France

Yannis Mylopoulos
Aristotle University of Thessaloniki
Civil Engineering Department
Hydraulics Laboratory
54124 Thessaloniki
Greece

Udaya Sekhar Nagothu
Norwegian Institute for Agricultural and
Environmental Research (Bioforsk)
Fr. A. Dahlsvei 20
1432 Ås
Norway

Miriam Ndini
Polytechnic University of Tirana
Institute of Energy, Water and
Environment
Durresi Street 219
Tirana
Albania

Benjamin Ngounou Ngatcha
University of Ngaoundéré
Faculty of Sciences
B.P. 454 Ngaoundéré
Cameroon

Elena Nikitina
Russian Academy of Sciences
Institute for World Economy and
International Relations
Prosouznaya st. 23
117997 Moscow
Russia

Dragana Ninković
'Jaroslav Cerni' Institute for the
Development of Water Resources
Jaroslava Cernog 80
11226 Belgrade
Serbia

Jožef Novak
Environmental Agency of the Republic
of Slovenia
Vojkova 1b
1000 Ljubljana
Slovenia

Petar Papic
University of Belgrade
Faculty of Mining & Geology
Department of Hydrogeology
Djusina 7
11000 Belgrade
Serbia

Marta Paris
National University of El Litoral
Faculty of Engineering and Water
Sciences
Ciudad Universitaria
Ruta Nacional 168-Km 472
S3000 Santa Fe
Argentina

Milana Pantelić
University of Novi Sad
Faculty of Natural Sciences and
Mathematics
Department of Geography, Tourism and
Hotel Management
Trg Dositeja Obradovica 3
21000 Novi Sad
Serbia

Didier Pennequin
BRGM
Water Division
3 avenue Claude Guillemin
BP 36009-45060 Orléans
France

Christian Perennou
Tour du Valat, Le Sambuc
13200 Arles
France

Marcela Perez
National University of El Litoral
Faculty of Engineering and Water
Sciences
Ciudad Universitaria
Ruta Nacional 168-Km 472
S3000 Santa Fe
Argentina

Andrej Perovic
University of Montenegro
Faculty of Natural Sciences and
Mathematics
20000 Podgorica
Montenegro

Sotiris Petropoulos
Harokopio University of Athens
Department of Geography
70 El Venizelou Str.
17671 Athens
Greece

Fotis Pliakas
Democritus University of Thrace
Civil Engineering Department
Engineering Geology Laboratory
Vas. Sofias 12
67100 Xanthi
Greece

Dušan Polomčić
University of Belgrade
Faculty of Mining & Geology
Department of Hydrogeology
Djusina 7
11000 Belgrade
Serbia

Irina Polshkova
Russian Academy of Sciences
Water Problems Institute
3 Gubkina Street
119333 Moscow
Russia

Joerg Prestor
Geological Survey of Slovenia
Dimičeva ulica 14
1000 Ljubljana
Slovenia

Samir Rhaouti
Sebou River Basin Organization
BP 2101 Rue Abou Alaa Al Maari
VN30000 Fes
Morocco

Lena Salame
UNESCO
Potential Conflict to Cooperation
Potential (PC-CP) Programme
Paris
France

Julio Sánchez Chóliz
University of Zaragoza
Faculty of Economics and Business
Administration
Department of Economic Analysis
Gran Vía 2
50005 Zaragoza
Spain

Samuel Sandoval-Solis
The University of Texas at Austin
Center for Research in Water Resources
10100 Burnet Rd., Bldg 119
Austin, TX 78703
USA

Spase Shumka
Agricultural University of Tirana
Faculty of Biotechnology and Food
Koder-Kamza
Tirana
Albania

Bach Tan Sinh
National Institute for Science and
Technology Policy and Strategy Studies
Science and Policy Studies Centre
Hanoi
Vietnam

Eva Skarbøvik
Norwegian Institute for Agricultural and
Environmental Research (Bioforsk)
Fr. A. Dahlsvei 20
1432 Ås
Norway

Stylianos Skias
Democritus University of Thrace
Civil Engineering Department
Engineering Geology Laboratory
Vas. Sofias 12
67100 Xanthi
Greece

Charalampos Skoulikaris
Aristotle University of Thessaloniki
Civil Engineering Department
Hydraulics Laboratory
54124 Thessaloniki
Greece

Alkis Stamos
Institute of Geology and Mineral
Exploration
Department of Geology and Geological
Mapping
Olympic Village, Entrance C
13677 Acharnae
Greece

Marianthi Stefouli
Institute of Geology and Mineral
Exploration
Department of Geology and Geological
Mapping
Olympic Village, Entrance C
13677 Acharnae
Greece

Raya Marina Stephan
Water Law expert
International consultant
38 rue du Hameau
78480 Verneuil sur Seine
France

Zoran Stevanović
University of Belgrade
Faculty of Mining & Geology
Department of Hydrogeology
Djusina 7
11000 Belgrade
Serbia

Pierre Strosser
ACTeon s.a.r.l., Le Chalimont
BP Ferme du Pré du Bois
68370 Orbey
France

Galina Stulina
Scientific Information Centre of
Interstate Coordination Water
Commission of Aral Sea Basin
(SIC ICWC)
Massiv Karasu 4, Building 11
100187 Tashkent
Uzbekistan

János Szanyi
University of Szeged
Department of Mineralogy,
Geochemistry and Petrology
Egyetem 2-6
6722 Szeged
Hungary

Peter Szucs
University of Miskolc
35152 Miskolc-Egyetemvaros
Hungary

Rebecca L. Teasley
The University Of Minnesota Duluth
Department of Civil Engineering
221 SCiv 1405 University Drive
Duluth, MN 55812
USA

József Török
ATIKOVIZIG
Directorate for Environmental
Protection and Water Management of
Lower Tisza District
Stefania 4
6701 Szeged
Hungary

Nikolaos Tsotsolis
Region of Central Macedonia
Thessaloniki
Greece

Ofelia Tujchneider
National University of El Litoral
Faculty of Engineering and Water
Sciences
Ciudad Universitaria
Ruta Nacional 168-Km 472
S3000 Santa Fe
Argentina

and

National Council of Scientific and
Technical Research (CONICET)
Av. Rivadavia 1917
C1033AAJ Buenos Aires
Argentina

Guido Vaes
HydroScan Ltd.
Tiensevest 26/4
3000 Leuven
Belgium

Anastasios Valvis
University of Peloponnese
Department of Political Science and
International Relations
Corinth
Greece

Jac van der Gun
IGRAC – International Groundwater
Resources Assessment Centre
3508 AL Utrecht
The Netherlands

Slavek Vasak
IGRAC – International Groundwater
Resources Assessment Centre
3508 AL Utrecht
The Netherlands

Evan Vlachos
Colorado State University
Department of Civil Engineering
Fort Collins, CO 80523
USA

Frank Wendland
Research Centre Jülich
Agrosphere Institute (ICG-4)
Leo-Brandt-Strasse
52425 Jülich
Germany

Rüdiger Wolter
Federal Environmental Agency (UBA)
Wörlitzer Platz 1
06844 Dessau
Germany

George Zalidis
Aristotle University of Thessaloniki
School of Agriculture
Laboratory of Applied Soil Science
Thessaloniki
Greece

1
Introduction and Structure of the Book

Jacques Ganoulis

This book is a practical guide that suggests methodological tools and answers to different questions related to Transboundary Water Resources Management (TWRM), including both surface and groundwater aquifer resources. Some of these questions may be formulated as follows:

- How could data and information from riparian countries be harmonized to better understand the physical characteristics of transboundary hydro-systems?
- Are hydrological and hydrogeological models available to predict different scenarios in TWRM?
- What methodology is available to delineate transboundary aquifers?
- What is the current status of international law in terms of sharing transboundary surface waters and groundwater aquifers between riparian countries and what are the main legal issues?
- How could international law improve the utilization and effective protection of shared water resources?
- How could public and stakeholder participation contribute to the implementation of integrated TWRM?
- What methodology is available for integrating different collaborative models of TWRM?
- How could potential conflicts in sharing transboundary waters be transformed into collaborative actions?

In this practical guide, different collaborative models and TWRM tools are identified and explained, not just theoretically or conceptually but through specific case studies from around the world. These case studies are grouped together in such a way that the wide range of tools available to effectively explain, address, assess, understand and resolve TWRM problems in the real world become apparent.

The book is organized in four parts, which are described below.

Transboundary Water Resources Management: A Multidisciplinary Approach, First Edition.
Edited by Jacques Ganoulis, Alice Aureli and Jean Fried.

1.1
Part I – A Global View

Part I is divided into two chapters (Chapters 2 and 3). Chapter 2 presents the importance of transboundary waters worldwide and the need for collaborative approaches to address global challenges of TWRM. The role of different disciplinary tools and regulatory instruments (technical, environmental, legal and socio-economical) for an effective collaborative approach is also explained.

Chapter 3 describes significant worldwide initiatives, such as the INBO (International Network of Basin Organizations) network, the UNECE (United Nations Economic Commission for Europe) Transboundary Waters Convention (1992), the UN Watercourses Convention (1997), the UN International Law Commission articles on shared natural resources (oil, gas and including shared groundwaters in 2002), UNESCO's International Hydrological Programme (IHP) components dealing with transboundary surface and groundwater resources, the UN CBD (Convention on Biological Diversity, 1992) and the European Union Water Framework Directive (EU-WFD, 2000). The importance of building international cooperation and management networks at the transboundary river catchment scale is emphasized (Chapter 3.1) and the role of international laws for transboundary water courses and aquifers is analysed (Chapters 3.2–3.4). The EU-WFD as a driving force for implementing the concept of Integrated Water Resources Management (IWRM) in transboundary regions is further explained (Chapters 3.5 and 3.6) and illustrated by characteristic case studies both from the EU and non-EU countries (Chapters 3.7–3.9).

1.2
Part II – Physical, Environmental and Technical Approaches

Part II is divided into two chapters, the first of which (Chapter 4) describes physical, environmental and technical approaches for transboundary aquifers, and the second (Chapter 5) covers transboundary lake and river basins. Chapters 4.1–4.3 are quite general and explain how hydrologic and hydrogeological approaches may be used to assess not only porous transboundary aquifers (Chapter 4.1) but also karst aquifers, which are globally very important sources for water supply (Chapter 4.2). The need to share information between neighbouring countries and to harmonize data is emphasized in these sections and the use of mathematical modelling as a tool for assessing groundwater hydrodynamics in transboundary aquifers is highlighted (Chapter 4.3).

Characteristic case studies from around the world, illustrating the application of the hydrogeological and scientific tools previously analysed, are reported in the second part of Chapter 4. In these case studies further details are given on how to assess and model transboundary aquifer systems, with examples from South America (Chapter 4.4), Africa (Chapter 4.5), Asia (Chapter 4.6) and Europe (Rhine Valley, Chapter 4.7), an aquifer shared by Hungary and Romania (Chapter 4.8), an aquifer shared by Serbia and Hungary (Chapter 4.9) and aquifers around Slovenia (Chapter 4.10).

For transboundary surface waters (Chapter 5), such as lakes and rivers, the hydrological monitoring data collected by individual countries are usually non-comparable, and even incomplete. This unfortunate situation is documented in (Chapter 5.1) and is also the case for the majority of hydrogeological data of groundwater aquifers. The non-comparability of monitoring data is a major obstacle in harmonizing information available from individual countries and applying global directives like the EU-WFD. International guidelines, such as those published by UN organizations like the WMO (World Meteorological Organization), could help remediate this situation.

However, despite the lack of systematic comparable monitoring systems, three case studies illustrating successful collaboration models are presented in Chapter 5: Lake Maggiore, shared between Italy and Switzerland (Chapter 5.2), Prespa Lakes, shared between Greece, Albania and FYR of Macedonia (Chapters 5.3 and 5.4), and the Kobilje River, shared between Slovenia and Austria (Chapter 5.5). This chapter also illustrates other problems in transboundary river basins, such as impacts from climate change (Chapters 5.6 and 5.7), identification of water bodies according to the EU-WFD (Chapter 5.8), sediment transport (Chapter 5.9) and river flow periodicities (Chapter 5.10).

1.3
Part III – Legal, Socio-Economic and Institutional Approaches

This part is also divided into two chapters (Chapters 6 and 7. Chapter 6 deals mainly with legal approaches; in Chapter 6.1 explanations are offered as to how international law on transboundary aquifers may be used. In Chapter 6.2 it is shown how adequate water policies may reduce over use of water in agriculture.

Regional and bilateral legal agreements can enhance effective cooperation between countries. Examples of this are given for the Aral Sea basin, Central Asia (Chapter 6.3), for the Kidron Valley, Middle East (Chapter 6.4) and for the Prespa Lakes basin in the Balkans (Chapter 6.5). A comparison between the rivers Mekong in SE Asia and Maritsa/Evros/Meriç in SE Europe illustrates how regional agreements can contribute to transform conflicts into cooperation (Chapter 6.6).

Adequate delineation of water resources regions adapted to specific regional conditions is also an important issue. The EU-WFD stipulates that water resources management should be performed on a river basin basis. Different criteria may be used to define water resources management regions in order to better promote the application of IWRM and contribute to transboundary water conflicts resolution. Examples are provided from the USA (Chapter 6.7) and Greece (Chapter 6.8).

Chapter 7 focuses on socio-economic and institutional approaches, which are very important for the implementation of technical and legal collaborative models in transboundary waters. Stakeholder participation, social learning and institutional design are important tools for achieving effective TWRM and reducing water insecurities and this is analysed in Chapters 7.1 and 7.2. Case studies from South

America (Chapter 7.3) and the Balkans (Chapters 7.4–7.6) demonstrate particular issues and problems in transboundary cooperation.

Economic governance, such as the model of common pool management of transboundary water resources (Chapter 7.7) and applications of game theory (Chapters 7.8–7.10), all important tools for facilitating negotiations in conflict resolution issues, is also discussed.

1.4
Part IV – Bridging the Gaps

To deal with the complexity of real world problems, where no distinction is made between different dependent physical and socio-economic processes, there is a need for the various approaches described in Parts II and III to be integrated. This process of integration could be facilitated in two main ways. Firstly, through education and capacity building, where special training programmes can show how multidisciplinary approaches can be coordinated to achieve an integrated view of a problem and solve it effectively in the real world (Chapter 8.1). Secondly, by taking into account a general framework for risk analysis in conflict resolution, where risks and benefits could be shared between riparian countries and "win-win" solutions to transbound-

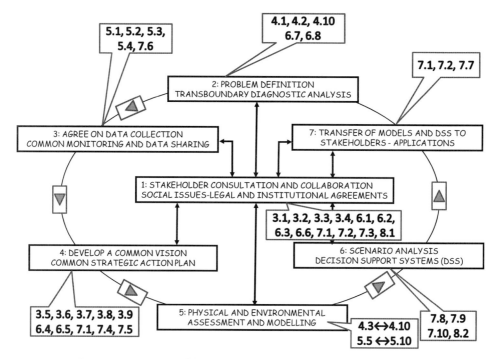

Figure 1.1 Seven steps of the TWRM collaborative model, showing the relevant chapter/section numbers of the individual contributions from this book.

ary disputes can be achieved (Chapter 8.2). Both these processes are based on specific programmes developed by UNESCO.

Figure 1.1 illustrates a collaborative model for TWRM based on the various contributions to this book. This uses the following seven steps and may be adapted to any particular case study of transboundary waters:

1) Stakeholder consultation and collaboration, social issues, legal and institutional agreements: this should interact with every one below;
2) problem definition: Transboundary Diagnostic Analysis (TDA);
3) agree on data collection, common monitoring and data sharing;
4) develop a common vision and common Strategic Action Plan (SAP);
5) physical and environmental assessment and modelling;
6) scenario analysis and Decision Support Systems (DSS);
7) transfer of models and DSS to stakeholders, applications.

Part One
A Global View

Transboundary Water Resources Management: A Multidisciplinary Approach, First Edition.
Edited by Jacques Ganoulis, Alice Aureli and Jean Fried.
© 2011 Wiley-VCH Verlag GmbH & Co. KGaA. Published 2011 by Wiley-VCH Verlag GmbH & Co. KGaA.

2
Transboundary Water Resources Management: Needs for a Coordinated Multidisciplinary Approach

Jacques Ganoulis and Jean Fried

2.1
Introduction

The global increase of population together with steady socio-economic development, especially of emerging economies, and the subsequent increase in water demand combined with the acceleration of water pollution from various point and diffuse sources, mean that transboundary water resources, located both on the surface (rivers and lakes) and in groundwater aquifers, are very important sources of water for different uses at global and regional scales, and form a significant part of the precious available water on earth.

Although the total amount of water on earth is substantial, only a very small fraction of it is not saline and can be directly used by man. According to the latest UN World Water Development Reports [1, 2] this amount is only 2.5% of the total water available on earth. When economically available renewable water resources are taken into account, global water availability is estimated at about $13\,500\,km^3$ per year that is only $2300\,m^3$ per person per year. This is approximately 37% less than in 1970.

About 60% of global river flow lies within transboundary river basins [3], the surface area of which amount to almost half of the world's land surface (Figure 2.1). The significance of transboundary waters may be seen from the following data [3, 4]:

- 40% of the world's population lives within these watersheds;
- 45% of the total land surface of our planet lies in this area;
- 263 major internationally shared basins are reported;
- approximately one-third of the 263 transboundary basins are shared by more than two countries;
- 145 countries have territory within transboundary river basins;
- 21 countries lie entirely within one transboundary river basin;
- more than 95% of the territory of 12 countries lies within one or more transboundary basins;
- 19 basins involve five or more different countries.

The distribution of transboundary basins per continent by number and as a percentage of the continent's surface, based on data revised in 1999 [3] and

Transboundary Water Resources Management: A Multidisciplinary Approach, First Edition.
Edited by Jacques Ganoulis, Alice Aureli and Jean Fried.
© 2011 Wiley-VCH Verlag GmbH & Co. KGaA. Published 2011 by Wiley-VCH Verlag GmbH & Co. KGaA.

Figure 2.1 Distribution of transboundary river basins worldwide [3].

2003 [4], is given in Figure 2.2. It can be seen that Europe has the greatest number of internationally shared basins (69), while in comparison with the others Africa's transboundary basins cover the greatest part of the continent (62%).

The actual number of transboundary basins may change not only because new political states emerge or in some cases, for example, Germany, become unified, but also because cartographic methods improve. As shown in Table 2.1, the number of transboundary basins varies over time in all continents. Owing to the collapse of the Former Republic of Yugoslavia and the Soviet Union, the number of transboundary basins in Europe as a whole increased from 48 in 1978 to 69 in 2002; in South East Europe (SEE) alone this meant that the number of transboundary basins almost doubled [4, 10]. Globally in 2002, there were 263 transboundary basins listed, compared to 261 in 1999 and 214 in 1978 (Table 2.1).

Maps of the world's transboundary aquifers were updated in 2007 by the German Federal Institute for Geosciences and Natural Resources (BGR)/Worldwide Hydrogeological Mapping and Assessment Programme (WHYMAP) [5], and, as shown in Figure 2.3, in 2009 by the International Groundwater Resources Centre (IGRAC) [6]. Transboundary fresh groundwater resources offer much higher volumes than transboundary river water flow. On a global scale, the importance of fresh groundwater resources is predominant. According to estimations by the United States Geological Survey, 99% of the available fresh water on the planet is stored in the ground. About 69% is stored in glaciers and permanent snow cover and is practically inaccessible for human use. Interestingly, while rivers and lakes hold only 0.3% of the total amount of the

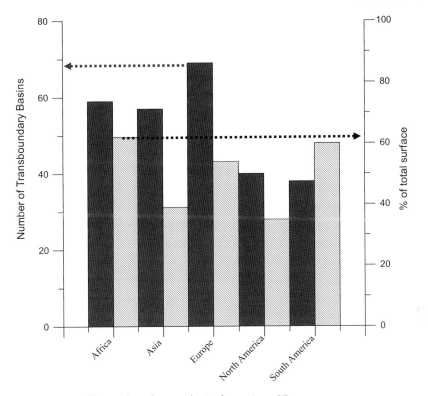

Figure 2.2 World's transboundary river basins by continent [4].

available fresh water, fresh groundwater represents about 30% of global fresh water, with the remainder being stored as soil moisture. This groundwater is located at depths up to 4000 m and half of this quantity is technically available at depths of less than 800 m.

The main characteristics of transboundary aquifers worldwide are not very well known because of the lack of joint monitoring systems, limited data sharing between

Table 2.1 Number of transboundary river basins by continent and as a percentage of the total surface [4].

	1978	1999	2002	Percentage of total surface
Africa	57	60	59	62
Asia	40	53	57	39
Europe	48	71	69	54
North and Central America	33	39	40	35
South America	36	38	38	60
Total	214	261	263	45

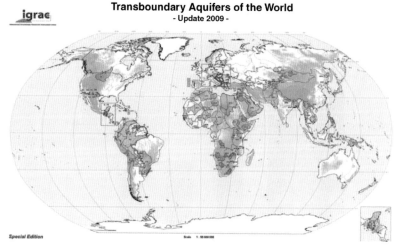

Figure 2.3 World's transboundary aquifers [6].

neighbouring countries and a low degree of political commitment. This is why UNESCO, and more specifically UNESCO's International Hydrological Programme (IHP) (Paris), having recognized that transboundary aquifer systems are important sources of fresh water in many regions of the world, decided in June 2002 to launch a new initiative to promote studies in regard to transboundary aquifers called the 'Internationally Shared Aquifer Resources Management' programme (ISARM) [7].

Since 2002 UNESCO has been implementing ISARM in different parts of the world. The first phase of the UNESCO/ISARM programme was initiated in Africa in 2002. In the same year, a project was prepared by the Economic Commission for Europe (UNECE) and the United Nations Economic and Social Commission for Western Asia (UN/ESCWA) and UNESCO's International Hydrological Programme (IHP) on the 'Sustainable Management and Protection of Internationally Shared Groundwater Resources in the Mediterranean Region'.

The second phase of ISARM was started in 2003 in the American continent in cooperation with OAS (Organization of American States). The first UNESCO/OAS ISARM-Americas Workshop was held in Montevideo, Uruguay from 24 to 25 September, 2003. Participation at the workshop was strong: 20 countries were represented, including Haiti and the Dominican Republic. After a series of annual workshops, UNESCO/OAS published in 2007 an inventory and a preliminary assessment of transboundary aquifers in America [8], followed in 2008 by the legal and institutional framework of these aquifers [9].

The third phase of ISARM was launched in South Eastern Europe (SEE – the Balkans) and the Mediterranean in 2004 by UNESCO/ISARM and the UNESCO Chair/International Network of Water-Environment Centres in the Balkans (INWEB) at the Aristotle University of Thessaloniki. In close cooperation with the International Association of Hydrogeologists/Transboundary Aquifer Resource Management Commission (IAH/TARM), INWEB held a workshop in Thessaloniki in October

2004 to present and assess its results. INWEB also cooperated closely with UNECE: Working Group on Monitoring & Assessment, Switzerland, to follow up the European inventory previously compiled by UNECE [10], as well as with UN/ ESCWA, and the Observatoire du Sahara et du Sahel (OSS), for the Mediterranean inventory. The inventory of transboundary aquifer resources in the Balkans and the Mediterranean was updated in 2007 and 2008 and is available, together with a preliminary assessment, on INWEB's Web site (http://www.inweb.gr/) [11].

The case of the SEE region is very particular because of the high number of transboundary aquifers, mainly due to the multitude of new borders that were created after the collapse of the Former Republic of Yugoslavia in 1990. As shown in Figure 2.4, two main types of aquifers may be distinguished: (i) alluvial or sedimentary, which are located along major river beds and especially along the Danube River and (ii) karst aquifers.

Karst aquifers are mainly located in the western part of the peninsula (Dinaric karst) and along the Central Karpathes (Serbo-Carpathian karst). Almost half of the water from the mountainous western area of SEE disappears underground in karst formations and flows in the shortest direction to the Adriatic. Karstification is the geologic process of differential chemical and mechanical erosion by water on soluble bodies of rock, such as limestone, dolomite, gypsum or salt, at or near the earth's surface. Karstification is exhibited best on thick, fractured and pure limestone in a humid environment in which the subsurface and surface are modified simultaneously. The resulting karst morphology is usually characterized by dolines (sinkholes), hums (towers), caves and a complex subsurface drainage system. Karst

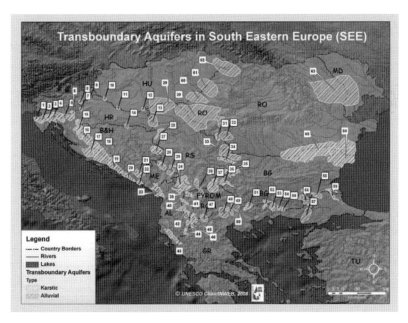

Figure 2.4   Transboundary aquifers in SEE (Balkan region) [11].

transboundary aquifers are very important for the region as the almost unique source for water supply in many cities and also for agricultural irrigation.

2.2
Assessment and Management of Transboundary Waters

Different scientific disciplines and professional groups have developed various methodologies and tools to assess, manage and share both transboundary surface and groundwater aquifer resources. In today's complex societies and global economy, transboundary water resources, as a topic and area of investigation, should be viewed from different scientific disciplines [12–14].

- **Hydrological sciences**: deal with water resources as part of the natural environment. Different aspects are studied mainly from a water quantity point of view: surface hydrology, hydrogeology.
- **Hydraulic engineering**: considers water resources management from a technical point of view. Structural or non-structural measures aim for the reliable satisfaction of water demand under specific water supply conditions, which may vary in time and space.
- **Environmental and ecological sciences**: focus on water quality and ecosystems.
- **Public health and toxicology**: address impacts on humans and ecosystems from various biological causes that may be found in water.
- **Economic sciences**: consider water to be an economic resource and recommend economic instruments, such as water pricing and economic incentives, to achieve a more competitive use of water in different sectors.
- **International law**: examines regulations, directives and international treaties for water sharing and water rights.
- **Social sciences**: focus on socio-political attitudes, public perception and relationships between individuals and the use of water.

Different conceptual models for TWRM have been suggested in the past. For example, there are models based on engineering approaches [12, 14], which underline the importance of assessing different hydrological uncertainties in order to assess and handle related risks [15]. A conceptual model showing the role of regional cooperative networks for developing sustainable monitoring and communication structures in TWRM was shown in Reference [13] for the Balkans.

2.2.1
Hydrological and Hydrogeological Approaches

The first important step towards assessing transboundary water resources is for riparian countries to share reliable data. Subsequent steps to be followed, including stakeholder consultation, data collection and suggesting effective TWRM plans are indicated in Figure 2.5 and specific issues in the hydrogeological approach to transboundary aquifers are illustrated in Figure 2.6.

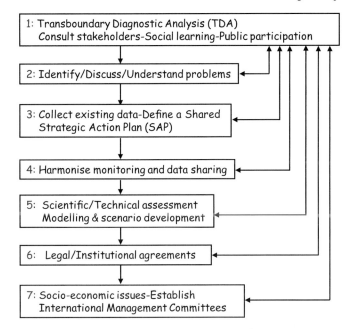

Figure 2.5 Recommended steps in hydrological and hydrogeological studies.

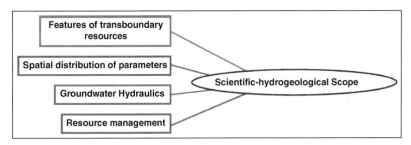

Figure 2.6 Issues in the hydrogeological approach of transboundary aquifers [7].

The framework document of the UNESCO-ISARM programme [7] recommends that when considering transboundary aquifers there should be good cooperation between various approaches, such as:

- **hydrogeology**: geophysical and geological prospecting, drilling techniques and mapping;
- **groundwater hydrodynamics**: quantitative aspects of flows, mathematical modelling, calibration and prediction scenarios;
- **groundwater management**: systems analysis, optimization techniques, risk analysis and multi-objective decision-making methods;
- **hydrochemistry**: chemical composition of the soil and water;
- **hydrobiology**: biological properties of groundwater systems.

Modern tools for groundwater development extensively use new information technologies, database development, computer software, mathematical modelling and remote sensing.

2.2.2
Environmental Issues

Preservation of water quality and ecosystem biodiversity should be an important objective for sustainability. Environmental protection should be realistically based on Environmental Risk Analysis (ERA) rather than on some precautionary principles that may not lead to any specific action [15]. ERA is a general and very useful approach for studying risks related to water overuse or water pollution in sensitive areas. The application of ERA consists of two main phases:

1) the assessment of risk,
2) risk management.

The assessment of risk is mainly based on modelling of the physical and social systems, including forecasting of its evolution under risk. The main objective of risk analysis is the management of the system; however, this is not possible if risk has not first been quantified.

The risk assessment phase involves the following steps [15]:

- **step 1**: identification of hazards and risks,
- **step 2**: assessment of loads and resistances,
- **step 3**: uncertainty analysis,
- **step 4**: risk quantification.

When it is possible to assess the risk under a given set of assumptions, then the process of risk management may begin. The various steps of the risk management phase are [15]:

- **step 1**: identification of alternatives and associated risks;
- **step 2**: assessment of costs in various risk levels;
- **step 3**: technical feasibility of alternative solutions;
- **step 4**: selection of acceptable options according to the public perception of risk, government policy and social factors;
- **step 5**: implementation of the best choice.

When applying ERA, different scenarios of socio-economic development, including possible climate change, should be taken into consideration. This is important in view of the natural and social vulnerability of transboundary water resources [15].

2.2.3
Legal Aspects

The main international laws on sustainable use and protection of transboundary water resources are (Chapter 3.3): the UN Watercourses Convention (1997) or the

Helsinki Rules, the UNECE (United Nations Economic Commission for Europe) Transboundary Waters Convention (1992), and in 2002 the UN International Law Commission articles on shared natural resources including transboundary aquifers (Chapter 3.4). Guidelines on monitoring and assessment of transboundary groundwaters have been issued already [10]; however, no international treaty yet exists for the use and protection of transboundary aquifers. The monitoring and assessment of surface waters are part of the 1999 Protocol on Water and Health to the UNECE Convention on the Protection and Use of Transboundary Watercourses and International Lakes. This Protocol contains provisions regarding the establishment of joint or coordinated systems for surveillance and early-warning systems to identify issues related to water pollution and public health, including extreme weather conditions. It also includes the development of integrated information systems and databases, the exchange of information and the sharing of technical and legal knowledge and experience.

The complexities of developing an international law on transboundary groundwaters have been described by many authors in the technical literature. Overpumping of groundwater in one country can endanger the future freshwater supplies of another country. In addition, groundwater overuse in one country can cause groundwater quality to deteriorate through salinity problems in another country, either by seawater intrusion or evaporation–deposition. The Bellagio Draft Treaty, developed in 1989, attempts to provide a legal framework for groundwater negotiations. The treaty describes principles based on mutual respect, good neighbourliness and reciprocity, which requires joint management of shared aquifers [16]. Although the draft is only a model treaty and not the result of accommodating actual state practice, it accepts that collecting groundwater data may be difficult and expensive and should rely on cooperation; and also provides a general framework for groundwater negotiations.

Only three bilateral agreements are known to deal with groundwater supply [the 1910 convention between Great Britain and the Sultan of Abdali, the 1994 Jordan–Israel peace treaty and the Palestinian–Israeli accords (Oslo II)]. In addition, the 1977 Geneva Aquifer Convention is also an important reference for internationalization of shared aquifer management and regulation by intra-State authorities for transboundary cooperation. Treaties that focus on pollution usually mention groundwater, but do not quantitatively address the issue. In 2008 the fifth report on shared groundwater resources was presented to the United Nations International Law Commission by the Special Rapporteur, who proposes a set of nineteen draft articles on the law of transboundary aquifers with commentaries (Chapter 3.4). As surface and groundwaters are interconnected, it is important that measures to protect ecosystems and surface water resources should also include the monitoring, assessment and protection of transboundary groundwaters.

2.2.4
Socio-economic Issues

It is widely accepted today that use of water resources, protection of the environment and economic development are not separate challenges. Development cannot be achieved when water and environmental resources are deteriorating, and similarly the environment cannot be protected and enhanced when growth plans consistently fail to consider the costs of environmental destruction. Nowadays it is clear that most environmental problems arise as 'negative externalities' of an economic system that takes for granted – and thus undervalues – many aspects of the environment. The integration of environmental and economic issues is a key requirement in the concept of sustainability, not only for the protection of the environment, but also for the promotion of sustainable long-term economic development, especially in water scarce areas (Chapter 7.1).

In the case of shared groundwater resources, the UNESCO/ISARM Framework Document [7] makes a preliminary overview of different socio-economic aspects of transboundary aquifer management. The main driving forces behind the over-exploitation of groundwater resources resulting in negative impacts are population growth, concentration of people in big cities and inefficient use of water for agricultural irrigation. The agricultural sector is most often mainly responsible for groundwater over-exploitation. The situation becomes particularly difficult when neighbouring countries share common transboundary water resources, as several differences arise in:

- socio-economic level;
- political, social, and institutional structures, including strict region-specific positions on national sovereignty;
- objectives, benefits and economic instruments;
- international relations, national legislation and regulation.

Competition for use of water resources for different purposes on one or both sides of the border may generate potential conflicts. Effective governance should consider specific hydrological and hydrogeological conditions, aquifer recharge rates and multiobjective use of renewable water resources involving stakeholders and multi-disciplinary regional working groups [13].

2.2.5
Institutional Considerations

For transboundary water courses and lakes, international commissions have proved to be the most effective institutional settings for transboundary surface water resources management. No such common institutions exist for transboundary groundwaters. Whether transboundary groundwater management should be a specific task of one or more specialized committees belonging to the same international river or lake committee, or whether a separate common institutional body should be created for this purpose, remains a question unanswered. In view of the physical interactions between surface and groundwaters, coordination between different specialized institutions is necessary for the overall sustainable management of water resources.

In the present situation national institutions dealing with groundwater are not sufficiently or effectively prepared to be able to undertake the joint management of transboundary groundwaters. Groundwater management units, when they exist, are often side-lined and invisible in surface-water dominated water administrations and groundwater is not explicitly addressed in national water legislations. Capacity building, especially on developing joint enabling capacity and consultation mechanisms at decision-maker level, including harmonization of domestic groundwater law, supported by common monitoring systems and sharing information and data, is essential. The role of regional partnerships between different decision makers, scientists from different disciplines, and other water stakeholders is also important for preventing conflicts and enhancing cooperation (Chapter 7.2) [13]. It is important to link and reconcile transboundary aquifer management with land management, and with regional political and social and economic regional cooperation and development policy.

2.3
The Integrated Water Resources Management (IWRM) Process

Effective management of transboundary water resources should be based on current best practices, which are grouped under the term Integrated Water Resources Management (IWRM). The term was first used in 1977 at the UN Conference in Mar del Plata and according to the Global Water Partnership (GWP) – an NGO based in Stockholm – IWRM is defined as [17]: 'a process which promotes the coordinated development and management of water, land and related resources to maximize the resultant economic and social welfare in an equitable manner without compromising the sustainability of vital ecosystems'.

If we were to analyse the different ways in which man uses water, such as for drinking, agricultural irrigation, hydropower production and industry, we would have to consider many activities and engineering structures that very often lead to conflicting functions. For example, industries producing large amounts of untreated wastewater may pollute groundwater in the surrounding aquifer, which in turn affects the quality of water pumped for drinking purposes. The increase of water pollution from industrial activities may also affect the quality of river water used for irrigation. When groundwater is over-pumped from a series of wells, the groundwater table is lowered and could affect agricultural production, as less water will be left for feeding crop roots. Lowering the water table in a coastal zone may also increase seawater intrusion and soil salinization, with negative impacts on agriculture and ecosystems.

Obviously, when actions are taken for different water uses, as can be seen in the examples above, there is a need for coordinating the related activities in various perspectives, such as between different:

- sectors of water uses (water supply, agriculture, industry, energy, recreation, etc.);
- types of natural resources (land, water and others);
- types of water resources (surface water, groundwater);
- locations in space (local, regional, national, international);

- variations in time (daily, monthly, seasonal, yearly, climate change);
- impacts (environmental, economic, social, etc.);
- scientific and professional disciplines (engineering, law, economy, ecology, etc.);
- water-related institutions (governmental, private, international, NGOs, etc.);
- decision-makers, water professionals, scientists and stakeholders.

In the past, traditional approaches for water resources management emphasized technical reliability versus the effective use of available economic resources in planning, construction and operation. Whilst still providing a reliable framework for water resources use, investment and maintenance costs were to be minimized. According to IWRM, apart from the above technical and economic criteria, at least two more additional general objectives should be considered, which are environmental security and social equity. In terms of an integrated approach, management issues should be considered at the basin scale and groundwater aquifers should be managed in relation to surface waters.

In view of the recent revival of the role of water resources for the sustainable development and protection of the environment, interest in analysing effective water management in the field has increased. IWRM has a multidisciplinary and interdisciplinary character involving many theoretical and applied fields of science. It is a traditional discipline in civil and agricultural engineering university curricula and in some countries it is considered as an independent engineering degree. Other disciplines involved in water quality and aquatic ecosystems are chemistry, biology and ecology. Law, economy, and also social and political sciences are important for implementing regulatory water policy, such as water allocation, water pricing and public participation. TWRM is characterized by the presence of a political boundary and in this view international law, socio-economic considerations and hydro-diplomacy also play an important role, mainly to promote cooperation between riparian countries and to prevent and alleviate potential water-related disputes.

As shown in Figure 2.7, IWRM could be achieved by coordinating different topics, areas, disciplines and institutions, which fall into two categories: natural issues (type of resources, space and time scales) and man-related (sectors, scientific disciplines, impacts, institutions, participants). There is no general rule about the optimum degree of integration and how to achieve it. Concerning the spatial scale that of the river basin is the most appropriate, taking into account the hydrological cycle and the water budget. The effect of possible climate change should also be taken into account, although large uncertainties still persist for quantifying such effects.

2.4
Capacity Building and Human Potential: The Role of Education

Transboundary water resources management has to face both the difficulties of traditional water resources management, such as correctly evaluating the resource, its variations with time, weather and climate, its quality and pollution and all its economic and financial dimensions, in particular the costs of infrastructures, the

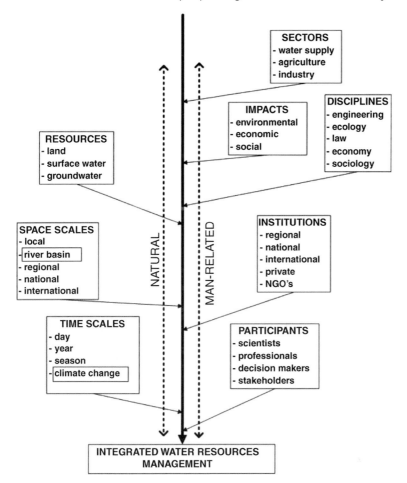

Figure 2.7 Areas and topics for IWRM.

risks of investments and many other management parameters, as well as specific issues on both sides of the boundary, such as different uses, different legal and administrative systems or even different scientific approaches to similar problems. Human potential is probably the key instrument of transboundary water resources management and specific capacity building can exploit this potential to its fullest to find solutions to management issues. Because of the specific difficulties related to groundwater compared to surface waters, a discussion focusing on transboundary groundwater resources management offers the most useful information towards understanding water resources management in general and highlights the need for education and training. Lessons on groundwater can then be usefully adapted to other kinds of water resources.

Although groundwater management has been included in higher education curricula in engineering and other related disciplines for decades, the subject of transboundary issues related to how this resource is managed across political boundaries has not emerged as an area of study to be covered in a comprehensive and global manner.

Practitioners and policy makers have not had the opportunity to benefit from specific education and training in the area of transboundary groundwater and its many complex factors. Such education and training would provide the added advantage of identifying best practices while also focusing on essential methodology and management. A comprehensive curriculum that extends from the scientific to the policy and legal aspects of managing transboundary groundwater is essential, not only for those who currently manage these resources or may do so in the future, but also for those who are studying or teaching in this field, as well as for those who are leaders in the community at local, national or international level.

It is necessary to conceive and design curricula that will teach participants how to have a very general overall view of transboundary groundwater problems, including issues such as differences in language, how to take part in negotiations, make political and managerial decisions, and create the institutions and instruments necessary for transboundary management, especially joint institutions and partnerships. This level of activity is known as the policy level. These curricula should be complemented by curricula aiming at training people who will work together in the field to implement the political, institutional and managerial decisions, on a scientific, legal and economic basis. This level of activity is known as the practitioners' level.

For the policy level, university graduate courses seem most suitable, while for the practitioner's level professional training courses, addressing already experienced specialists, seem to be the answer. As it can be very difficult and time-consuming to adapt existing university courses, and even more difficult to create new courses, UNESCO decided to stimulate the design and organization of short, intensive, professional training courses, held directly under its own auspices at institutions willing to host them. These training courses would deal with both the policy and the practitioners' levels.

For both levels, a multidisciplinary approach has to be adopted, involving hydro-geology, water management, economy, law, engineering and political science. This approach should emphasize communication between the different future actors of transboundary groundwater management, and facilitate the introduction of a common language or, at least, the use of concepts and technical words understood by all actors. This multidisciplinary approach should be supported by existing practical examples, provided by UNESCO-sponsored institutions such as ISARM, WHYMAP and IGRAC. These institutions identify the many internationally shared aquifers, and also those within a nation state that may have regional impacts, but are not necessarily managed under international law or treaties.

UNESCO decided that two pilot courses would be organized to test the possible formats of the training courses and these courses are briefly presented in Chapter 8.1. Although the results of the second course are still being analysed as at the time of writing, it already appears that multidisciplinarity is a key concept for the training of personnel able to manage transboundary groundwater, and that it should be understood as follows:

Multidisciplinary capacity building means the training of personnel willing and able to communicate and work as a team with specialists of disciplines different from their own. Examples such as those presented in the present book form useful support.

Chapter 8.1 examines how multidisciplinarity was achieved in the pilot courses presented above, attended by professionals from different backgrounds all sitting in the same class. Firstly general examples of transboundary groundwater cases were presented by a generalist, then experts from the various disciplines concerned in the case in question were invited to stimulate interest, comment and define the vocabulary of their particular discipline, clarify the concepts that were used to address the various issues involved, and open the minds of all participants to the idea of an integrated approach to the case. A subsequent general discussion allowed the participants to question the experts and test their understanding of the issues.

2.5
Conclusions

While the management of water resources is in itself a complicated process from the scientific, legal, social and economic points of view, requiring a good knowledge of the concerned disciplines, often to their very limits as use of the most up to date tools is made, it becomes even more complicated when the resource extends on both sides of a political boundary, either national within one country or international between two countries. The increased complexity is essentially related to the human factor, as different sociological parameters may exist on each side of the boundary, as well as different needs, uses, languages, even scientific tools, approaches and interpretations.

Management of these transboundary water resources cannot be neglected, as recent surveys, often performed by UNESCO and its various Programmes, have identified more than 400 international basins, which means that a vast quantity of world water is actually internationally shared, without even taking into account the national water resources shared by different administrative regions of the same country.

The management of such resources will involve not only the classical tools of water resources management, especially scientific and technical, but also the specific legal, economic, cultural, anthropological and socio-historical tools of transboundary water resources management, including communication instruments and methods for an efficient joint management. Among others, joint institutions, common monitoring networks, information and data sharing and a common vision for sustainable development of the entire river catchment can be listed. To summarize, both scientific and political dimensions play specific roles in transboundary water resources management, which requires innovative approaches especially in terms of regional cooperation and also in terms of ethics [18].

Especially in the case of internationally shared groundwater resources, a very particular management approach in legal and socio-economic terms is necessary, as well as specific education and training. This book addresses some of these issues.

For both surface and groundwater, regional partnerships and networks involving decision makers and stakeholders, as well as specialists of different scientific disciplines, are important driving forces behind the promotion of innovative approaches and the development of effective action plans. In this sense UNESCO's ISARM and PC-CP worldwide programmes have provided important tools, allowing methodological approaches to be designed, tested and demonstrated, and illustrating how correctly implemented multidisciplinarity has become a key to sustainable transboundary water management.

References

1 United Nations (2006) *Water: A Shared Responsibility: The United Nations World Water Development Report 2*, UNESCO Publishing, Paris; ISBN: 92-3-104006-5. http://www.unesco.org/water/wwap/wwdr/ (accessed 26 January 2011).

2 United Nations (2009) *Water in a Changing World: The United Nations World Water Development Report 2* UNESCO Publishing, Paris; ISBN: 978-9-23104-095-5. http://www.unesco.org/water/wwap/wwdr/ (accessed 26 January 2011).

3 Wolf, A., Natharias, J., Danielson, J., Ward, B. and Pender, J. (1999) International river basins of the world. *International Journal of Water Resources and Development*, **15**(4), 387–427.

4 Water Academy , France (2003) Proposal for a Strategic Guide to Form International Management Committees of shared Waters (in French). http://www.academie-eau.org/article.php?id=34 (accessed 26 January 2011).

5 German Federal Institute for Geosciences and Natural Resources (2007) Worldwide Hydrogeological Mapping and Assessment Programme (WHYMAP), Bundesanstalt fur Geowissenschaften und Rohstoffe, Hannover, Germany; http://www.whymap.org/whymap/EN/Home/whymap_node.html?_nnn=true (accessed 26 January 2011).

6 IGRAC (2009) Transboundary Aquifers of the World; http://www.igrac.net/publications/323# (accessed 26 January 2011).

7 UNESCO/ISARM (2001) A Framework Document, Paris, UNESCO, Non-Serial Documents in Hydrology.

8 UNESCO (2007) Preliminary Assessment: Transboundary Aquifer Systems in the Americas, Series ISARM Americas No 1 (in Spanish) http://www.isarm.net/publications/314 (accessed 26 January 2011).

9 UNESCO (2008) Legal and Institutional Framework in the Management of the Transboundary Aquifer Systems of the Americas, Series ISARM Americas No 2. http://www.isarm.net/publications/314 (accessed 26 January 2011).

10 UN/ECE (2000) *Guidelines on Monitoring and Assessment of Transboundary Groundwaters*, UN/ECE, Lelystad, UNECE Task Force on Monitoring and Assessment, under the Convention on the Protection and Use of Transboundary Watercourses and International Lakes (Helsinki 1992). ISBN 9036953154. http://www.unece.org/env/water/publications/documents/guidelinesgroundwater.pdf (accessed 24 March 2011).

11 INWEB (2008) Internationally Shared Aquifers in South Eastern Europe (Balkan Region): a Preliminary Assessment, UNESCO Chair and Network INWEB, Thessaloniki, Greece. http://www.inweb.gr/ (accessed 26 January 2011).

12 Ganoulis, J., Duckstein, L., Literathy, P. and Bogardi, I. (eds) (1996) *Transboundary Water Resources Management: Institutional and Engineering Approaches*, Vol. 7, NATO ASI Series, Partnership Sub-Series 2: Environment, Springer-Verlag, Heidelberg, 478 pp.

13 Ganoulis, J., Murphy, I.L. and Brilly, M. (eds) (2000) *Transboundary Water Resources*

in the Balkans: Initiating a Sustainable Co-operative Network, Vol. 47, NATO ASI Series, Partnership Sub-Series 2: Environmental Security, Kluwer Academic, Dordrecht, Boston, London, 254 pp.

14 Ganoulis, J. (2006) *Water Resources Management and Environmental Security in Mediterranean Transboundary River Basins. Environmental Security and Environmental Management: The Role of Risk Assessment* (eds B. Morel and I. Linkov), Springer, pp. 49–58.

15 Ganoulis, J. (2009) *Risk Analysis of Water Pollution*, 2nd edn, Wiley-VCH Verlag, Weinheim, 311pp.

16 Hayton, R. and Utton, A. (1989) Transboundary groundwaters: the Bellagio Draft Treaty. *Natural Resources Journal*, **29**, 663–722.

17 Global Water Partnership (GWP) (2000) Integrated Water Resources Management, Technical Committee Background Paper No 4, Stockholm, Sweden. www.gwpforum.org/gwp/library/Tacno4.pdf (accessed accessed 26 January 2011).

18 Fried, J. (2008) Water governance, management and ethics: new dimensions for an old problem. *Santa Clara Journal of International Law*, **6**, 1.

3
Global Challenges and the European Paradigm

3.1
Towards Integrated Management of Transboundary River Basins over the World

Jean-François Donzier

3.1.1
Introduction: Towards a Worldwide Water Crisis?

Climate change, floods, droughts, pollution, wastage, water-related diseases, food shortages and destruction of ecosystems all indicate the seriousness of the situation regarding water in many countries. It is essential that comprehensive, integrated and consistent management policies for water resources, aquatic ecosystems and the territories that form their supply area are implemented to prepare for the future and to meet the quickly increasing needs for water.

Global warming now seems to be unavoidable, and freshwater resources in particular will be directly and quickly affected, with the following consequences:

- increase in extreme hydrological phenomena, such as droughts and floods, with the risk of huge human losses, displacement of populations, destruction and catastrophic economic damage;
- reduction of the snow cover in mountains, which thus will not be able to play their part of 'water towers of the planet', by regulating the flows of the large rivers that originate there;
- modification of the vegetation species and soil cover, which will result in increased erosion;
- increase in the ocean water level modifying the flow of rivers at their coastal mouth and increasing the salinity of the aquifers.

At the level of large river basins, it is then necessary to develop or to increase the means for observing changes and for modelling their probable effects and to assess the resources available in the long term, so as to more effectively manage the reserves, wetlands and soil cover, to evaluate existing or planned hydraulic works, to control water demand and the various uses, and to protect agglomerations, collective infrastructures, areas of activity and arable lands against the damage caused by water.

The catchment areas of rivers, lakes and aquifers, which are a sole water system leading to the same mouth in the sea, are the relevant natural geographical territories to organize integrated freshwater resources management and to cope with the inevitable adaptation to the consequences of climate change. It is, indeed, on this geographical scale that there is physical, economic and political interdependence, which is expressed in resource sharing to meet the various uses or in the effects of anthropogenic pollution from upstream to downstream, and the effect of floods and droughts, hydropower production or continuity of waterways navigation.

True common management policies between upstream and downstream areas must be established in these basins to prevent natural or accidental risks, to fight against pollution, to optimize the uses of the available resources and their consequences on the economy and development, to protect ecosystems, to organize navigation, and thus avoid conflicts and to share the benefits of coordinated joint management.

3.1.2
Water has no Boundary

It is especially necessary to take into account the specific situation of the 276 rivers or lakes and several hundreds of aquifers (274) whose basins are shared by at least two riparian countries or sometimes many more (e.g. 18 in the case of the Danube). On a worldwide basis, 15% of countries depend on the water resources of other upstream countries for more than 50% of their water: some countries, such as Botswana, Bulgaria, Congo, Egypt, Gambia, Hungary, Iraq, Luxembourg, Mauritania, Niger, Paraguay, the Netherlands, Romania, Sudan and Syria exceed the threshold of $^2/_3$ of their water resources coming from outside of their own borders.

After Europe and Central Asia, Africa, in particular, is characterized by very large hydrological systems, which originate in wet tropical areas and flow into arid or semi-arid areas, carrying large volumes of water that have a direct effect on development. There are 59 transboundary basins in Africa, including 28 in West Africa, covering 80% of all the territory of the area. Except for Cape Verde and Madagascar, all the African States share at least one river with a neighbour, and there are also several tens of large transboundary aquifers. In Africa, transboundary water resources account for 80% of surface water. This results in a very strong sub-regional interdependence. Thus, Niger, Gambia, Botswana, Mauritania, Sudan, Chad and Egypt have a very significant share (exceeding 75%, and even up to 98%) of their resources coming from other countries. The Congo, Nile, Zambezi, Niger, Volta and Lake Chad basins are shared between six and ten countries. The Gambia, Senegal, Limpopo, Orange and Okavango rivers are each shared by three or four countries. This interdependence also exists for groundwater resources: the large aquifer of the Northern Sahara ('continental terminal' essentially formed by fossil groundwater) is shared by Algeria, Libya and Tunisia.

It is imperative that on a worldwide basis cooperation agreements are initiated or consolidated between the riparian countries of transboundary resources, to create an essential common strategy for the basin and to develop a common vision for the future.

3.1.3
Transboundary Cooperation should be Strengthened

Many cooperation agreements were signed by riparian countries several centuries ago, mainly on issues concerning the freedom of navigation or sometimes on the sharing of flows or the prevention of floods. Since the end of the nineteenth century, there have also been agreements on the building of hydropower dams. However, even today, there are still very few agreements, conventions or treaties on pollution control, environmental protection and integrated management of these shared basins.

The existing agreements are very disparate: some only organize quantitative water sharing in the dry season (this is the case, for example, of the Joint Commission created by India and Bangladesh for the Ganges), while others have already created water management and planning bodies on the scale of the concerned basin (the Mekong River Commission, and the Organization for the Development of the Senegal River).

At the multilateral level, awareness only advances rather slowly. Europe appears to be the furthest ahead: the UNECE Convention, known as the Helsinki Convention of 17 March 1992, lays down a framework for cooperation in this field in Europe and applies it in a positive way. The European Water Framework Directive, for its part, lays down an objective of good ecological status in the national or international river basin districts of the 27 current Member States and the countries applying for accession to the European Union, Norway and Switzerland: for the first time in history, 29 countries are committed to jointly managing all their freshwater resources on a basin scale.

Elsewhere efforts still need to be made. There are elements for the definition of the concept of Integrated Water Resources Management (IWRM), which aims to be applied to national and transboundary water resources, in Chapter 18, devoted to freshwaters, of Agenda 21, adopted in Rio in June 1992. However, the Convention adopted by the United Nations General Assembly on 21 May 1997 on the uses other than navigation of international rivers has not yet come into effect 13 years later and only 18 out of the necessary 36 countries have ratified it.

The G8 Heads of State and Government who gathered in Evian in 2003 retained the stakes of better governance of transboundary basins among the priorities of their actions in the field of water, firstly in Africa.

Various factors, indeed, have already emerged and could in the future worsen tensions between States of the same basin in Africa. These come from strong hydrological interdependence, problems linked to access to water, a general reduction in the availability of surface waters, a decrease in groundwater level, and an increase in the number of infrastructure projects such as for large dams, irrigation canals and transfers between basins. Nigeria, to take only one example, is concerned with the building of the Kandadji dam in Niger and the Tossave dam in Mali, which could reduce the flows of the Niger River and affect the huge energy and hydro-agricultural investments downstream of this basin.

The risk of conflicts between the riparian states of a river has recently become much more obvious with the increase in water demand in the area and the additional pressure of climate change. Before 2025, it is expected that African water consumption will increase five times compared to current levels. The dams planned in Africa will worsen the pressures on river ecosystems. This requires better coordination between riparian states and the creation of suitable mechanisms for conflict resolution and prevention for all the African transboundary basins, especially through a vision on sharing the benefits drawn from the increase in hydropower production. Since most of the African surface or ground water resources are located in transboundary basins, an approach at this level is indeed essential.

During the World Water Forum in Istanbul in March 2009, issues such as the 'international' status of transboundary waters, the methods for financing and implementing common infrastructures, the ratification of the United Nations Convention of 1997 and the management of transboundary aquifers saw divergent positions clashing, sometimes vehemently. This shows that it is still difficult to achieve real consensus, although most of the participants agreed on the benefit of the basin approach, either on a national or transboundary basis, to face the challenges of water resources management.

The ministerial declaration of the Forum supports 'the implementation of integrated water resources management (IWRM) at the level of river basins and groundwater systems, within each country, and, where appropriate, through international cooperation, to equitably meet economic, social and environmental demands and, inter alia, to address the impact of global change.' The ministers also declared that 'they resolve to develop, implement and further strengthen transnational, national or/and local plans and programmes to anticipate and address the possible impacts of global changes,... that they will strive to improve water related monitoring systems and ensure that useful information is made freely available to all concerned populations, including neighbouring countries.' Finally, they also declared 'that they will take, as appropriate, tangible and concrete steps to improve and promote cooperation on sustainable use and protection of transboundary water resources through coordinated actions of riparian states, in conformity with existing agreements and/or other relevant arrangements, taking into account the interests of all riparian countries concerned. They will work to strengthen existing institutions and develop new ones, as appropriate and if needed, and implement instruments for improved management of transboundary waters.'

Of course, some people will point out that these formulations are subject to interpretation and obviously all the problems will not be miraculously solved, as some positions still remain too different, but unmistakably basin management and transboundary cooperation was a success during the 5th World Water Forum in Istanbul.

3.1.4
Basin Management is Essential Everywhere in the World

It is thus now widely recognized that water resources management should be organized:

1) on the scale of local, national or transboundary basins of rivers, lakes and aquifers;
2) based on integrated information systems, allowing knowledge on resources and their uses, polluting pressures, ecosystems and their functioning, and the follow-up of their evolutions and risk assessment; these information systems will have to be used as an objective basis for dialogue, negotiation, decision-making and evaluation of undertaken actions, as well as coordination of financing from the various donors;
3) based on management plans or master plans that define the medium and long-term objectives to be achieved;
4) through the development of Programmes of Measures and successive multi-year priority investments;
5) with the mobilization of specific financial resources, based on the 'polluter-pays' principle and 'user-pays' systems;
6) with the participation in decision-making of the concerned governmental administrations and local authorities, the representatives of different catego-ries of users and associations for environmental protection or of public interest; notably, at present in most countries of the world the decision-making processes in water management suffer from a strong democratic deficit.

Legal and institutional frameworks should allow the application of these six principles. It seems especially necessary to support the creation of international commissions or similar organizations, such as basin authorities, and to reinforce those already existing. Such international commissions or authorities allow better dialogue, the exchange of useful information, the solving of potential conflicts and the sharing of the benefits of better joint management and the reinforcement of transboundary cooperation.

Depending on the needs, local situations and history, various formulas have been adopted to organize water management at the basin level and there is a great diversity in the mandates and selected options. One can quote:

- **Administrative International Commissions, with or without a permanent secretariat**, in which participants are mainly representatives of the 'ministries' concerned, aiming to coordinate various projects on the same river or aquifer, to exchange information or data, formalized or not, in particular on emergency situations, to define common rules (navigation, etc.), and whenever necessary to allocate the available resources between the countries, the categories of uses, especially in periods of crisis or when regulation structures do exist, and so on.

- **Arbitration Authorities** to which the interested 'parties' refer for decision-making on conflicts that arise; this is the case of the International Joint Commission (IJC) between the USA and Canada, for example.
- **Organizations or Basin Authorities in charge of contracting large structures or combined installations**; this is the case for navigation, flood control, water transfers, the building of reservoirs, in particular for irrigation, hydropower production, and so on. These organizations, often created as public or private 'companies', usually have the concession of community infrastructures for which they are responsible for their construction and long-term management, generally by providing services, electricity and raw water or by levying specific taxes on waterways transport in particular.
- **Basin Committees or Councils**, or specific working groups which gather, together with the administrations, representatives of the local authorities, economic sectors, water users, civil society, and so on, who can be advisers or decision-makers, in particular as regards planning, the fixing of taxes, the allocation of available resources, and so on.
- **Projects**, which are usually temporary and initiated by a bi or multilateral donor for specifically implementing an action plan with specific financing.

It is especially recommended that support is given to the creation of tools necessary for coordination between bordering countries, such as:

- water monitoring, information and observation systems;
- prevention and fight against floods and droughts, through better information exchange and harmonization of action plans between the upstream and downstream parts of the basins;
- systems for warning against floods, droughts and pollution and prevention and action mechanisms for facing natural disasters caused by water and for protecting human lives and properties;
- practices of long-term planning and programming of priority investments;
- adapted financing mechanisms;
- adequate measures to prevent the introduction and dissemination of invasive aquatic species, which cause considerable ecological and economic damage and of which new specimens are continuously found;
- methods and means for consultation and mobilization of the populations concerned.

In particular, the signing of transboundary agreements for aquifer management should be sought, taking into account their fragile nature, especially fossil groundwater, and the time needed for restoring degraded situations, from a quantitative and qualitative viewpoint. It is also recommended that practical experience sharing and the comparison of approaches and methods between the managers and technicians concerned be promoted, in particular by supporting the action of specialized cooperation networks in this field. This is why the International Network of Basin Organizations (INBO), whose secretariat is ensured by the International Office for Water in Paris, has undertaken, for the past 15 years, to strengthen and develop effective basin organizations over the world, for transboundary rivers, lakes and aquifers.

3.1.5
The European Union is a Pioneer

Europe is the continent where there is the greatest number of transboundary basins shared between at least two countries. The European Water Framework Directive of October 2000 strengthens transboundary basin management by introducing the concept of 'International Basin Districts'. Riparian states have the same obligations towards these international basins as they do for their own national basins. Existing international commissions will be reinforced, and new ones will be created. In Europe, most of these international commissions have a similar organization, which is based firstly on the plenary assembly of the international commission itself, made up of official representatives of the States, which makes the decisions committing the Member States and its permanent secretariat, and secondly on many official geographical, sectoral or technical Working Groups, which are the places of associations between economic partners, local authorities and the civil society of the basin and where the decisions are prepared, the plans and programmes developed or the common tools designed for observation, monitoring and warning in particular.

The Framework Directive requires each of these International River Basin Districts to draft a master plan of the basin data and an overall Management Plan and Programme of Measures to achieve the objectives of good ecological status for water and the ecosystems of the basin, before 2015, 2021 and 2027. These 'Management Plans', defining the objectives to be reached, and these 'Programmes of Measures', defining the actions necessary, must be elaborated and operational by 2012 at the latest. The results obtained will be the subject of an evaluation that will be made public, and the European Commission will be able to call the failing Member States to appear before the European Court of Justice for non-respect of the Directive and possibly to make them subject to heavy financial obligations.

The current European effort shows that a suitable and constraining integrated management of the resources of basins of rivers, lakes or aquifers shared between several bordering countries is considered both necessary and feasible. The first results obtained are positive and encouraging.

3.1.6
Implementing Integrated Transboundary Water Resources Management Requires Political Will and Long-Term Commitments

Water is a possible cause of tensions but also, and more importantly, a powerful incentive for cooperation. The management of these essential and shared resources is crucial for poverty alleviation strategies, economic and social development and environmental conservation worldwide.

Transboundary cooperation is essential to effectively face the effects of climate change on the freshwater cycle. The creation of transboundary basin organizations is successful in many cases, but many institutions of transboundary basins have not yet sufficient power, capacities or resources. Worse still, there is no inter-state institution to manage water in most transboundary river basins and the joint management of shared aquifers has hardly started to be considered. There is thus still an enormous need for reinforcing governance in this field.

In particular, it would be appropriate if donors made it a condition of their support of large projects for hydropower, navigation, irrigation or any other project implying the abstraction or diversion of significant water flows in transboundary rivers, lakes or aquifers, for prior agreements on a vision for a common future between all the riparian countries of the same basin to be made, which could be then formalized in a Basin Management Plan. An assessment of the long-term effects of these projects and its feasibility under different development scenarios should also be made. The creation of coordinated observation systems allowing the exchange of information between riparian countries, not only on the flows but also on pollution, abstractions, quality of the ecosystems and more generally on the economy of the water sector in the basin concerned, should also be a prerequisite for the implementation of any large project, since information transparency is the key to trust. Finally, the existence of mechanisms for involving the concerned populations in decision-making is a real need.

Remarkable progress has been made since the 1990s, thanks to the reforms undertaken in many areas and countries worldwide. The experience gained illustrates that integrated water resources management in river and aquifer basins has real advantages. This experience should guide countries wishing to implement effective water resources management and to strengthen international cooperation.

Further Reading

International Network of Basin Organizations (November 2009) INBO Newsletter (no 18), http://www.inbo-news.org/IMG/pdf/inbo18.pdf (accessed 12 January 2011).

International Network of Basin Organizations (December 2008) INBO Newsletter (no 17), http://www.inbo-news.org/IMG/pdf/inbo17.pdf (accessed 12 January 2011).

International Network of Basin Organizations (January 2008) INBO Newsletter (no 16),

http://www.inbo-news.org/IMG/pdf/inbo16.pdf (accessed 12 January 2011).

International Network of Basin Organizations (January 2007) INBO Newsletter (no 15), http://www.inbo-news.org/IMG/pdf/inbo15.pdf (accessed 12 January 2011).

International Network of Basin Organizations (December 2005 – January 2006) INBO Newsletter (no 14), http://www.inbo-news.org/IMG/pdf/inbo14.pdf (accessed 12 January 2011).

3.2

Antarctic Subglacial Lakes and Waters: The Challenge to Protect a Hidden Resource

Lilian Del Castillo-Laborde

3.2.1

Introduction

The Antarctic continent is governed by the Antarctic Treaty of 1959 and other conventions concerning living resources in the Antarctic, and the Protocol on Environmental Protection of 1991 (in force since 1998). From this last instrument stems the Committee for Environmental Protection (CEP), which has dealt with subglacial lakes and waters since it was set up in 1998, paying special attention to the environmental protection of the resource, while the Operational Matters Working Group, established by the Antarctic Treaty Consultative Meetings (ATCM), has focused on the scientific data of the subject.

3.2.2

Antarctic Environmental Regime

The 1991 Protocol on Environmental Protection establishes the general guidelines for the protection of the Antarctic Environment, while Annex V of the Protocol deals with area protection and management, and is in charge of the establishment of Antarctic Specially Protected Areas and Antarctic Specially Managed Areas (ASMAs) such as McMurdo Dry Valleys ASMA. Special standards for the protection of the environment of subglacial waters have not been established, although the microbial organisms living in this environment could be significantly affected by the intrusion of devices for exploration of subglacial lakes, which includes drilling and sampling. The adoption of such specific guidelines and the establishment of ASMAs seem highly necessary.

Under the 1959 Antarctic Treaty and Protocols, the 28 Consultative Parties to the Antarctic Treaty are able to participate in the decision-making process, and although there are seven State parties that claim territorial sectors of Antarctica and consider those sectors as national territories, they can neither exercise exclusive rights over the resources nor exclude other member States' activities in those sectors.

The 1991 Antarctic Protocol on Environmental Protection has banned the exploitation of Antarctic mineral resources for 50 years. However, whether subglacial water is a mineral resource remains uncertain, since underground water is not considered as such. The Protocol establishes that in the event of a dispute arising regarding the interpretation of its provisions, there is the possibility to submit the issue to an arbitral tribunal or to the International Court of Justice. Taking into account that the Continent has a specific legal framework, there is an exclusive competence of Antarctic Treaty Parties and bodies to deal with the regime of renewable and non-renewable Antarctic resources below 60°S, a special competence entailing a special responsibility.

Research activities have been to a large extent at the core of Antarctic explorations and initiatives. In 1958 the International Council for Science (ICSU), a non-governmental scientific organization, established the Scientific Committee on Antarctic Research (SCAR) to promote and coordinate scientific research in

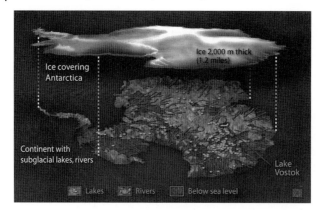

Figure 3.2.1 Antarctic subglacial hydrology. (Courtesy of Professor Priscu)

Antarctica and, once the Antarctic Treaty came in force in 1961, to provide scientific advice to the Antarctic Treaty member States. SCAR organized specific programmes for the assessment of subglacial lakes and basins and convened a Group of Specialists on Subglacial Antarctic Lake Exploration (SALEGOS).

3.2.3
Subglacial Lakes and Waters (Figure 3.2.1)

The basic standards for the exploration of subglacial Antarctic lakes were agreed upon by the *ad hoc* group of experts, and were included in its 2001 Report [1]. Taking into account the extreme vulnerability of the resource, these activities should be developed in compliance with the Environmental Impact Assessment (EIA) requirement established in Annex I of the Protocol on Environmental Protection.

The physical and biological characteristics of these waters are as yet not known, and the protection of such a particular environment is the responsibility of the Antarctic Treaty system member States. In 2004 the SCAR launched a programme for subglacial lakes research, SALE_UNITED [2]. Within the Treaty system, member States are entitled to carry out research and scientific activities in the Antarctic continent and in the surrounding Southern Ocean, and they are also accountable for the protection of the Continent's environment.

3.2.4
The Vostok Subglacial Lake

One of the most important subglacial lakes is the Vostok Lake, which stretches over an area approximately 250 km long and 50 km wide, lying beneath the Russian Vostok research station (Figure 3.2.2). Between 1989 and 1998 Russian scientists carried out explorations including deep drilling of the ice sheet, under the Russian Antarctic Programme. These explorations, though, are carried out in a legal vacuum regarding the proper devices and prevention measures for the protection of the untouched environment lying beneath the Antarctic ice cover. The ice core from the Vostok

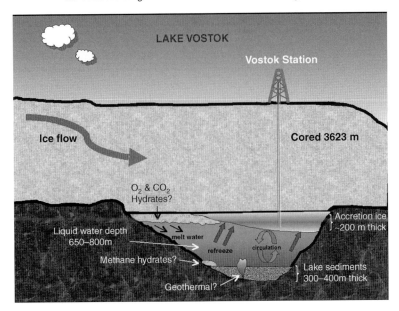

Figure 3.2.2 Lake Vostok.

borehole at 3623–3650 m is still structured as large (up to 1.5 m diameter) crystals with insignificant intercrystal spaces, which largely prevents rapid leakage of the drill liquid towards the water body.

In 1996, the XX(a) Antarctic Treaty Consultative Meeting (ATCM) 'urged Russia to take the necessary steps to ensure that the planned ice coring is stopped at a safe distance above the reported lake, so that there is no risk of polluting it' (para. 108). The Subglacial Antarctic Lake Exploration Group of Specialists (SALEGOS), a special working group set up by the Scientific Committee on Antarctic Research (SCAR), expressed the interest of the scientific community in Lake Vostok and in the activities that Russia had under implementation. In its Information Paper (XXV ATCM/IP55), 2002, the SCAR stated that 'there is as yet no international consensus among the scientific community on appropriate lake sampling or on drilling methods to penetrate into the lake' and 'recommended that additional studies should be carried out before further drilling towards Lake Vostok is undertaken using the existing Russian drill hole'. In 2003 Russia circulated its revised draft Comprehensive Environmental Evaluation (CEE) for Water Sampling of the subglacial Lake Vostok (ATCM XXVI/WP01) which was reviewed by the CEP. In its comments, the CEP VI meeting (2003) Final Report expressed that the draft CEE was incomplete, that is, the possibility of spilling of drilling fluid (60 m^3 of kerosene) was not envisaged.

At the 2006 meeting of the Committee on Environmental Protection (CEP) the SCAR 'noted that it was aware of recent scientific literature which suggested that if one subglacial lake was contaminated, contamination may spread downstream to connected lakes' as a result of the possible interconnection between the different Antarctic lakes. At the Kyiv meeting in June 2008, Russia recalled several incidents that occurred in borehole 5G-1 during 2007 that delayed progress with further drilling of the ice core and penetration of the subglacial Lake Vostok. However, the Antarctic advisory scientific bodies, or the other consultative members of the Treaty, cannot prevent the continuation of the drilling even if the proposed conditions are not met.

Not only Russia on Lake Vostok but also other countries have projects for the investigation of subglacial lakes and waters. In 2009 the UK started a five year long project on the exploration of subglacial Lake Ellsworth with a team of nine universities and the British Antarctic Survey (www.nerc.ac.uk, accessed 5 October 2010).

3.2.5
Conclusion

Subglacial Antarctic freshwaters are a large and important water reserve, mostly unknown, which deserve the attention of the international community at large and of the specialized water experts of the world.

At present, States and the scientific community are only at the stage of discussing the development of research programmes. In case benefits could be derived from these subglacial resources, their equitable apportionment should be agreed upon by means of special agreements according to the overall rules of the 1959 Antarctic Treaty, the Antarctic Treaty Consultative Meetings (ATCM) Recommendations, and the Protocol on Environmental Protection and its Annexes.

Acknowledgements

This chapter was prepared with the bibliography and documentation filed with the Antarctic Treaty Secretariat, Buenos Aires, Argentina, for which appreciation is expressed.

References

1 Report of the Subglacial Antarctic Lake Exploration Group of Specialists (SALEGOS) Meeting – I, Bologna, Italy, 29–30 November 2001. Available at http://salegos-scar.montana.edu/ (accessed 17 May 2011). Zotikov, I.A. (2006) *The Antarctic Subglacial Lake Vostok: Glaciology, Biology and Planetology*, Springer Praxis Books, p. 10.

2 Subglacial Antarctic Lake Environments: Unified International Team for Exploration and Discovery (SALE-UNITED) Programme. Available at http://www.sale.scar.org/ (accessed 5 November 2010). Priscu, J.C. *et al.* (2003) An international plan for Antarctic subglacial lake exploration. *Polar Geography*, **27** (1), 69–83.

Further Reading

Committee on Principles of Environmental Stewardship for the Exploration and Study of Subglacial Environments, National Research Council (2007) *Exploration of Antarctic Subglacial Aquatic Environments: Environmental and Scientific Stewardship*, The National Academies Press, Washington D.C., 162 pp. ISBN: 0-309-10635-4.

Siegert, M.J., Carter, S.P., Tabacco, I., Popov, S. and Blankenship, D.D. (2005) A revised inventory of subglacial lakes. *Antarctic Science*, **17** (3), 453–460.
Reports of the Committee for Environmental Protection – CEP – I a XI.

3.3
Progressive Development of International Groundwater Law:
Awareness and Cooperation

Raya Marina Stephan

3.3.1
Introduction

Groundwater represents between 98 and 99% of our planet's reserves in freshwater [1]. Like surface water, groundwater knows no boundaries. The number of transboundary aquifers is not well known. To date, in the frame of the ISARM project of UNESCO IHP, 270 transboundary aquifers have been identified. Despite their importance, transboundary aquifers have until recently received limited consideration in international law, and have been considered as subsidiary to surface water. This is the case in the only global convention on freshwater resources, the UN Convention on the law of the non-navigational uses of international watercourses (1997) (UN Watercourse Convention) [2]. The Convention applies to the 'uses of international watercourses and of their waters for purposes other than navigation and to measures of protection, preservation and management' (article 1§1). In its article 2§a on the use of terms, the Watercourse Convention defines a watercourse as: 'a system of surface waters and groundwaters constituting by virtue of their physical relationship a unitary whole and normally flowing into a common terminus'. It appears from this definition that groundwaters fall within the scope of the Convention only when they are related to a surface water system, *and* they flow to the same terminus. In reality this condition is rarely satisfied, and therefore the Convention excludes a great number of transboundary aquifers. And even if a transboundary aquifer satisfies the conditions of the Convention, its rules were tailored for surface waters; the topic of groundwater was introduced at a late stage in the debates at the International Law Commission. Groundwater is more vulnerable to pollution and depletion than surface water; it needs more protective obligations and principles.

The large majority of interstate treaties and agreements concluded on transboundary waters are concerned with international rivers and rarely address transboundary aquifers. Transboundary groundwaters are addressed only when they interact with the surface water body concerned. Below are a few examples of treaties and agreements addressing a surface water body that give some consideration to the related groundwaters:

- Convention on the Sustainable Development of Lake Tanganyka, 12 June 2003, (Burundi, Congo, Tanzania, Zambia).
- The Protocol for Sustainable Development of Lake Victoria basin, 29 November 2003, (Kenya, Tanzania, Uganda).
- Convention and Statutes relating to the development of the Chad basin, 22 May 1964 (Cameroon, Chad, Niger, Nigeria).

- Agreement concerning the equitable sharing in the development, conservation and use of the common water resources, 18 July 1990 (Nigeria, Niger).
- Convention on Cooperation for the Protection and Sustainable Use of the River Danube, 29 June 1994.
- Convention on the Protection of the Rhine, 12 April 1999 (Germany, European Community, France, Luxemburg, Netherlands, Switzerland).

The result of this dominating practice is the exclusion of a great number of transboundary aquifers from interstate agreements, and the absence of adequate and adapted provisions to cover the specific characteristics of aquifers. A few exceptions exist for transboundary aquifers. One major and remarkable exception is the Convention on the protection, utilization, recharge and monitoring of the Franco-Swiss Genevese aquifer. This Convention replaced the previous agreement of 1977, which lasted 30 years. The case of the Genevese aquifer is the only example of a treaty dealing exclusively with the management of a transboundary aquifer. There are two other examples to be mentioned here:

- **The case of the Nubian Sandstone Aquifer System (Chad, Egypt, Libya and Sudan)**: In 1992, Egypt and Libya signed an agreement on the 'Constitution of the Joint Authority for the study and development of the Nubian Sandstone Aquifer Waters'. They were joined later on by Chad and Sudan. The Authority is responsible *inter alia* for collecting and updating data and conducting studies.
- **The case of the North Western Sahara Aquifer System (Algeria, Libya and Tunisia)**: In 2008 the three countries established among themselves a permanent consultation mechanism. The main role of the coordination mechanism is to enforce and update the common database by exchanging data and information, and developing and managing common monitoring networks of the aquifer system.

However, this trend is changing with recent development at the global level, and with increased cooperation and awareness.

3.3.2
Development, Evolution and Cooperation at the Global Level

In its programme of work the UN International Law Commission (ILC) included in 2002 the topic of 'Shared Natural Resources', which covers transboundary groundwaters, oil and gas. Six years later, the UN General Assembly adopted Resolution 63/124 on the law of transboundary aquifers including as an annex the draft articles prepared by the ILC.

3.3.2.1 The Process at the UN ILC
The Special Rapporteur submitted two reports, the first in 2003 ('First report on outlines') and the second in 2004 ('Second report on shared natural resources:

transboundary groundwaters')[1]. In 2006, after completing the work on the third report, the ILC adopted the draft articles at first reading. The draft articles were then transmitted to Governments for comments and observations by 1 January 2008. At the meetings of the sixth Committee of the UN General Assembly (Legal Committee), the delegates, representatives of their governments, have the opportunity to comment on the work of the ILC. In 2006 and 2007, 45 oral comments were presented on the draft articles on the law of transboundary aquifers. Nineteen written comments were received. All these comments were generally favourable. In fact, the delegates at the sixth Committee have the opportunity to comment every year on the work of the ILC. Between 2003 and 2007, the reactions expressed at the sixth Committee showed a global support of the project: the delegates generally acknowledged the importance of groundwater and transboundary aquifers, and showed their awareness of the need for specific rules for transboundary aquifers. They also gave general support to the approach adopted by the Special Rapporteur in the reports and to the scientific consultation with groundwater experts, which was organized by UNESCOs International Hydrological Programme upon the request of the ILC.

After reviewing all the comments received, the Special Rapporteur submitted in 2008 his fifth report at the ILC including a revised version of the draft articles. At second reading the ILC subsequently adopted a preamble and a set of 19 0draft articles on the law of transboundary aquifers with commentaries thereto [3]. This marks the end of the process at the ILC. The draft articles were then transmitted to the UN General Assembly with the following recommendations:

- to take note of the draft articles on the law of transboundary aquifers in a resolution and to annex the articles to the resolution;
- to recommend to States concerned to make appropriate bilateral or regional arrangements for the proper management of their transboundary aquifers on the basis of the principles enunciated in the draft articles;
- to consider at a later stage the elaboration of a convention on the basis of the draft articles.

On 11 December 2008, the UN General Assembly adopted Resolution 63/124 on the law of transboundary aquifers including in its annex the draft articles prepared by the UN ILC.

3.3.2.2 Cooperation of Two UN Bodies: The UN ILC and UNESCO-IHP

Since the beginning of its work on transboundary aquifers, the UN ILC has sought scientific cooperation from UNESCOs International Hydrological Programme (IHP).

---

1) The reports of the special Rapportear are available at www.un.org/law/ilc

The UN International Law Commission was established by the General Assembly in 1947. Its role is to promote the progressive development of international law and its codification. It is composed of 34 members elected by the General Assembly. The members serve in their individual capacities as persons of recognized competence in international law, and not as representatives of their respective States.

The International Hydrological Programme of UNESCO is the only global intergovernmental scientific programme on water resources in the UN system. The programme was created in 1975. Its member States define its needs and the plans of its phases with growing emphasis on management and social aspects.

During the preparation phase of the draft articles on the law of transboundary aquifers, UNESCO IHP provided scientific and technical advice on the issues related to hydrogeology to the Special Rapporteur and to the members of the ILC. It invited, coordinated and supported contributions from international experts, international and national institutions, and centres on groundwater resources, such as the International Association of Hydrogeologists (IAH), Food and Agriculture Organization (FAO), United Nations Economic Commission for Europe (UN ECE), International Groundwaters Assessment Centre (IGRAC) and so on.

The role played by UNESCO IHP has always been acknowledged by the ILC in its annual reports, and it was also acknowledged by the UN General Assembly in the Resolution 63/124.

3.3.3
Cooperation and Awareness

Some existing legal instruments had already given some consideration to transboundary aquifers. However, the legal aspects of the management of transboundary aquifers are gaining in importance.

3.3.3.1 Existing Legal Frameworks

The UN ECE Convention on the protection and Use of Transboundary Watercourses and International Lakes (1992) [4] applies to the 37 States that have ratified it. The Convention applies to all transboundary waters, which are defined as 'any surface or ground waters which mark, cross or are located on boundaries between two or more States' (article 1§1). The Convention is guided by the equitable and reasonable use principle, the precautionary principle and sustainable development. It includes provisions related to procedural rules such as developing monitoring.

Under the Convention, an assessment of transboundary aquifers in Europe and in Central Asia was undertaken (Figures 3.3.1 and 3.3.2).

With the objective of assisting the Parties and joint bodies in developing harmonized rules for the setting up and operation of systems for transboundary groundwater management, 'Guidelines on monitoring and assessment of transboundary groundwaters' were published [5]. The Guidelines include indications on:

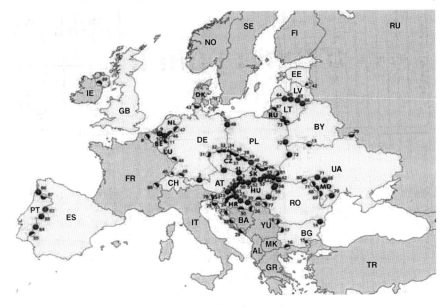

Figure 3.3.1 UNECE survey of European transboundary aquifers.

- monitoring programmes
- data management
- quality management
- joint action and institutional arrangements.

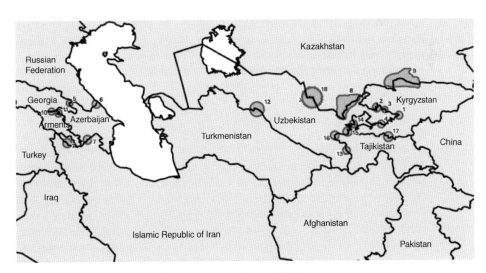

Figure 3.3.2 Transboundary aquifers in Caucasia and Central Asia.

The EU Water Framework Directive (WFD) [6] includes provisions on transboundary waters among EU Member States, and between a Member State and a non-Member State. The WFD requires cooperation on such waters (article 3§3). Member States sharing groundwater bodies have to coordinate their activities in respect of monitoring, setting of threshold values and identifying relevant hazardous substances. Coordination for establishing groundwater threshold values is made even more specific in Article 3 (criteria for assessing good chemical status), for both transboundary groundwater bodies within the Community and bodies shared with non-Member States.

3.3.3.2 Legal Component in Projects

Various projects on transboundary aquifers are emerging, which include a legal component aimed at establishing mechanisms for cooperation and eventual joint management.

In the project on the Iullemeden aquifer system (IAS) (Niger, Nigeria and Mali), Managing Hydrogeological Risk in the Iullemeden Aquifer System (2004–2008), a joint legal and institutional cooperative framework was prepared. However, it has still not been adopted by the riparian States.

One of the objectives of the project on the Guarani aquifer system (Argentina, Brazil, Paraguay and Uruguay) (2003–2009) is to support the four countries in jointly elaborating and implementing a common institutional and technical framework for managing and preserving the aquifer system. While the project did not come out with any legal and institutional framework for the Guarani aquifer system, the riparian countries signed on 2 August 2010 the Guarani Aquifer Agreement. The agreement refers to Resolution 63/124 and is on the equitable and reasonable use principle, the no harm rule, the exchange of data and other principles of international law. The project has been a catalyst for cooperation on the regional, national and local levels.

Another project that could also be mentioned is the project on the Dinaric karst aquifer system (Albania, Bosnia-Herzegovina, Croatia and Montenegro) that has recently been initiated. The project includes a component on 'establishing cooperation among countries sharing the aquifer', and one of its outcomes is the establishment of a multi-country consultative body.

3.3.4
Conclusion

The development of international law is a slow process. In the case of transboundary aquifers, the process has started, supported by a codification activity that took place at the UN ILC and which resulted in the adoption of the UN General Assembly (GA) Resolution on the law of transboundary aquifers. The UN GA is certainly a non-binding text; however, it can provide guidance for States on their transboundary aquifers. This process is also supported by an emerging cooperation on transboundary aquifers among riparian States.

References

1 Margat, J. (2008) *Les Eaux Souterraines Dans Le Monde*, BRGM Éditions, Orleans, ©UNESCO/BRGM.

2 United Nations (1997) Convention on the Law of the Non-navigational uses of International Watercourses, www.un.org/law/cod/watere.htm (accessed 01 November 2010).

3 United Nations International Law Commission (2008) Report on the work of its sixtieth session (5 May to 6 June and 7 July to 8 August 2008), General Assembly, Official Records, Sixty-third Session, Supplement No. 10 (A/63/10), http://untreaty.un.org/ilc/reports/2008/2008report.htm (accessed 01 November 2010).

4 UNECE (1992) UN ECE Convention on the protection and Use of Transboundary Watercourses and International Lakes, (http://www.unece.org/env/water/text/text.htm (accessed 01 November 2010).

5 UN/ECE Task Force on Monitoring & Assessment (2000) Guidelines on Monitoring and Assessment of Transboundary Groundwater, http://www.unece.org/env/water/publications/documents/guidelinesgroundwater.pdf (accessed 01 November 2010).

6 EU (2000) Directive 2000/60/EC of the European Parliament and of the Council of 23 October 2000 establishing a framework for Community action in the field of water policy. http://eur-lex.europa.eu/LexUriServ/LexUriServ.do?uri=CELEX:32000L0060:EN:NOT (accessed 01 November 2010).

3.4

The Role of Key International Water Treaties in the Implementation of the Convention on Biological Diversity

Sabine Brels, David Coates and Flavia Rocha Loures

3.4.1
Introduction

The Convention on Biological Diversity (CBD) was adopted in 1992 as a global policy framework aiming to promote the conservation and sustainable use of Earth's biological resources. When ecosystems stretch across two or more countries, interstate cooperation becomes essential for the long-term sustainability of such ecosystems and dependent human communities.

Recognizing this, CBD Decision VII/4, on inland waters biodiversity, underscores the need for parties to address transboundary water issues, especially with regard to the allocation and management of shared freshwater resources.

Furthermore, that same decision reiterates a call, made three years earlier at the 8th Conference of the Parties to the CBD, in Decision VIII/27, for parties to join and implement the 1997 United Nations Convention on the Law of Non-Navigational

Uses of International Watercourses [1], as well as related regional and watercourse agreements, as a means to strengthen the governance of transboundary inland water ecosystems.

Both CBD decisions add considerable legal and political weight to ongoing efforts to improve regulatory frameworks for transboundary water cooperation. The decisions in question also broaden the arguments for such cooperation by highlighting the linkages between transboundary freshwater management, biodiversity conservation and sustainable use, and human well-being.

Within this context, this chapter draws from an earlier study [2] containing an in-depth analysis of the role and relevance of the UN Watercourses Convention in supplementing and strengthening the regulatory framework under the CBD with respect to water allocation and management, as well as other transboundary water issues. This earlier paper also considered the relationship between the CBD and the Convention on the Protection and Use of Transboundary Watercourses and International Lakes [3].

Here, we summarize and highlight our main points and conclusions.

The UN Watercourses Convention was adopted at the UN General Assembly in 1997 by an overwhelming majority. It is the only global legal instrument codifying, clarifying and developing minimum substantive and procedural standards for transboundary water cooperation. Counting now 24 contracting states – 11 short of the required for entry into force, the convention is open for accession by all states and regional economic organizations [4].

The UNECE Water Convention is a regional instrument adopted in 1992 among the states that are members of the UN Economic Commission for Europe (UNECE). In force since 1996, the convention creates a consistent, stringent and detailed legal framework for transboundary waters, and has fostered cooperation and environmental sustainability within Europe and neighbouring regions. As it stands today, the convention is only open to accession by UNECE member states, and counts 37 state parties, plus the European Community (EC). In 2003, its Parties adopted amendments opening it for accession by states beyond the UNECE region. Currently, 16 parties have accepted those amendments out of the 23 that are necessary for entry into force. However, the Meeting of the Parties will not consider requests for accessions by non-UNECE members until those amendments become effective for the EC and all states that were parties to the convention at the time of their adoption.

3.4.2
Water Conventions and the Protection and Sustainable Use of Aquatic Biodiversity

Globally, freshwater ecosystems are in serious decline due largely to rapid population and economic growth. In fact, the rate of aquatic biodiversity loss already surpasses that from other major biomes by a considerable margin. Climate change will add further pressure on those ecosystems, resulting in major impacts on the global hydrological cycle. In so doing, climate change will aggravate significantly the challenge of establishing and sustaining cooperation between the states that share watercourses. Yet, the equitable and sustainable allocation and management of water across international borders are crucial requisites for sustaining aquatic ecosystems and maintaining their ecological functions and services, such as clean water, food and flood control, in support of human well-being.

It is thus imperative that states pursue and achieve much closer levels of transboundary cooperation in the management, protection, and sustainable use of freshwater resources. Several important international policy statements, undertakings or commitments have recognized this need, including, for example, the Hashimoto Action Plan II [5].

In the process of transboundary water cooperation, the conservation and sustainable use of biodiversity is not just an end in itself, but a means to sustain ecosystem service provision to the benefit of all the states concerned in an equitable manner. For example, all states sharing lake systems are likely to suffer from unsustainable land use practices that lead to land erosion and siltation, no matter where they occur in the catchment. For rivers that cross international borders, dams built downstream, if not planned properly, may harm key fisheries upstream and thus the communities that depend on those resources for food and livelihoods. Similarly, riparian states in contiguous rivers might suffer the impact of pollution originating across the border. Where states cooperate, on the other hand, they can share in the benefits from joint water planning and management. For example, an upstream state can improve clean water service provision to a lower riparian country by rehabilitating wetlands, with some of the costs incurred by the latter [3]. Naturally, there are complex economic and political issues regarding interstate payments for these services, but such challenges can best be met through improved cooperation in the context of solid regulatory frameworks.

Against this background, we have looked at the potential role of international water law in aiding states in the transboundary cooperation process, taking into account relevant CBD provisions and decisions. In particular, we examined the roles of the UN Watercourses Convention and the UNECE Water Convention in supporting and strengthening the implementation of the CBD towards the conservation, sustainable use and equitable sharing of biological resources within or dependent upon transboundary waters.

This analysis becomes particularly relevant when one considers that most transboundary river systems and aquifers are not subject to specific management frameworks to supplement and regulate the relevant CBD provisions. Even where treaties governing transboundary waters exist, most do not involve all the states concerned or fail to incorporate key elements of international law, such as information exchange and obligations in the case of emergencies. In particular, most existing watercourse agreements do not incorporate climate change and variability considerations.

The UN Watercourses Convention and the UNECE Water Convention share common goals with the CBD. All three conventions promote international cooperation as a crucial prerequisite for parties to achieve their goals. While the CBD broadly addresses water resources management and use, particularly through the programme of work on inland waters biodiversity, that convention lacks specific rules and principles governing cooperation between watercourse states and promoting the equitable and reasonable use and management of international watercourses.

In this sense, the sound implementation of that work programme could benefit from strengthened regulatory frameworks focusing specifically on transboundary water cooperation. In the absence of other specific management frameworks, therefore, the UN Watercourses Convention and the UNECE Water Convention could help address the relevant regulatory gaps in the CBD, further supporting the sound management of aquatic ecosystems and biological resources within transboundary watersheds.

For example, Article 14(1) (c) of the CBD simply requires parties to notify and consult with other states that might be significantly affected by major unilateral planned measures. The UN Watercourses Convention takes a step further and establishes a detailed procedure, with specific timetables and obligations, for states to exchange information, consult and negotiate with each other on planned developments, with a view to reaching mutually beneficial outcomes. The UN Watercourses Convention also supplements the CBD with respect to pollution flowing across international borders, which represents a major threat to biodiversity. While the CBD does not specifically address the problem, the UN Watercourses Convention goes as far as to: include a broad definition of the term 'pollution of an international watercourse'; require states to individually and, where appropriate, jointly, prevent, reduce and control the pollution of an international watercourse that may cause significant harm to other watercourse states or to their environment; call on states to take steps to harmonize their policies in this connection; and create a duty on states to, at the request of any of them, consult with a view to arriving at mutually agreeable measures and methods of pollution prevention, reduction and control, such as joint water quality standards. The UNECE Water Convention governs pollution in even greater detail; for example, the convention specifically refers to the polluter-pays principle, promotes the concept of best available technology and offers specific guidelines for parties to develop water quality objectives and criteria. The same applies to an obligation pertaining to data sharing, which the CBD simply enunciates. Both the UN Watercourses Convention and the UNECE Water Convention would add regulatory value to the CBD by clarifying the content and scope of that duty and related obligations in relation to transboundary waters (Table 3.4.1).

Having established the complementarities between the CBD and the two water conventions in question, the next step would be to consider what would be necessary for those legal instruments to make a meaningful contribution to the achievement of CBD goals. In our view, parties to the CBD and the UNECE Water Convention should cooperate more closely to assess synergies and develop harmonious and mutually supportive approaches for implementing their relevant obligations under each convention. The parties to the UNECE Water Convention should also accept the 2003 amendments opening the convention for

Table 3.4.1 Examples of issues on which the water conventions would add regulatory value to the CBD.

Similarities and Common goals	CBD	UNECE Water Convention	UN Watercourses Convention
Scope			
– Subject matter	Art.2(1): Biodiversity: variability among 'living organisms from... aquatic ecosystems'	Arts.1(1), 2(2)(b): Transboundary waters and 'water resources'	Arts.2(b), 6(f), 21(2): International Watercourses 'water/ecosystems'
– Main objectives	Art.1: Conservation, sustainable use and equitable sharing of biodiversity	Arts.2(2)(b), 2(2)(c):	Arts. 5(1), 6(f):
		– 'conservation of water resources';	– 'conservation ...of the water resources'
		– 'reasonable and equitable use of transboundary waters'	– utilization ...in an 'equitable and reasonable manner'
Obligations			
– Prevention, control and reduction of pollution	Art.3-1(a): 'The emission of pollutants is prevented, controlled and reduced at source'	Preamble: obligation to 'anticipate, prevent and attack the causes of significant reduction or loss of biological diversity at source'	Art.21(2): Obligation to 'prevent, reduce and control the pollution of an international watercourse that may cause significant harm to ...the living resources'
		– Art.8 (l): 'regulate or manage the relevant processes and categories of activities'	
– Conservation and/or restoration of aquatic ecosystems	Arts.1, 8(f): 'conservation... of... Ecosystems'; states to 'rehabilitate and restore degraded ecosystems'	Art.2(2)(d): Obligation to 'ensure conservation and, where necessary, restoration of ecosystems'	Art.20: Obligation to 'protect and preserve the ecosystems of international watercourses'
– Environmental Impact Assessment	Art.14(1)(a): Obligation to 'Introduce appropriate procedures requiring *environmental impact assessment* of its proposed projects that are likely to have significant adverse effects on biological diversity'	Art. 3(1)(h): '*Environmental impact assessment* and other means of assessment are applied'	Art.12: 'Before a watercourse State implements or permits the implementation of planned measures which may have a significant adverse effect upon other watercourse States, it shall provide... *environmental impact assessment*'

(Continued)

Table 3.1 (Continued)

Similarities and Common goals	CBD	UNECE Water Convention	UN Watercourses Convention
– Prevention of Invasive Alien Species introduction	Art.8(h): Obligation to 'Prevent the introduction of, control or eradicate those alien species which threaten ecosystems, habitats or species'	Art.2: Obligation to 'prevent, control and reduce any transboundary impact' (1) as 'significant adverse effect on the environment resulting from a change in the conditions of transboundary waters caused by human activity' (art.1–2)	Art.22: Obligation to 'prevent the introduction of species, alien or new, into an international watercourse'
Principles – International cooperation	Art.5: Obligation to 'cooperate... for the conservation and sustainable use of biological diversity'	Art.9: Cooperation through adopting 'agreements' (1) and establishing 'joint bodies'(2)	Art.8(1)–(2): Cooperate for the 'optimal utilization and adequate protection' through, where appropriate, the 'establishment of joint mechanisms or commissions'
– Not cause transboundary harm/damage (good-neighbourliness)	Art.3: States shall 'not cause damage to the environment of other States or of areas beyond the limits of national jurisdiction'	Art.2: Obligation to 'prevent, control and reduce any transboundary impact'	Art.7(1)–(2): Obligation to 'prevent the causing of significant harm' to other watercourse States' and 'eliminate or mitigate such harm'
– Sustainable development	Preamble: 'Determined to conserve and sustainably use biological diversity for the benefit of present and future generations'	Art.2(5)(c): 'Water resources shall be managed so that the needs of the present generation are met without compromising the ability of future generations to meet their own needs'	Preamble: 'Expressing the conviction that a framework convention will ensure the utilization, development, conservation, management and protection of international watercourses and the promotion of the optimal and sustainable utilization thereof for present and future generations'

accession by all UN member states. Moreover, all CBD parties should join the UN Watercourses Convention, actively contribute to its entry into force and engage in its implementation, to enable the effective and consistent implementation of those two multilateral agreements. Entry into force would also allow the UN Watercourses Convention to serve as a global framework for the development of international water law – a major contribution to the protection of transboundary inland water ecosystems and the need to address emerging challenges, such as climate change, the global food crisis and the overexploitation of transboundary aquifers.

3.4.3
Conclusions

Biodiversity considerations add significant weight to the case for the wider adoption and implementation of the UN Watercourses Convention and the UNECE Water Convention. Once in force and widely implemented, the UN Watercourses Convention will reinforce interstate cooperation at the basin level and significantly improve global freshwater governance.

Similarly, the good progress made so far under the UNECE Water Convention illustrates its potential value to support CBD implementation at the global level once the 2003 amendments become effective, and if parties collaborate more closely in implementing those two instruments.

Through their widespread implementation, therefore, the water conventions at hand would support and strengthen the CBD regulatory framework governing biodiversity within transboundary freshwater ecosystems, including, for example, in areas such as information exchange, major planned measures and transboundary pollution, as discussed above.

References

1 United Nations (1997) Convention on the Law of Non-Navigational Uses of International Watercourses, UN Doc. A/51/869,New York City, 21 May 1997, reprinted in 36 Int'l Legal Materials 700 ("UN Watercourses Convention"), http://untreaty.un.org/ilc/texts/instruments/english/conventions/8_3_1997.pdf (accessed 27 September 2010).

2 Brels, S., Coates, D. and Loures, F. (2008) Transboundary Water Resources Management: the Role of International Watercourse Agreements in Implementation of the CBD, CBD Technical Series No. 40, http://www.cbd.int/doc/publications/cbd-ts-40-en.pdf (accessed 27 September 2010).

3 United Nations Economic Commission for Europe (2006) UNECE Water Convention, Recommendations on Payments for Ecosystem Services in Integrated Water Resources Management, http://www.unece.org/env/water/publications/documents/PES_Recommendations_web.pdf (accessed 27 September 2010).

4 United Nations Economic Commission for Europe (1992) UNECE Convention on the Protection and Use of Transboundary Watercourses and International Lakes, Helsinki, http://www.unece.org/env/water/pdf/watercon.pdf (accessed 27 September 2010).

5 United Nations (UN) Secretary General's Advisory Board on Water and Sanitation (UNSGAB) (2010) Hashimoto Action Plan II, http://www.unsgab.org/HAP-II/HAP-II_en.pdf (accessed 27 September 2010).

Further Reading

United Nations (1992) Convention on Biological Diversity, Rio de Janeiro, 5 June 1992, http://www.cbd.int/doc/legal/cbd-en.pdf (accessed 27 September 2010).

United Nations (2010) Convention on Biological Diversity, Global Biodiversity Outlook 3, http://www.cbd.int/doc/publications/gbo/gbo3-final-en.pdf (accessed 27 September 2010).

Iza, A. (ed.) (2004) International Water Governance: Conservation on Freshwater Ecosystems, Vol. 1, International Agreements, Compilation and Analysis, Environmental Policy and Law Paper No. 55, http://data.iucn.org/dbtw-wpd/edocs/EPLP-055.pdf (accessed 24 March 2011).

International Union for Conservation of Nature (IUCN) (2006) *Pay: Establishing Payments for Watershed Services* (ed. M. Smith, D. de Groot, D. Perrot-Maître and G. Bergkamp), IUCN, Gland, Switzerland. Available at http://data.iucn.org/dbtw-wpd/edocs/2006-054.pdf (accessed 24 March 2011).

Loures, F., Rieu-Clarke, A. and Vercambre, M.L. (2009) Everything you need to know about the UN Watercourses Convention, World Wildlife Fund, http://assets.panda.org/downloads/wwf_un_ watercourses_brochure_for_web_july2010_ en.pdf (accessed 27 September 2010).

Tanzi, A. (2000) The Relationship between the 1992 UNECE Convention on the Protection and Use of Transboundary Watercourses and International Lakes and the 1997 UN Convention on the Law of the Non Navigational Uses of International Watercourses, Report of the UNECE Task Force on Legal and Administrative Aspects, http://www.unece.org/env/water/publications/documents/conventiontotal.pdf (accessed 27 September 2010).

United Nations Development Programme (UNDP) (2004) Protecting International Waters – Sustaining Livelihoods, http://www.undp.org/gef/documents/publications/intlwaters_brochure2004.pdf (accessed 27 September 2010).

United Nations Development Programme (UNEP) (2002) Atlas of International Freshwater Agreements, http://www.transboundarywaters.orst.edu/publications/atlas/ (accessed 27 September 2010).

3.5

The European Union Water Framework Directive, a Driving Force for Shared Water Resources Management

Elpida Kolokytha

3.5.1
Introduction

. . . and they must act in a changing world, a world driven by forces that they often do not control – forces of demography, the global economy, changing societal values and norms, technological innovation, international law, financial markets and climate change.

(WWAP 3, 2009)

Water management is, by definition, conflict management as it has to consider the competing interests of different users. On a national level these interests include urban use, agriculture, industry, hydropower generators and tourism. The possibility of satisfying all of these by adopting mutually acceptable solutions drop exponentially as more stakeholders are involved. On an international level the problem becomes even more difficult and the chances of finding mutually acceptable solutions decrease exponentially yet again [1, 2].

In this sense, the management of shared water resources is indeed conflict management and is exceptionally important since it is directly related to the safety and the maintenance of world peace. In recent decades the problem of competing demands, depleting groundwater resources and degrading water stocks has increased both in national and international river basins. Today there are 263 international basins compared to the 214 listed in a 1978 UN report. This increase is a result of political changes, such as the break up of the Soviet Union and the former Yugoslavia, as well as due to better mapping sources and technology. Even more striking than the total number of basins are the figures showing how much of each nation's land surface falls within these watersheds. A total of 145 nations include territory within international basins. Twenty-one nations lie in their entirety within international basins, and a total of 33 countries have more than 95% of their territory within such basins. The number of countries sharing each international basin is impressive, varying from 3 to 17 riparian countries (The Danube basin) [3].

In assessing the sources of tension between co-riparian states over shared water systems, the hydropolitical literature has largely focused on the issues of scarcity and inequitable allocation of available water stocks [3–5], while water quality issues are often overlooked in the context of international freshwater management [6].

In Europe 20 countries depend on neighbouring countries for more than 10% of their water resources. Clearly, there is an urgent need for concerted management and harmonization of policies. International cooperation for the management of shared water bodies is a priority for the European Commission. To further develop this issue the 'Improvement of transboundary cooperation' was identified as a key activity of the Work Programmes for 2005–2006 and 2007–2009 of the Common Implementation Strategy for the Water Framework Directive (CIS WFD).

3.5.2
The EU WFD

The Water Framework Directive (WFD) sets new common objectives in EU water policy and establishes a coherent legal and administrative framework for their achievement. Through the WFD, EU policy has moved from the protection of particular waters of special interest (such as drinking water, a nature area, coastal water, etc.) to protection and use, based on an overall approach and extended to all waters, both surface and groundwater.

The overall system provided by the WFD is based on a central concept, present also in the Helsinki Convention, which is *integration*. Integration is a key notion for all water resources, environmental and ecological objectives, water uses, functions and values, interdisciplinary analyses and expertise and different decision-making levels within a common policy framework. The ultimate goal is 'the good ecological status' for water bodies.

Physical rather than administrative boundaries in the sense of management at river basin level represent a major innovation in procedure. Managing the river basin as a whole is the best way to ensure the integrity of the ecosystem. In addition, the economic analysis of water use, the provision of river basin management plans and programmes of measures and public participation provide the main framework of the water resources management process. Table 3.5.1 summarizes all the information needed and the procedure that should be followed to fulfil such an integrated approach.

3.5.3
Shared Water Management According to the EU WFD

According to the Water Framework Directive Member states are obliged to identify their river basins and assign them to 'river basin districts'. For all districts, national and international, six-yearly river basin management plans and programmes of measures need to be developed. To ensure the necessary national coordination, member states need to identify a 'competent authority'. A description of the legal status of the competent authority and a detailed description of its responsibilities as well as of its role within each river basin district are required. For international basins EU member states have to coordinate their activities and 'endeavour' to coordinate with non-EU members in the basin (art. 3.5). Finally, public participation plays a crucial role in the WFD. In the implementation of the WFD it is stipulated that public consultation and 'active involvement' in the planning process have to be promoted and encouraged three times [8, 9].

3.5.4
WFD: A Driving Force for Cooperation

The WFD could act as a guideline for international cooperation as it promotes the sustainable management of shared watercourses among both EU and non-EU countries through the design and implementation of joint management plans, joint river authorities, transboundary river basin units and coordinated national measures at a river basin scale.

The WFD states explicitly the requirement for joint management in the case of a transboundary aquifer. All programmes should be coordinated for the whole of the river basin even if that basin extends beyond the boundaries of the Community. At the substantive level, the process of harmonization can be basically facilitated by a set of legal elements, respective criteria and standards based on and deriving from the

Table 3.5.1 Summary of the information needed and the procedure to be followed to fulfil for an integrated approach. Source: Reference [7].

Selected items related to an integrated approach	Some examples of information needed
Integration of environmental objectives	Information on water-quality, water-quantity and ecological status as well as objectives/targets for protecting highly valuable aquatic ecosystems and ensuring a general good status of other waters
Integration of all water resources	Inventories of fresh surface water and groundwater bodies, wetlands, coastal water resources at the river basin scale and their interactions
Integration of all water uses, functions and values	Knowledge on the links between functions and uses in a specific river basins (e.g. water for the environment, role of the ecosystems such as forests and wetlands in the water cycle, water for health and human consumption, water for industry, transport, leisure (see Table 3.5.2 below) and their societal valuation.
Integration of disciplines, analyses and expertise	Uses of resources, hazardous activities (e.g. manufacturing industries), water-construction works and other economic activities to assess current pressures and impacts on water and water-related ecosystems by combined knowledge on hydrology, hydraulics, morphology, ecology, chemistry, soil sciences, technology, engineering, social and political sciences and economics.
Integration of all significant management and ecological aspects	Knowledge on the extent of floodplains; inventories of land use (e.g. forests, agricultural areas, urban areas); inventories of protected areas and their relationship with, or influence on, the ecological situation; link between policy measures and ecology to assess effectiveness of measures.
Involvement of stakeholders and the civil society in decision-making	Local experience and traditional knowledge on water issues; identification of stakeholders and their information needs; reporting requirements/information need of the stakeholders and general public (e.g. level of detail, frequency of reports)
Integration of different decision-making levels that influence water resources and water status	National legislation: transboundary agreements; decisions taken at international forums (e.g. MDGs): functions and mandates of local, provincial, national and transboundary authorities and other bodies.

principle of equitable utilization, the environmental protection rules and from sustainable development considerations [9, 10].

The WFD allows continuous improvement during the three planned management cycles: 2009–2015, 2015–2021 and 2021–2027. In fact, it urges the Member States to adopt a common process, with the same objectives, the same methods and tools. Valuable tools for the assessment of transboundary cooperation are treaties, cooperative arrangements and databases combining hydrological, geographic, socioeconomic, environmental and political data relating to transboundary waters.

This procedure is very important for international river basins. The main tool is the formulation of a common River Basin Management Plan (RBMP) [11]. Formulation of different scenarios for sustainable development in river basins, focusing in particular on the potential benefits that could be gained from joint river management, is provided by the shared RBMP. This way emphasis is placed not on the water itself but on benefit sharing that is allocating the benefits derived from the various uses (and non-uses) of water.

Table 3.5.2 illustrates the requirements of the WFD and the information needed to fulfil each requirement in the management of shared water resources.

3.5.5
International Cooperation – Transboundary Cooperation

An analysis based on a total of 1831 events connected to transboundary 'basins at risk' has shown that the riparian countries in fact tend to cooperate, rather than enter into conflicts [12, 13]. Since in certain international river basins it seems hard to reach an agreement on water rights and water allocation for each riparian country, the introduction of the benefit sharing concept promotes an alternative approach based on benefits that may be derived from the use of the water rather than on the actual sharing of the water itself. In other words, this approach focuses on optimizing the values (economic, social, cultural, political and environmental) generated from water in its different uses and equitably distributing the benefits amongst water users and suppliers [14, 15].

More emphasis is now placed on the dynamics of cooperation at all levels of planning and management. The decision on cooperation is based on perceptions of greater benefits. These benefits can be environmental (benefits to the river), economic (benefits from the river), political (benefits because of the river) or even broader (benefits beyond the river) [16]. The costs of cooperation include financial, institutional, political and any costs of unilateral opportunities (benefits) foregone [17].

Cooperation, and in particular benefit and cost sharing, can promote more efficient and more equitable river basin management by separating the physical location of river development (where activities are undertaken) from the economic distribution of benefits/costs (who profits from/pays for those activities) [9, 15].

Table 3.5.2 WFD requirements and the information needed to fulfil each requirement in the management of shared water resources. Source: Reference [7].

Requirement in Annex VII of the WFD		Where does the information come from?
1)General description of the characteristics of the river basin district(RBD)		Article 5(characteristics of the RBD, Review of environmental impact of human activity and economic analysis of water issues)
		Annex II(characterisation of water bodies and assessment of impact)
2)Summary of significant pressures and impact of human activity on water status		
3)Identification and mapping of protected areas		Article 6(Register of protected areas)
		Article 7(Waters used for the abstraction of drinking water)
		Annex IV(Protected areas)
4)Map of monitoring networks		Article 8(Monitoring of surface water status, groundwater status and protected areas)
		Annex V(Surface water status and ground water status)
5)List of environmental objectives		Article 4(Environmental objectives)
6)Summary of the economic analysis of water use		Article 9(Recovery of costs for water services) Annex III(Economic analysis)
7)Summary of the programme of measures		Article 11(Programme of measures
		Annex VI(List of measures to be included within the programmes of measures)
8)Register of any more detailed programmes and management plans for the RBD dealing with particular sub-basins, sectors, issues or water types and a summary of their contents		Article 13(RBMPs)
9)Summary of public information and consultation measures taken		Article 14(Public information and consultations)
10)List of competent authorities		Article 3
11)Contact points and procedures		Annex 1

For an effective transboundary agreement the following questions should be raised and an answer satisfactory to all riparian sides found: *What benefits exist* in joint river basin management, and *how best may such benefits be shared?* It is important to shift the discussion from benefits of water to benefits from the use of water.

Transboundary institutional and regulatory frameworks are the backbone of cooperative management systems. While they will not prevent all disputes, they are essential in clarifying the 'rules of the game' and thus enhancing legal security and reducing the likelihood of water disputes among sharing states [15].

All major stakeholders should be involved in order to maximize the chances of an agreement that actually contributes to mutual benefit and minimizes the chance of national opposition to its implementation. It is very important to identify the major stakeholders and moreover to stress the need to basically enable not only 'national governments' but also regional and local governments, civic society and individual water users [18].

3.5.6
Improvement of the Nestos/Mesta GR-BUL Shared Basin Agreement

The current international agreement for the management of the Nestos/Mesta shared river basin between Greece and Bulgaria signed in 1995 as a result of previous negotiations that started back in 1964, followed by a series of efforts in 1971, 1982 and 1988, concerns mainly water resource allocation issues. In Mylopoulos *et al.* [19] the bilateral agreement is explained and discussed in detail. The most important concern of the Greek side was to secure a standard amount of water. The text of the agreement includes some of the main principles set by the WFD but in practice there are two negative aspects. On the one hand, the articles of the agreement are characterized by generalities and, on the other hand, implementation of the agreement has not been activated yet.

According to the WFD, for the situation to be improved, the integrated river basin approach should be promoted rather than the focus being only on water quality and quantity, which is the case now.

There is no cross-border cooperation concerning the monitoring of the river and the exchange of information is problematic due to lack of an effective organizational background. The installation of a common monitoring system would provide reliable information and data about the status of the river basin. Especially where water quality issues and environmental protection are concerned, the agreement should be enhanced with the integration of quality parameters (e.g. the specific pollution loads and their respective limits) and specific environmental objectives together with the characterization of water based on ecological criteria in accordance with the aims of the EU WFD.

The two countries completely disregard the need for a common management plan and fail to set the essential background conditions for future implementation of such a plan. Public participation and involvement in the formulation of a joint RBMP is also absent. Economic analyses of the water uses as well as responsibilities of the competent authorities are not mentioned at all. Obviously, the agreement in its present form cannot serve environmental protection, economic development and social prosperity of the people living in the river basin.

For the agreement to be effective three basic notions should be incorporated:

- Envisioning of both countries to share the dream and the goals. The negotiations should be supported by reliable methodological tools for conflict resolution.
- Empowerment to provide joint decision making and to share power based on the 'right' of each country to water for the benefit of all.
- Enactment in order to proceed to implementation and civil engagement, which are the basic components of successful shared water management.

3.5.7
Conclusions

The key activity for transboundary water resources management is to promote the development and implementation of formal or informal transboundary agreements. Knowledge and participation should guide the process of implementation, as they build trust, ownership and common understanding among stakeholders. Effective implementation requires real commitment from governments and stakeholders, and a difficult balance between maintaining the spirit and applying the specific obligations of agreements, while developing operational modalities and adapting to changing circumstances. Adaptive management allows flexibility in transboundary water management. To act for the future vision, lessons of the past and an understanding of the present conditions should be combined. Common river basin management plans are critical. The WFD can play an important role in the direction of effective transboundary agreements as it incorporates all the basic requirements.

References

1 MacQuarrie, P., Viriyasakultorn, V. and Wolf, A. (2008) Promoting cooperation in the Mekong region through water conflict management, regional collaboration, and capacity building. *GMSARN International Journal*, 2, 175–1842.

2 Wolf, A.T. (2007) Shared waters: conflict and cooperation. *Annual Review of Environment and Resources*, 32, 3.1–3.29.

3 Wolf, A.T. (1999) Criteria for equitable allocations: the heart of international water conflict. *Natural Resources Forum*, 23, 3–30.

4 Duda, A.M. and La Roche, D. (1997) Sustainable development of international waters and their basins: implementing the GEF operational strategy. *Water Resources Development*, 13(3), 383–1997.

5 Wouters, P. (2000) National and international water law: achieving equitable and sustainable use of water resources. *Water International*, 25 (4), 499–512.

6 Giordano, M. (2002) International river basin management: global principles and basin practice. Dissertation thesis, Oregon State University.

7 UNECE, Background paper on information management in transboundary water cooperation http://www.unece.org/env/water (accessed 15 Oct 2010).

8 European Commission IDA programme (2002) Guidance on public participation in relation to the water framework directive; active involvement, consultation, and public access to information, Prepared in the Framework of the Common Implementation Strategy of the European Commission and the EU Member States, http://forum.europa.eu.int/Public/irc/env/wfd/library (accessed 16 June 2011).

9 Barraque, B. and Mostert, E. (2006) Transboundary river basin management in Europe, Thematic paper for Human Development Report 2006/21.

10 Mylopoulos, Y. and Kolokytha, E. (2008) Integrated water management in shared water resources. The EU Water Framework Directive implementation in Greece. *Physics and Chemistry of the Earth*, **33/5**, 347–353.

11 Kolokytha, E. and Mylopoulos, Y. (2005) Hydrodiplomacy: a ground for action transboundary cooperation and economic development in Nestos River basin, Greece. in E-proceedings (F04-4 – Sustainable Water Resources, 4544–4552), IAHR Congress, 11–16 Sep 2005, Seoul, Korea.

12 Yoffe, S. and Larson, K. (2001), in Basins at risk: water event database methodology, Department of Geography, Oregon State, University, Corvallis, USA.

13 Wolf, A.T., Yoffe, S.B. and Giordano, M. (2003) International waters: identifying basins at risk. *Water Policy*, **5**(1), 29–60.

14 White, D., Wester, F., Huber-Lee, A., Hoanh, C.T. and Gichuki, F. (2008) Water benefits sharing for poverty alleviation and conflict management: topic 3 synthesis paper, CGIAR Challenge Program on Water and Food, Colombo.

15 Sadoff, C., Greiber, T., Smith, M. and Bergkamp, G. (eds) (2008) *Share – Managing Water Across Boundaries*, IUCN, Gland, Switzerland, ISBN 978-2-8317-1029-7.

16 C.W. Sadoff and D. Grey (2002) Beyond the river: the benefits of cooperation on international rivers. *Water Policy*, **4**, 389–403.

17 Qaddumi, H. (2008) *Practical Approaches to Transboundary Water Benefit Sharing*, Working paper 292, Overseas Development Institute, London. ISBN 978 0 85003 877 4.

18 Trottier, Julie (2003) The Need for Multiscalar Analysis in the Management of Shared Water Resources in (eds F.A. Hassan, M. Reuss, J. Trottier, C. Bernhardt, A.T. Wolf, J. Mohamed-Katerere and P.v.d. Zaag), pp. 1–10.

19 Mylopoulos, Y., Eleftheriadou, E. and Kabragou, E. (2004) The transboundary catchment of River Nestos and the bilateral agreement between Greece and Bulgaria. Presented at the Proceedings of the Eco-Geowater Conference, GI for International River Basin Management, 3–5 June, Budapest, Hungary.

Further Reading

Eleftheriadou, E. and Mylopoulos, Y. (2006) A methodological framework supporting transboundary water agreements: the case of the Nestos/Mesta river basin. *Water Policy*, **10**(3), 239–257.

Phillips, D.J.H., Daoudy, M., Öjendal, J., McCaffrey, S. and Turton, A.R. (2006) *Transboundary Water Cooperation as a Tool for Conflict Prevention and Broader Benefit Sharing*, Swedish Ministry for Foreign Affairs, Stockholm.

United Nations (1978) *Register of International Rivers*, Pergamon, New York.

Viessman, W. Jr. (1998) Water policies for the future: an introduction, water resources update, issue no. Ill, Universities Council on Water Resources, Carbondale, Illinois.

Wolf, A., Natharius, J., Danielson, J., Ward, B. and Pender, J. (1999) International river basins of the world. *International Journal of Water Resources Development*, **15** (4), 387–427.

WWAP3 (2009) Facing the challenges, The United Nations World Water Development Report 3.

3.6
Transfer of Integrated Water Resources Management Principles to Non-European Union Transboundary River Basins

Katharina Kober, Guido Vaes, Natacha Jacquin, Jacques Ganoulis, Francesca Antonelli, Kamal Karaa, Samir Rhaouti and Pierre Strosser

3.6.1
Litani River Basin within Lebanon

In EU countries, where transboundary water management has been obligatory since the Water Framework Directive (WFD) came into force, a large quantity of knowledge about related issues has been accumulated, which should be fruitfully passed on to benefit transboundary water management in non-EU countries. The approach to transferring EU knowledge to Third States, described as the WFD/EUWI Joint Process, is the base of the EU external policy regarding water management. This process was established to be able to pass on the impressive quantity of knowledge acquired during the implementation of the WFD to non-EU countries, and support them in their efforts to implement Integrated Water Resources Management (IWRM). Efforts to improve the Science–Policy interface in EU water management as carried out in the SPI-Water project [1] will have a direct impact on the speed of knowledge transfer to Third Countries. One of the working groups in the SPI-Water project focused especially on this issue and tried to find an effective way to facilitate the transfer of IWRM results to non-EU countries. Of course the existing knowledge has to be adapted to fit the needs of these countries. This was shown in phase 1 of the WFD/EUWI JP in the Mediterranean area.

Although various tools and mechanisms are known and used to enhance transboundary water cooperation (i.e. international commissions, basin authorities, arbitration commissions, etc.) it is still rare to find well developed joint water management between neighbouring countries in many regions. The SPI-Water project shows Third Countries how to identify efficient tools for transboundary river basin (RB) management, how to adapt them to their own specific needs and how to organize transboundary cooperation.

3.6.2
Objectives of the Working Group on IWRM Knowledge Transfer

The objective of the working group on knowledge transfer to non-EU countries was to facilitate the transfer of IWRM principles, as considered in the WFD, which includes transboundary river management. For this purpose the main goal of the involved partners was to further adapt and develop the EUWI/WFD Joint Process and to help non-EU countries to benefit in the best possible way from the knowledge accumulated by EU-countries during the implementation of the WFD.

To reach this goal four tasks were carried out: (i) bridging inputs and needs for non-EU countries: a review of recent EUWI/WFD joint process activities and related practices was completed [2]; (ii) detailed description of selected RBs and identification of water management needs: reports on the two non-EU piloted river basin organizations (PRBO) Sebou (Morocco) and Litani (Lebanon) were written [3, 4]; (iii) matching WFD RTD solutions to the PRBOs needs [5] and (iv) development of recommendations [6].

3.6.3
Description of Results

3.6.3.1 Review of Recent EUWI/WFD Joint Process Activities and Related Practices

A review on water policy experiences in non-EU countries, and especially the recent practices on knowledge transfer acquired in EU countries towards non-EU countries, was carried out [2]. This activity aimed to propose a mechanism for the implementation of IWRM/WFD principles in non-EU countries. A desktop review of knowledge transfer processes and direct interviews with several stakeholders were realized so that an updated picture and a general overview of the strengths and weaknesses of the past knowledge transfer activities could be gained.

The main lesson learnt regarding the implementation of the IWRM/WFD principles in non-EU countries was that a bottom-up and participatory approach needs to be promoted, both enhancing the interface between the local and national level and improving the dissemination of results and capacity training. More initiatives should be financed to set up expert networks and to transfer technologies, analysis tools and so on. EU cooperation instruments should be developed and maintained for a sufficient period of time, taking into account the fact that many non-EU countries are currently in a phase of transition.

3.6.3.2 Detailed Description of Pilot River Basin Organizations and Identification of Weaknesses

Based on the WFD methodology, a description of each Pilot River Basin (PRB) (Litani in Lebanon and Sebou in Morocco, see Figures 3.6.1 and 3.6.2, respectively) was produced by the PRBOs with the support of several project partners [3, 4]. The reports give a general overview of the administrative framework and the physical, climatic and biotic situation in the PRBs. Additionally they include a description of the status of IWRM implementation. The WFD Article 5 Reports served as example for these reports. The documents focus on the identification of pressures, characterization of the actual status of water bodies and identification of the main water management issues. Based on the results it was possible to establish a definition of environmental objectives. The authors of both reports put special emphasis on weaknesses in the PRBs, because the main aim of the report was to be able to match major water management problems and needs of the PRBs with solutions from EU experiences. Considering the identified problems and needs, the working group partners searched in the WISE-RTD Web portal for tools, initiatives and research that could offer possible solutions.

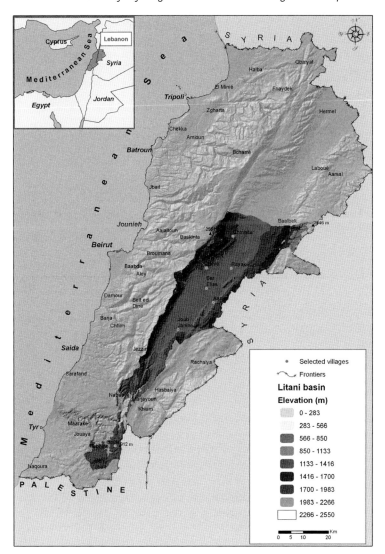

Figure 3.6.1 Litani River Basin within Lebanon [4].

3.6.3.3 Matching WFD RTD Solutions to the River Basins' Needs

Once specific problems and needs have been identified in the PRBs, a link was made to those RTD projects that propose specific solutions via browsing the WISE-RTD Web portal (www.wise-rtd.info). This activity translated into facilitating access to already available information on technical projects and research. The results of this activity were summarized in a special report [5]. This document includes: (i) a synthesis of the main water management issues in the PRBOs; (ii) a first

Figure 3.6.2   Sebou River Basin [3].

identification of the potential needs of water authorities in terms of tools and methods that would help improving decisions; (iii) a description of the search in WISE-RTD to identify existing guidance, research projects, reports and tools relevant to the practical areas selected; and (iv) a conclusion on the WISE-RTD Web portal, in which the non-EU partners give a summary of their experiences of using the portal.

3.6.3.4 Development of Recommendations
Finally detailed and specific recommendations were published. Among those recommendations, none can be considered as being 'brand new', as the stake is more to promote a comprehensive policy for transferring EU IWRM and WFD knowledge, while taking stock of existing tools and ongoing initiatives, which are sometimes too isolated, inadequately publicized or lacking in political backup to be identified as part of a global approach offering practical solutions. The recommendations serve as guidelines for EU and non-EU countries, which need to focus on IWRM principles, for example, in transboundary waters management, and on how the EUWI/WFD joint process activities can best facilitate the implementation. The guidelines concentrate on the transfer of WFD knowledge from the field of research

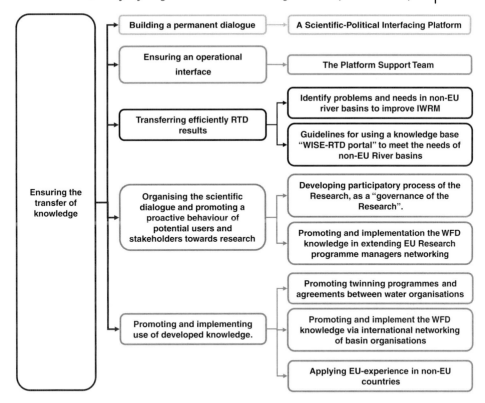

Figure 3.6.3 A set of nine recommendations to enforce five pillars.

to actual water managers and on the promotion of exchanges between EU and non-EU researchers, between researchers, stakeholders and water managers in non-EU countries. The recommendations aim to uphold five pillars (Figure 3.6.3). The nine recommendations are:

1) building a permanent dialogue through a multi-stakeholder platform (scientist, decision makers from non-EU countries, donors);
2) ensuring an operational interface open to technological and social innovations: a support team for the Science–Policy Interfacing;
3) Identification of problems and needs in non-EU RBs;
4) extension of WISE-RTD Web portal to meet the needs of non-EU RBs;
5) develop a participatory process of research as a governance of researches;
6) extending water research programme managers networking (IWRM.Net) to Mediterranean countries;
7) develop twinning programmes and agreements between RBOs;
8) promotion and implementation of WFD knowledge via international networking of RBOs;

9) applying EU-experience in non-EU countries: concrete transfer of knowledge between EU and non-EU countries [6].

To facilitate the transfer of knowledge and information exchange, especially in the case of the two PRBOs of the SPI-Water project, four technical visits to EU RBs were organized as an example of transboundary cooperation. They gave experts from both PRBOs the possibility to learn on-site from European partners facing similar challenges about issues of specific interest to them. As a result of these visits concrete project proposals on pressing water management issues were elaborated for both PRBOs.

3.6.4
Conclusions

The SPI-Water project shows that there is a great need for transfer of the knowledge gained from research and experience to the daily practice of water managers. This is not only the case for Europe with respect to the implementation of the WFD, but also relevant to countries outside Europe. The good management of water bodies does not stop at the EU borders and transboundary cooperation outside Europe is necessary. For that reason a better transfer of knowledge to these countries is needed. The SPI-Water recommendations on how to improve this process will facilitate the successful transfer of IWRM knowledge to non-EU countries.

More information on the SPI-Water project and all mentioned reports can be found on the web site www.spi-water.eu.

Acknowledgement

SPI-Water was a Scientific Support Priority project developed under the sixth Framework Programme (sponsored by the EC, DG-RTD, contract 044357). It started on 1st November 2006 and ran for two years. Sixteen partners were involved: HydroScan (Coordinator, Belgium), QualityConsult (Italy), Centre for Water Management (Netherlands), Mediterranean Network of Basin Organizations (working group leader, Spain), Environment Agency (UK), XPRO Consulting (Cyprus), National Technical University of Athens (Greece), Katholieke Universiteit Leuven (Belgium), Potsdam Institute for Climate Impact Research (Germany), International Water Office (France), 2Mpact (Belgium), Aristotle University of Thessaloniki (Greece), WWF European Policy Programme (Italy), Litani River Authority (Lebanon), Sebou River Basin Organization (Morocco) and ACTeon (France).

References

1 Vaes, G., Willems, P., Swartenbroekx, P., Kramer, K., de Lange, W. and Kober, K. (2008) SPI-water: science-policy interfacing in support of the water framework directive implementation. Paper presented at the 11th International Conference on Urban Drainage, 31 Aug–5 Sep, 2008, Edinburgh, Scotland, UK.

2 OIEAU, MENBO, AUTh, WWF, LRA, ABHS & ACTeon (2007) IWRM needs in non-EU countries: review of recent EUWI/WFD joint process activities and related practices, report for the EC contract 044357.

3 ABHS, WWF, MENBO & ACTeon (2007) Description of the river basin Sebou (Morocco): state of the art in the frame of pilot establishment of WFD tools, report for the EC contract 044357.

4 LRA & MENBO (2007) Description of the river basin Litani (Lebanon), report for the EC contract 044357.

5 WWF, MENBO, OIEAU, AUTh, LRA, ABHS & ACTeon (2008) Matching WFD solutions in non-EU pilot river basins, report for the EC contract 044357.

6 OIEAU, MENBO, AUTh, WWF, LRA, ABHS & ACTeon (2008) Guidelines and recommendations on the transfer of IWRM and WFD knowledge to non-EU countries, report for the EC contract 044357.

3.7
Implementation of the Water Framework Directive Concepts at the Frontiers of Europe for Transboundary Water Resources Management

Didier Pennequin and Hubert Machard de Gramont

3.7.1
The EU Water Framework Directive

The October 2000 European Water Framework Directive (WFD) [1] sets up as an objective that all European water bodies should reach a good water status by 2015, provided that they are not under one of the derogation regimes that allows this deadline to be extended. This ambitious objective applies to all soft water bodies, including continental surface water and groundwater.

To reach good water status by 2015, the WFD gives a step by step approach, fixes deadlines for each step, proposes a common framework to apply in all European countries, and asks for the designation of national entities to be in charge of its implementation.

The WFD considers water resources as a patrimony that should be used in a balanced and equitable manner, while at the same time being preserved for future generations. It is based on the concept of sustainable management and development. It recognizes the need to consider the scale of the catchment area as a natural working unit to manage water resources, and uses this as a basic principle for many of its recommendations. At the same time, the 'water body' concept is introduced as a management unit of which the contours are defined by each Member State (MS) according to water resources management criteria.

The WFD also states that all water resources must be thoroughly characterized and monitored, and in the case of poor water status or upward trends in contaminant concentration, programmes of measures must be established to decrease levels of contamination and to improve water quality, in addition to the elaboration of basin management plans.

In this framework, water resources management must be viewed as being an action driven by an integrated reasoning process: in each catchment area, both surface water and groundwater must be addressed together and along with socio-economic parameters to ensure sustainable management of water resources. As such, it introduces both the concept of (i) 'public participation' – water management must result from a concerted action involving representatives of all categories of stakeholders – and (ii) the 'economic analysis' – as a support tool to determine measures to be implemented to reach the environmental objectives set up by the WFD.

These are only a few of the concepts and constraints brought about by the WFD, but clearly they suggest that a new approach to water resources management is being implemented in Europe. In practice, this is not easy to apply in the field as it requires (i) a multidisciplinary approach to water resources management, often involving scientists, engineers, stakeholders, administrative bodies and water resources managers, and (ii) to acquire a good knowledge of the characteristics and functioning modes of the water resources . A spirit of dialogue and cooperation are mandatory in order to succeed.

3.7.2
The EU Water Framework Directive and Transboundary Water Resources

The WFD also applies to transboundary water resources, and the situation is even more complicated in these cases. Indeed, extending the water resources natural management units all the way to the boundaries of the catchment areas, and creating water bodies, led to many situations in Europe where single water resource entities now spread over the political boundaries of two or more countries. In such situations, the WFD clearly states that concerned Member States must ensure that the prescribed requirements for achieving the good water status conditions by 2015 and compliance with existing regulation be coordinated over the whole catchment, just as is the case for water bodies contained within one single country. In fact, the WFD says that for international catchments or water bodies, the 'Member States concerned shall together ensure this coordination and may, for this purpose, use existing structures stemming from international agreements'.

Transboundary water resources must therefore be managed in the same manner as the national water resources, but adding the international dimension, which means calling for the appropriate reciprocal political will, proper empowered supra-national institutional mechanisms and transnational concerted actions involving stakeholders from all countries concerned. This can, however, be somewhat facilitated by the EU Commission in certain cases ('at the request of the Member States involved, the Commission shall act to facilitate the establishment of the programmes of measures').

The problem increases further in complexity at the boundaries of Europe. Indeed, the WFD requires that Member States located at the frontiers of Europe, which share water resources with non EU members, must 'endeavour to establish appropriate coordination with the relevant non-Member States, with the aim of achieving the objectives of this Directive' (WFD) throughout the entire basin.

In theory, two European countries that share a common water resource, guided by the same objective of reaching good water status, with a common working framework and common deadlines, can reasonably hope to agree on a concerted approach to implement in order to reach their common goal (*even though several difficulties may often arise*). However, at the frontiers of Europe, the situation is not as 'naturally straightforward', and it may even reach a high degree of complexity, as the non European neighbouring countries abide by their own regulations, which generally do not impose a deadline for good water status. In addition, the perception of the value of water resources may be quite different outside of Europe, and the principles of water resources management may therefore significantly deviate from those applied today within the European Union.

3.7.3
Applying the WFD Concept at the Frontiers of Europe for a Transboundary Lake and Underlying Aquifer System

The Lake Peipsi water complex and the underlying hydraulically connected aquifer system is one example of a transboundary water resource that extends across the boundaries of Europe [2, 3]. It is shared by Estonia and Russia. Being the largest transboundary water body and the fourth largest lake in Europe, Lake Peipsi extends along a north–south direction over a distance exceeding 150 km, and covers an area of approximately 3555 km^2, which spreads over both sides of the Russian–Estonian border (Figure 3.7.1). It is underlain by a series of five major aquifers (Figure 3.7.2). The shallowest of these consist of glacial continental deposits and usually provide good quality water. Underneath lies the Devonian limestone and sandstone complex, and, below, the Ordovician limestone aquifer (upper part of the cambro-ordovician series), which contains layers of oil shale deposits. The two deeper aquifers, the Voronka and the Gdov aquifers, consist of sandstones and are generally separated from each other and from the overlying aquifers by thick clay horizons.

Hydraulic connections and flux exchanges between the Quaternary glacial aquifers, the Devonian complex, the Ordovician limestone and the lake system do exist to a different extent according to the area. Hydraulic connections between these and the two deeper aquifers are probably poor, due notably to the Lontova and Kotlini clay-rich aquitards.

The Lake Peipsi region has a population of about 800 000 inhabitants, 51% of whom are on the Estonian side. It has undergone extensive oil shale mining activity in the northern part in both countries for the last seven decades or so, reaching a peak of 20 MT on the Estonian side alone in 1991. Today, oil-shale production has decreased and stabilized at around 10–14 MT per year in Estonia and less on the Russian side. Over the years this mining activity generated millions of tons of waste (cracking residues, ashes, etc.) that was piled up as spoil heaps in various areas. These often reach lengths exceeding 1 km or more, widths of several hundreds of meters and can be from 30 to more than 100 m high, as is the case in the area of Kohtla-Järve (Figure 3.7.3).

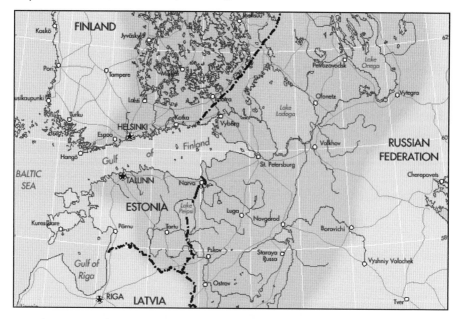

Figure 3.7.1 Lake Peipsi at the eastern border of Estonia. Source: Map Design Unit – World Bank.

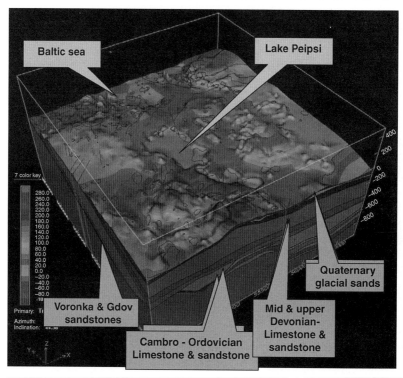

Figure 3.7.2 Hydrogeological diagram of the Lake Peipsi area. Source: BRGM–Viru-Peipsi-Camp project.

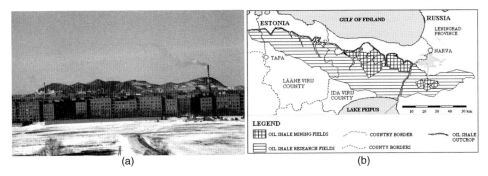

(a)   (b)

Figure 3.7.3 (a) Kohtla-Järve ash heaps. The highest summit is the third from left (122 m) (photograph: Riina Vaht); (b) oil shale mining area north of Lake Peipsi.

The mining activity creates a significant pollution hazard to local water resources. In fact, pollution from hydrocarbon compounds, sulfates and salts derived from oxidation of the sulfates has already reached the Ordovician limestone in the mining areas. It is feared that rising water levels where mining activity ceased might still amplify this problem and come to contaminate the shallow quaternary aquifer. Surface water bodies close to the spoil heaps and in contact with ashes became highly alkaline in some areas (pH 13). They are sometimes also contaminated by phenols (Figure 3.7.4) produced from cracking residues.

South of the mining area, a more rural socio-economic system prevails, with well set agricultural activity which, as in many developed countries, generates diffused pollution mainly consisting of nitrates and pesticides.

Figure 3.7.4 A phenol polluted pond near the ash plateau in Kohtla-Järve (photograph: Tõnis Saadre).

In addition, averted pollution or the fear of potential pollution often brought local authorities to turn to the two deeper aquifers for drinking water supply, leading to groundwater overexploitation with declining water levels and water quality degradation in certain areas. To date, this primarily affects the mining zones. Further details can be found in the literature [4–6].

This situation and the obligation for Estonia to comply with European legislation, and in particular with the WFD, prompted several projects in the last few years with the aim to help (i) in improving the water resource status and water management in the area and (ii) implement different key aspects of the WFD on the Estonian side, and if possible on the Russian side too . One such project dealt with the Lake Peipsi system (including surface and groundwater) and was funded by the European Union (LIFE), FFEM (French Fund for Global Environment), BRGM (French Geological Survey), GTK (Finnish Geological Survey) and the two respective countries, Estonia and Russia (the EU-FFEM Lake Peipsi project). Designed to enhance collaboration between the two countries, it focused on selected pilot areas to establish the necessary corrective actions, and contributed in providing the ingredients and skills to Estonia and Russia to trigger similar parallel actions in the remaining portions of the Lake Peipsi water resource system, in order to elaborate the River Basin Management Plan and the Programmes of Measures as required by the WFD.

In particular, three major key aspects of water resources management were addressed: (i) shared monitoring – collecting, assessing and organizing common data to be shared, (ii) building common tools for transboundary water resources management and (iii) adding the socio-economic dimension to water resources management in the mining areas.

Concerning the data aspect, emphasis was placed on optimization processes in data collection and monitoring network configuration. The importance of meta-data was stressed as it is essential that each party understands the physical meaning and limitations of the data, and appreciates their reliability [7]. A common data base and a Geographic Information System (GIS) prototype were built to allow for data organization and assessment for various applications related to the protection and management of the shared water resources. This task also aimed at initiating the data exchange process between the two sides.

In addition, numerical models were built for pilot areas in the northern region of the Lake Peipsi catchment, including the mining area; one example is given in Reference [8]. These models were intended to give both parties common transboundary water resources management tools to progressively establish programmes of measures and management plans based on reliable data and knowledge of the water resources. The exercise was also meant to enhance discussion between both parties and give them common techniques and converging points of view for water resources management.

The third aspect of the project aimed at adding the sustainable development dimension to water resources management as is advocated by the WFD. In this respect, the work carried out included economic analyses focusing both on cost-efficiency and cost benefit aspects, which were applied to pilot areas where the impact assessment and the cost of remediation measures were being determined. One example concerned the costs and benefits of important works that needed to be built to protect the Vasavere glacial aquifer from the polluted Ordovician layer after the closing of oil shale mines in the area. Stakeholder involvement and more generally public participation in water resources management issues were also addressed in the framework of this project, as requested by the WFD [9].

Capacity building was carried out with the purpose of preparing local authorities to trigger similar actions over the remaining portions of the Lake Peipsi transboundary water resource system. As such, workshops and various bilateral or group meetings were held in Estonia, Russia, Finland and France. Fieldwork was also an important component of the project. Further details can be found in the literature [10, 11].

3.7.4
Discussion and Conclusion

Projects like the Lake Peipsi project obviously cannot pretend to solve all problems and ensure that upon completion the Lake Peipsi transboundary water resource system will be entirely in line with the requirements of the WFD. First of all, this project can only focus on selected key aspects and a limited number of demonstration actions. Extending this experience to other areas takes time. Secondly, when it comes to transboundary water resources management a common and concerted effort is needed from all the parties involved. This is already complicated inside of Europe, and it is even more so in the case of lake Peipsi as it is located at the frontiers of Europe and beyond, in countries with different legislation, different working methods, different obligations and a different culture on the value of 'water resources' on either side of the border.

The important point, however, is that through this kind of project the transboundary gap with respect to water resources management definitely starts to narrow and a convergence of viewpoints between the two neighbouring countries can begin. The EU-FFEM Lake Peipsi project (i) initiated a sustainable discussion and data exchange process between both parties, (ii) raised awareness on environmental matters, (iii) started providing a shared vision of the problems and of the solutions to implement and (iv) helped in building common tools and methodologies for sample areas that can be replicated in other places to eventually build up a sound river basin management plan and the necessary programmes of measures, acceptable to all stakeholders, on both parts of the border. As such, it clearly constitutes a step forward in the implementation of the WFD.

In fact, it is, notably, through this type of concrete transboundary project that the ideas conveyed by the WFD may slowly percolate outside of Europe. In this way progressive acceptance of these new concepts is enhanced. This will eventually be of benefit to all, including Europe itself, as management of shared water resources at its frontiers is facilitated, thereby ensuring at the same time better compliance with the internal regulations of each country.

In the short term, it is probably an illusion to believe that the full application of the WFD – including reaching good water status by 2015 – is possible in most of the shared water resources located at the frontiers of the European Union. However, at the same time, it is also believed by many specialists that the objective of reaching good water status by 2015 in many European water bodies stems from utopia, for example, just due to the great inertia of many aquifer systems.

However, in the long run the principles of the WFD will most probably be adopted by many countries, as this legislation is in line with today's needs and with the perception that most educated citizens have on the issues involving water resources, even in the southern countries [12]. The WFD is indeed one of the most advanced pieces of legislation in this respect, and is regarded by many countries as a model to follow. Applying the WFD concepts at the frontiers of Europe for transboundary water resources management has started in several places, and will eventually become a completely fulfilled reality, but it will take some time to be fully effective everywhere. The iterative nature of the WFD implementation should help in this respect.

Acknowledgement

We wish to express our gratitude to the whole CAMP-Life Environment Project team, and particularly to our Estonian partners MM. Ain Laane and Peter Marsoö who led the Estonian project team and participated in the socio-economical inquiries with the BRGM specialists; and to M. Rein Perens, Chief Hydrogeologist at the Estonian Geological Survey who proved to be the key scientist able to provide the project with a clear vision of the transboundary aquifer conceptual model.

We also wish to express our gratitude to the Saint Petersburg Expedition Russian team which took the lead of the project on the Russian part of the Chuskoye catchment, and particularly to M. Boris Anechkine, its project Director, and to Olga Kruglova.

References

1 EU (2000) Water Framework Directive – WFD. Directive 2000/60 EC of the European Parliament and of the Council of 23 October 2000.

2 Machard de Gramont, H., Houix, J.P., Pennequin, D. and Pinard, F. (2004) Réhabilitation et développement durable de la ressource transfrontalière du Lac Peipsi. *L'Eau, L'industrie, Les Nuisances*, 11, 31–44.

3 Guillaneau, J.C., Pennequin, D., Machard de Gramont, H., Pinard, F. and Houix, J.P. (2005) Sustainable water resources management and international cooperation: the example of Russia, Estonia and other countries. *Géoscience*, (2) 44–53.

4 MAVES (1999) Water Resources Assessment for Ida Viru County, Estonia, EU Phare.-REDOS2, Draft report: 92 pp.

5 Perens, R., Andresmaa, E., Antonov, V., Roll, G. and Sults, U. (2001) Groundwater Management in the northern Peipsi-Narva River Basin – Background Document of the Tartu Seminar: 45 pp.

6 Alexeeva, N. (2004) Integrated Transboundary Water Management in the Lake Peipsi/Chuskoye Basin, Publication of the Peipsi Centre for Transboundary Cooperation Project, 4 pp.

7 Mardhel, V., Breton, L., Rolland, N. and Théry, S. (2005) Report on GIS and Metadata in Peipsi Lake catchment, Life

Environment CAMP Project. Final Report – Appendix 2, 234 pp.

8 Le Nindre, Y.M. and Amraoui, N. (2005) Hydrogeological modelling of the northern catchment of Peipsi Lake in Estonia, Life Environment CAMP Project. Final Report – Appendix 1, 63 pp.

9 Rinaudo, J.D., Houix, J.P., Laane, A. and Marksö, P. (2005) Stakeholder consultation in the Viru-Peipsi water basin (Eastern Estonia), Life Environment CAMP Project. Report BRGM RP 53905-FR, 30 pp.

10 EU EU LIFE TCY/ROS/000049 (2006–2008) Narva Groundwater Management Plan, Progress Reports 1 to 6.

11 EU EU LIFE LIFE00/ENV/000925 (October 2003–September 2005) Viru Peipsi CAMP project – Implementation of a Basin Management plan on the Russian-Estonian border, Progress Reports 1 to 4.

12 Pennequin, D., (2010) Management of transboundary aquifer systems: a worldwide challenge, a need for increased concertation and political support. *Managing Shared Aquifer Resources in Africa, Third International Conference Tripoli, Libya, May 25–27, 2008*, UNESCO, Paris, pp. 50–58. Available at http://unesdoc.unesco.org/images/0018/001884/188462m.pdf.

Further Reading

Web site concerning water data in the Viru-Peipsi catchment area. Ministry of Environment – Environment Information Centre–Estonia. Karin Pachel, Water. http://grida.no/enrin/htmls/estonia/env2001/content/soe/water_3-3.htm (accessed 16 December 2010).

3.8
Implementation of the European Union Water Framework Directive in Non-EU Countries: Serbia in the Danube River Basin

Dragana Ninković, Marina Babić Mladenović, Miodrag Milovanović, Milan Dimkić, and Dragana Milovanović

3.8.1
Introduction

This chapter aims to show the challenges and difficulties of implementing EU legislation in a non-EU member state. It is based on the example of Serbia within the scope of the implementation of the EU Water Framework Directive in the international Danube river basin.

EU Directive 2000/60/EC – Water Framework Directive – came into force in October 2000 [1]. The Directive addresses EU countries and relates to establishing a framework for the protection of all kinds of waters. The main goal of the Directive is to achieve a good status of all waters by 2015 through the development of River Basin Management Plans.

The managing unit of the WFD implementation is a river basin district defined as '... the area of land and sea, made up of one or more neighbouring river basins, together with their associated groundwater and coastal waters'. Having in mind that a river basin district usually covers more than one country's territory and sometimes extends beyond the territory of the EU, appropriate coordination with relevant non-member states should be established by member states concerned.

The Danube river basin (DRB) is the second largest and the most international in Europe. It covers territories of EU member states, accession countries and other states that have not applied for EU membership. The territory of the Republic of Serbia accounts for approximately 10% of the DRB, both by state territory and population. For this reason, Serbia's involvement in the WFD implementation activities in the DRB was necessary, although it was neither an EU member state nor an EU accession country.

The implementation process of the WFD in such political, territorial and legal surroundings is described here.

3.8.2
Territorial and Institutional Scope

The Danube river basin (Figure 3.8.1) covers more than 800 000 km^2. It encompasses the territories of 19 countries with 81 million people and has a rich history and a strong cultural heritage. Since more than 90% of its territory lies within the Danube river basin, the Republic of Serbia is strongly committed to being a part of all joint activities related to water management in the basin.

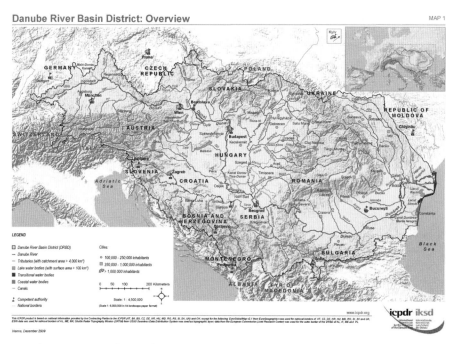

Figure 3.8.1 Danube river basin extent. ICPDR and IKSD (2009) Danube River Basin District: Overview. Available at http://www.icpdr.org/participate/sites/icpdr.org.participate/files/ DRBMPmap01_Over-view_5.pdf (accessed 10 November 2010).

The body responsible for implementing the WFD in the Danube river basin is the International Commission for the Protection of the Danube River (ICPDR). It serves as a platform for coordinating development and for establishing the Danube River Basin Management Plan. The ICPDR was established under the 'Convention on Cooperation for the Protection and Sustainable Use of the Danube River' (Danube River Protection Convention, DRPC, signed in Sofia in 1994 and came into force in 1998).

The ICPDR is formally consists of the Delegations of all DRPC Contracting Parties: Austria, Bosnia and Herzegovina, Bulgaria, Croatia, Czech Republic, Germany, Hungary, Moldova, Montenegro, Romania, Slovakia, Slovenia, Serbia, Ukraine and the European Union. As well as the above the following countries also cooperate with the ICPDR under the EU Water Framework Directive are Italy, Switzerland, Poland, Albania and the Former Yugoslav Republic of Macedonia. The contracting parties are also committed to making every effort to draw up a coordinated international River Basin Management Plan for the Danube river basin.

The ICPDR's main goals are the protection of water resources, healthy and sustainable river systems and damage-free floods. The different bodies of the ICPDR are the Ordinary Meeting Group, in charge of making political decisions, the Standing Working Group, which provides political guidance and the Technical Expert Groups. The latter two are the backbone of the operation of the ICPDR. They were formed by national experts from the contracting parties and representatives of the observer organizations. They deal with various issues – from policy measures to reduce water pollution, to the implementation of the EU-WFD.

3.8.3
Political Background and Legal Basis

At the time of ratification of the DRPC and active involvement in ICPDR activities (2003), Serbia was one of the two republics in the State Union of Serbia and Montenegro (CS). The Council of Ministers of the State Union designated the Serbian Ministry of Agriculture, Forestry and Water Management/Directorate for Water as the competent authority and body responsible for the synchronization of efforts in relation to the activities of the ICPDR and its working bodies.

In 2006 the State Union separated into two independent countries, making Serbia competent only for waters on its state territory. However, the southern Serbian province of Kosovo and Metohija stayed under the competence of the United Nations and this part of the Danube river basin is not covered by the WFD analyses and activities.

Serbian water management activities are legally based on the Water Law of the Republic of Serbia (1991), which addresses use of water, water protection, flood control, pollution control, water regime/quality and quantity, and so on, but is not harmonized with EU legislation. This means that the implementation of the WFD was not based on national legislation but on obligations related to the DRPC and relevant activities of the ICPDR.

Currently, a new Water Law compliant with the provisions of the WFD is under adoption procedure.

3.8.4
The WFD as a New 'Water Philosophy' – Challenges and Difficulties

In general, water management in Serbia encompasses three main issues: use of water, protection of water and protection from any adverse impact of water. Having that in mind, the WFD, which is based mainly on ecological rather than on quantitative aspects of water, came as a quite new water philosophy in Serbia. Starting from new basic definitions, with phrases that did not even exist in the Serbian language, for example 'water body', it was not easy to pass from traditional to new approaches in water management.

A lot of data and information necessary for the implementation of the WFD were not available. Good examples of this are data and methods for pressures and impacts analyses and related risk assessment, and GIS basics and techniques. Biological monitoring was not developed in a way that could be used for the implementation of the WFD.

Moreover, in comparison with other EU members or accessing countries in the Danube river basin that had the advantage of several years for preparation, Serbia had only a few months to get acquainted with the provisions of the WFD and to start the process of implementation.

Lack of capacities in certain sectors was a large obstacle too. Different kinds of experts from various institutions had to be involved with the process of the implementation of the WFD. It can be seen that the conditions for Serbia to start implementing the WFD on its territory were somewhat complicated.

3.8.5
Steps in Implementation of the WFD – Participants, Step-by-Step Approach, Capacity Building

To join other Danube countries in the implementation process of the WFD, the Serbian Directorate for Water called on the expertise of several important organizations. The organization with the highest level of involvement in this process is the 'Jaroslav Cerni' Institute for the Development of Water Resources, which is acting as coordinator, but a significant contribution also comes from the Public Water Management Companies 'Srbijavode' and 'Vode Vojvodine', as well as from some other institutes (Institute for Biological Research 'Siniša Stanković', Belgrade), faculties (Faculty of Mining and Geology, Belgrade) and public institutions (Republic Hydrometerological Service, Statistical Office). Other ministries contribute as well (e.g. Ministry of Environmental Protection).

The first step was the preparation of data and information for the DRB Characterization Report in accordance with Article 5 of the WFD. Serbia developed a detailed timetable to complete the necessary work within the deadline (March, 2005). Activities consisted of the Danube River and its tributaries with a catchment size larger than $4000 \, km^2$ and all transboundary groundwater bodies larger than $4000 \, km^2$. The 13 largest Serbian rivers and one important transboundary groundwater body were analysed in the report.

Over the next two years (2006, 2007) Serbia continued with Article 5 analyses for smaller rivers and groundwater bodies. Most of the tasks were completed for rivers with a catchment area of $100\,km^2$ or more.

Capacity building was mostly carried out by actively participating in ICPDR Expert groups. Different related projects (e.g. Sava CARDS Regional Project, Twinning Project with German Ministry of Environment, and Nature Protection and Nuclear Safety) and participation in training seminars provided the opportunity to exchange and share information, experiences and knowledge among related experts from both EU and non-EU states, which was also of great benefit.

In the meantime, involvement in the activities of the International Sava River Basin Commission (established in 2005 by the Framework Agreement on the Sava River Basin), which convenes former Yugoslav riparian republics, provided opportunity for further development of know-how among experts in the region.

3.8.6
Serbia's Contribution to the First Danube River Basin Management Plan

The first Danube River Basin Management Plan (DRBMP) was developed under the competences of the ICPDR. It focuses on the main transboundary problems, so-called Significant Water Management Issues (SWMI) that can directly or indirectly affect the quality of rivers and lakes as well as transboundary groundwater bodies. Such problems include pollution by organic substances, pollution by nutrients, pollution by hazardous substances and hydromorphological alterations. The plan includes visions and objectives for each SWMI as well as a joint programme of measures.

The activities of the expert groups and the Danube countries ensured that all preparatory measures were undertaken. Data collection and respective analyses were based on the DanubeGIS database, which provides a platform for exchanging, harmonizing and viewing geo-information and related issues.

Serbia's contribution is given in accordance with available data and information. The reason for unavailability of some data is mostly based on the fact that there is no relevant legal basis (e.g. data on different kinds of protected areas are not available because Serbia still has no related legislation).

At the same time the competent Serbian authorities, together with experts, are continuously taking action to fill existing gaps in data, methodologies and capacities. They are making efforts to prepare the Serbian water management authorities and relevant institutions for the EU integration procedure, when all obligations related to the WFD will be institutionalized and officially based on EU legislation.

3.8.7
Final Considerations

The Republic of Serbia is a non-EU member state with more than 90% of its territory lying in the international Danube river basin. It has no legal obligations as regards the implementation framework of the WFD, but relevant Serbian authorities and institutions have made a strong effort to actively participate in the preparation of the Danube River Basin Management Plan through the work of the International Commission for the Protection of the Danube River Basin.

Serbia has faced many difficulties due to lack of data, capacities and appropriate funds. However, relevant experts responded to the challenge by making significant efforts and have made Serbia an equal partner in the Danube River Basin Water Management process.

Moreover, after the adoption of the new Water Law, the previous activities of Serbia in the implementation process of the WFD and the experience gained will certainly make any similar tasks related to EU integration procedures much easier in the future.

References

1 European Parliament (2000) Directive 2000/60/EC of the European Parliament and of the Council, 23 October 2000.
2 Danube River Basin District Management Plan, Part A – Basin-wide overview, International Commission for the Protection of the Danube River (ICPDR) in cooperation with the countries of the Danube River Basin District, 2009.

3.9
Basic Problems and Prerequisites Regarding Transboundary Integrated Water Resources Management in South East Europe: The Case of the River Evros/Maritza/Meriç

Stylianos Skias, Andreas Kallioras and Fotis Pliakas

3.9.1
Introduction: The Way Towards Implementing IWRM

Integrated Water Resources Management (IWRM) has been defined as a process that promotes the coordinated development and management of water, land and related resources in order to maximize the resultant economic and social welfare (efficiency) in an equitable manner without compromising the sustainability of vital ecosystems. The term first appeared in the 1930s as a new paradigm that reinforces the importance of considering the world's complexities, which are due to the interactions between environment, society and technology [1]. In this context, geographic integration is an important dimension in a range of water related activities, such as planning, controlling and monitoring and resource allocation.

The declaration of the four Dublin principles, Agenda 21 of the Rio Earth Summit (1992) and the World Summit on Sustainable Development, Johannesburg 2002, are considered milestones towards shaping and implementing IWRM, since they established a new way of thinking based on three key strategic objectives: *efficiency, equity* and *environmental sustainability*. The statement of the Dublin Conference equates the term 'integrated' (IWRM) to 'holistic'. Besides engaging 'sustainability' into the concept of IWRM, the time dimension is activated. Sustainability refers directly to resource use that can be sustained over time for generations to come. Two international bodies were set up to address these issues: the Global Water Partnership (GWP) and the World Water Council (WWC).

A river basin is internationally considered to be the most suitable geographic unit for IWRM. Dourojeanni *et al.* [2], have justified this on the grounds that a river basin (i) corresponds to the principal terrestrial form of the hydrologic cycle; (ii) reflects the interrelationship and interdependence between water uses and users; (iii) represents the region where water and physical and biotic systems interact; and (iv) is strongly linked to the socio-economic system of the region. However, some countries suggest other criteria for defining IWRM units, including historic development, cultural and environmental aspects and strategic water uses, representing the 'problem shed' concept, as defined by Vlachos and Mylopoulos [3] and Allan [4]. In addition, political boundaries, which generally do not coincide with the hydrological limits, can represent a strong barrier to using river basin areas as territorial units for IWRM. Political boundaries can even exist between different regions in the same country [5]. Internal issues within national borders and external issues between riparian countries regarding water sharing [5] can be reduced by defining IWRM units and by ensuring that there is a comprehensive institutional structure [6] with sufficient power to lessen the boundary effects.

3.9.2
The Transboundary Dimension of IWRM: Problems, Principles and Goals

Worldwide more than 45% of the land surface is located within international river basins and many groundwater aquifers are shared by more than one country [7]. Unilateral action by one country concerning these resources is often ineffective (e.g. fish ladders in an upstream country only), inefficient (e.g. hydropower development in a flat downstream country) or simply impossible (many developments on boundary stretches). Moreover, unilateral action (e.g. change in water flow regime) can significantly harm the other countries and may result in serious international tension.

Perhaps the biggest problem in sharing an international water system is its sheer scale and the opaqueness of system interactions over large distances (upstream and downstream). This opaqueness may result in unforeseen negative consequences of human interventions (engineering-structural or/and policy measures), which are difficult to correct and may give rise to tensions between riparian countries sharing the water system. In addition, within the same international river basin, national interests usually differ; thus riparian nations may develop diverging policies and plans that are not compatible. This is a manifestation of the sovereignty dilemma: that is to what extent may individual countries develop and use resources found within their territories and to what extent do they have to consider interests of riparian countries, and the 'common interest' of the river basin as a whole? One of the biggest challenges in sharing international rivers is to identify development strategies whereby all riparian countries eventually benefit from an equitable allocation of costs and benefits.

There are many reported principles of transboundary IWRM, but the guiding principles that are recognized by international conventions, treaties and judicial decisions are: the theory of limited territorial sovereignty; the principle of equitable and reasonable utilization; an obligation not to cause significant harm; the principles of cooperation, information exchange, notification and consultation; and the peaceful settlement of disputes. These principles form the basis of the 1966 Helsinki Rules on the Uses of the Waters of International Rivers and the 1997 UN Convention on Non-Navigational Uses of International Watercourses. The inclusion of these principles in an agreement between two or more countries offers common ground and a window of opportunity to foster coordinated and sustainable water resources development and management. The clarified acceptance and practical application of these principles is a very difficult task but can greatly facilitate synergy towards an operational water sharing process. Equitable water sharing among all stakeholders must always be the ultimate goal of IWRM. The practical application/achievement of equitable water sharing in an international basin necessitates (as a prerequisite) the formation and operation of a proper International River Commission (IRC) as a formal interstate institutional body, which will have as a basic task the recommendation (and monitoring upon implementation) of appropriate (in space and time) decisions (regarding plans, projects and policies engaged to IWRM) to decision makers of the participating countries. The formulation of this IRC should be based on the above-mentioned guiding principles and its structure and functioning on three basic supporting pillars: the *operational* (technical cooperation), the *political* (responsible for an enabling environment) and the *institutional* (responsible for laws and institutions). The operational pillar is central to the success of any IRC tasks. It may support most of the load if one of the two outer pillars is weak, cracked or in the process of being repaired or restructured [8].

3.9.3
Transboundary IWRM, EU WFD and Cooperation in SE Europe

Some 90% of the area of the SE European (Balkan) countries falls within transboundary river basins, including those of Danube, Drin, Martisa/Meriç/Evros, Neretva and others. These and other transboundary rivers flow into the Adriatic, the Aegean, the Ionian and the Black seas. More than half of the transboundary basins are shared by three or more riparian states. Shared lake basins include Doiran, Ohrid, Prespa and Shkoder. The SE European region is also characterized by a large number of transboundary aquifers that are often karstic in their nature. Prior to 1992 there were six major transboundary rivers crossing the sub-Danubian geographical area which consists of territories belonging to SE European countries. These rivers are the Aoos/Vjosa, Drim, Axios/Vardar, Strymon/Struma, Nestos/Mesta and Evros/Maritza/Meriç. With the emergence of new states (Croatia, Slovenia, Bosnia and Herzegovina, the Former Yugoslav Republic of Macedonia, Serbia, and Montenegro) in the SE European-Balkan region, the number of shared rivers in the area has more than doubled. In fact, several other rivers (e.g. Sava, Kupa/Colpa, Cetina, Una, Drina, Neretva and Trebišnjica) are now listed as transboundary ones.

There have been numerous initiatives regarding cooperation for sharing transboundary waters among SE European countries, but the existing formal agreements are very limited and they are almost exclusively of a bilateral nature. These bilateral agreements do not cover all existing country pairs and some of them are rather problematic in their implementation. Taking as an example Greece, cooperation among riparian countries is of vital importance, since roughly 25% of the country's renewable water resources are 'imported', as Greece is the downstream country in four out of the five transboundary rivers. The lack of the necessary, functional water agreements between Greece and its neighbouring countries (there are only a few in existence with Bulgaria and more recently with Albania), negatively affects regional cooperation and the state of the water resources in the respective transboundary basins. In addition, certain existing water related agreements between Greece and Bulgaria are either entirely (river Nestos/Mesta) or partly (river Evros/Maritza) problematic and do not cover certain important issues (e.g. protection from flooding in the river Evros basin).

The potential for international conflicts as a result of water scarcity, quality degradation or even flooding regarding shared waters poses a risk to stability and development in SE Europe. The international community (including the EU, donor countries, international organizations, inter- and non-governmental organizations) has undertaken a series of initiatives, many of which are complementary. Particular reference is made to the St. Petersberg Process (1998) and the Athens Declaration Process (2003). Regrettably, no sound or formal, water related agreements have yet been drawn up as a result of the above-mentioned initiatives and processes.

Transboundary rivers in SE Europe currently cross but a few EU member countries and mostly non-member states. The latter have obviously no obligation to implement the European Directives. The EU Water Framework Directive (WFD), 60/2000, is based on a holistic management approach and in the case of international basins requires each of them to be assigned to an international River Basin District (RBD). The Directive further specifies that member countries shall ensure cooperation for producing one single River Basin Management Plan for an international RBD falling within the territories of the EU; however, somewhat confusingly, the Directive at the same time indicates that if not produced then plans must be set up for the part of the basin falling within each country's own territory. If the basin extends beyond the territories of the EU, the directive encourages Member States to establish cooperation with non-Member States and, thus, manage the water resource on a whole basin level (Articles 3 and 13). The guidance document 'Best Practices in River Basin Management Planning', produced as a part of the Common Implementation Strategy, touches upon international RBDs but does not actually go any further than the Directive in specifying how to designate international RBDs. Thus, the rather vague formulations in the WFD may result in multiple interpretations by Member States as to how to actually implement it. The international dimensions are more explicit in the WFD than in other Directives, potentially requiring member States to move towards close cooperation in managing shared river basins. The strict legal requirements to actually achieve joint management are weak. This fact has already created cooperation problems, as in the case of the rivers Nestos/Mesta and Evros/Maritza/Meriç. Thus a prerequisite for implementing IWRM in transboundary rivers, especially within SE Europe, is the formulation of a clear, strict and rational set of legal requirements by the EU (under the Common Implementation Strategy).

Problems and challenges regarding shared waters in SE Europe include: (i) surface and ground water quantity/quality, (ii) navigation, (iii) balancing conflicting interests to ensure ecosystem/biodiversity conservation and (iv) management of flood risk. All of the above obviously require information networking and sharing.

3.9.4
The River Evros/Maritza/Meriç Case: Problems and Recommendations

The transboundary river Evros basin [about $53\,000\,km^2$, shared by Bulgaria (66%), Turkey (28%) and Greece (6%)] represents a very complicated case. The numerous water related initiatives and agreements that have been undertaken and signed so far, mostly on a bilateral basis, have been proved inadequate for establishing a cooperative and contemporary management status in the basin of the river as a whole, for various political, cultural and other socio-economic reasons, which exhibit varying content and intensity over time. The present geographically and operationally fragmented management status facilitates the occurrence of political tension, especially during crisis periods (flooding, water scarcity, pollution, etc.). As far as the repeated and devastating flood events are concerned, key-factors manifesting poor water management are represented by the operational status of a series of old dams in the Bulgarian territory, the functioning of the protection dikes and the land use regime in the flood plains of the river network of all three riparian countries [9].

Thus, the three riparian countries (Greece, Bulgaria and Turkey) must urgently identify and appreciate all the existing cross border water (quantity and quality) problems and look at the relevant lessons learned the hard way by other European countries. Then, a framework agreement (on a win-win basis) has to be reached, where (i) immediate (data sharing, monitoring, databases), (ii) intermediate (IWRM) and (iii) final (reasonable and equitable water sharing) tasks and goals are clearly defined, as well as the prerequisites (as mentioned in ii) needed to achieve them. The kick-off for establishing this framework agreement requires strong political commitment from all three countries and perhaps pressure from an international political body (third party's involvement) such as the EU. The water-related scientific and academic communities can play a decisive role in creating an enabling environment, which facilitates cooperation and agreement at the top political level in their countries.

Greece, the oldest EU member in SE Europe, and as a receptor, downstream country the one that suffers the most with floods and pollution from the negative impacts of poor transborder cooperation regarding water management, should commit to use all available European experience and support to ensure that initiatives are taken towards achieving a rational tripartite agreement as already mentioned. Bulgaria, being the upstream country with more than 50% of the Evros river basin's area on its territory, has the greatest economic and ecological incentives to facilitate the formulation of an agreement on a long-term, multi-criteria, cost–benefit basis. The tripartite agreement for the river Evros basin should be in accordance with the Helsinki Rules and refer to an IRC (equivalent to that of the Danube or Rhine rivers) based on the previously mentioned three supporting pillars. A first priority scientific task of the IRC should be the compilation of a Master Plan for implementing IWRM on a 'win-win' basis in the whole transboundary basin. The EU water related Directives (60/2000, 60/2007, etc.) should be the guiding documents for the compilation of the Master Plan. This agreed process (IRC and Master Plan) for the river Evros may well act as a paradigm for establishing similar processes for other transboundary river basins in SE Europe [9].

References

1 Vlachos, E. (2008) *Lecture Notes. Seminar on Technology Assessment and Social Forecasting – CIVE639*, Colorado State University, Fort Collins, Colorado.

2 Dourojeanni, A., Jouravlel, A. and Chavez, G. (2002) *Gestión del Agua a Nivel de Cuencas: Teoría y Práctica*, Serie Recursos Naturales e Infraestructura, No. 47, CEPAL Division de Recursos Naturales e Infraestructura Santiago, Chile.

3 Vlachos, E. and Mylopoulos, Y. (2000) The status of transboundary water resources in the Balkans: establishing a context for hydrodiplomacy, in *Transboundary Water Resources in the Balkans: Initiating a Sustainable Co-operative Network* (eds J. Ganoulis et al..) NATO Science Series, Vol. 74, Kluwer, Dordrecht.

4 Allan, J.A. (2005) Water in the environment/socio-economic development discourse: sustainability, changing management paradigms and policy responses in a global system. *Government and Opposition*, 40, 181–199.

5 Ganoulis, J., Duckstein, L., Literathy, P. and Bogardi, I. (1996) *Transboundary Water Resources Management: Institutional and Engineering Approaches* (eds J. Ganoulis et al..), NATO ASI Series, Vol. 7, Springer Verlag, Heidelberg.

6 Waterstone, M. (1996) A conceptual framework for the institutional analysis of transboundary water resources management: theoretical perspectives, in *Transboundary Water Resources Management: Institutional and Engineering Approaches*, (eds J. Ganoulis et al.), NATO ASI Series, Vol. 7, Springer Verlag, Heidelberg.

7 Wolf, A.T. (1998) Conflict and cooperation along international waterways. *Water Policy*, 1 (2), 251–265.

8 Savenije, H.H.G., van der Zaag, P. and Wolf, A.T.(guest editors) (2000) The management of shared river basins, *Water Policy*, 2 (Special issue 1–2), 149 pp.

9 Skias, S. and Kallioras, A. (2007) Cross border co-operation and the problem of flooding in the Evros Delta, in *Many Rivers to Cross. Cross Border Co-operation in River Management* (eds J. Verwijmeren and M. Wiering), Eburon Academic Publishers, pp. 119–143.

Part Two
Physical, Environmental and Technical Approaches

Transboundary Water Resources Management: A Multidisciplinary Approach, First Edition.
Edited by Jacques Ganoulis, Alice Aureli and Jean Fried.
© 2011 Wiley-VCH Verlag GmbH & Co. KGaA. Published 2011 by Wiley-VCH Verlag GmbH & Co. KGaA.

4
Transboundary Aquifers

4.1
Towards a Methodology for the Assessment of Internationally Shared Aquifers

Neno Kukuric, Jac van der Gun and Slavek Vasak

4.1.1
Introduction

The issue of shared international waters is as old as the national borders that make those waters international. During the last century, significant progress was made in the regulation of joint management of surface watercourses; many international river-, lake- or basin commissions were set up and legal treaties signed. Although some of these activities address 'a groundwater component' as well, major comparable efforts related to the invisible groundwater only started a few years ago.

Much can be learned from surface water experiences with respect to the socio-economical, legal and institutional aspects of international water management. Moreover, when surface and groundwater are hydraulically connected, integrated water resources management should be exercised. However, compared to surface water, the assessment of groundwater is much more demanding and less certain. This chapter discusses the assessment of internationally shared groundwater in an attempt to further clarify the assessment specifics and to elaborate the procedure to be followed.

4.1.2
Principles and Basis Steps

The assessment needs to include all relevant aspects or facets of transboundary groundwaters. The framework document of the Internationally Shared Aquifer Resources Aquifer (ISARM) initiative (www.isarm.net) [1] distinguishes five aspects of transboundary aquifers, namely hydrogeological, legal, socio-economical,

Transboundary Water Resources Management: A Multidisciplinary Approach, First Edition.
Edited by Jacques Ganoulis, Alice Aureli and Jean Fried.
© 2011 Wiley-VCH Verlag GmbH & Co. KGaA. Published 2011 by Wiley-VCH Verlag GmbH & Co. KGaA.

institutional and environmental. In the document, the development of guidelines for each of the aspects listed above was suggested. In the meantime, the only substantial progress in preparing the guidelines has been made by developing a legal instrument – the articles on the law of transboundary aquifers. Numerous activities [particularly Global Environment Facility (GEF) projects and ISARM inventories] conducted in the last couple of years have yielded precious information, particularly on the hydrogeological and institutional aspects of transboundary aquifers. Yet, there are still no guidelines to assist these activities, nor an agreed overall methodology for assessment.

The original ISARM suggestion of having a separate guideline for each aspect of transboundary aquifers was apparently very ambitious. Besides, there is a strong link amongst the various aspects in practice. Assessment of transboundary aquifers is often mostly limited to the assessment of their hydrogeological situation, but if properly conducted it should also incorporate sufficient information on other aspects. This also needs to be reflected in the assessment procedure, which should guide the user through the assessment steps, pointing out the specifics of various aquifer aspects in an international context.

The assessment of the shared groundwater could be seen as being composed of the following steps:

- delineation and description,
- classification, diagnostic analysis and zoning,
- data harmonization and information management.

The first two activities (delineation and description) could be clustered as 'inventory' or 'characterization', depending on the stage and the scale of activities. In any case, delineation and description are chiefly about collecting, combining and interpreting the field information.

The second set of activities provides the stakeholders with information necessary for decision-making, such as on problems that may develop and opportunities that will be missed in the absence of coordinated groundwater resources development and management. Later on, the stakeholders need to know which aquifers are likely to be most the responsive ones to transboundary aquifer management, and which zones within such aquifers should be targeted for the greatest positive impacts.

One could argue whether the activities mentioned in the third step should be addressed separately; data harmonization takes place in the previous steps and the information management is a management measure. Nevertheless, data harmonization and information management have an additional dimension and importance in the international content; they are more difficult to carry out, and are more elaborate and politically sensitive. At the same time, they are also an opportunity for building trust and mutual understanding amongst the involved parties.

The following section contains a brief elaboration of the basic assessment steps.

4.1.3
Elaboration of the Basic Steps

4.1.3.1 Delineation

Delineation of aquifer geometry (lateral extent and depth) is the most essential and usually the most difficult part of the assessment. An aquifer may be formed by an alluvial strip along a river, a single hard rock formation or a complex of various hydraulically interconnected formations. Though the regional hydrogeological settings are usually known and the approximate boundary of aquifers might be shown on existing hydrogeological maps, information on the exact boundaries of transboundary aquifers is often lacking. For the sharing countries a question may arise: Can we delineate the aquifers, and if yes to what level of detail should this information be shared with other parties? There are several concepts and levels of delineation encountered in on-going transboundary initiatives (from lines through circles to detailed delineation). Figure 4.1.1 shows a detail from a transboundary aquifer map of the world [2] with various levels of delineation and aquifer recognition. The map shows the aquifer extent (if known) of aquifers with an area larger than 6000 km². Smaller aquifers are represented with squares. If the exact aquifer boundaries are known and acknowledged by all sharing countries they are delineated with solid red lines. If not, they are delineated with dashed red lines. Filled and half-filled circles are used to depict aquifers whose extent is not known. A filled circle represents an aquifer whose occurrence is confirmed by all countries involved; if an aquifer is not recognized by all countries, it is depicted by a half-filled circle.

4.1.3.2 Description

Once the lateral extent and thickness of a transboundary aquifer is defined (or approximated), its main properties have to be described. The recharge/discharge

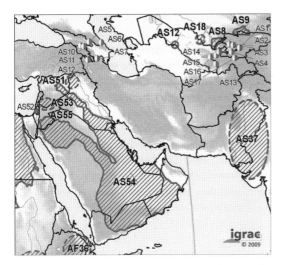

Figure 4.1.1 Detail from a transboundary aquifer map of the world.

mechanism and the hydraulic properties of the aquifer are needed to determine the direction and velocity of groundwater flow and its interaction with other water bodies (rivers, lakes, seas). These characteristics are also necessary to assess the aquifer's vulnerability to overexploitation and pollution. Superimposed on these hydrogeological characteristics are anthropologic influences, such as abstraction and pollution from various sources. An impression of the current status of transboundary aquifer descriptions can be obtained from the overview of ISARM's transboundary activities [3]. The general lack of data, the main bottleneck of any aquifer description, can be tackled by the proper design of inventory forms (clear formulation of data type, units, etc.) and by use of proxy information from various sources. A comprehensive inventory form is used for instance in the inventory of TBAs (transboundary aquifers) in Southern Africa [4].

4.1.3.3 Classification
Classification is simplification, but its intention is to deepen knowledge by revealing patterns, by highlighting certain features and/or by facilitating comparison between objects. In the case of aquifers, it allows aquifers to be compared with each other according to different characteristics. Many different characteristics can be chosen, depending on available information and the context of the analysis. Examples of characteristics for classification that may be relevant in the context of transboundary aquifer management are aquifer size and hydraulic properties, vulnerability, current functions, observed or perceived stresses, need for transboundary aquifer management and so on.

4.1.3.4 Diagnostic Analysis
This step concerns interfacing assessment with transboundary aquifer management planning. Diagnostic analysis can be carried out in the first place as a screening step at the regional level, covering a certain number of transboundary aquifers. Its objective is ranking the inventoried transboundary aquifers according to criteria related to priority for transboundary management and so help water resources managers to decide which ones to select for inclusion in the transboundary action planning.

Transboundary diagnostic analysis (TDA) can also be applied to a single transboundary aquifer system, in analogy to TDA applied to lakes, inland seas and river basins. Published TDA results in the surface water domain usually contain the following elements: (i) inventory of major perceived issues and problems; (ii) overview of possible actions in response to the perceived issues and problems; (iii) details on the main proposed actions and related aspects (stakeholders, institutions, expected impacts, etc.). Although there is no significant experience yet of TDA application to groundwater systems, it is expected that this approach can be used successfully. After concluding TDA, a strategic action plan can be developed.

4.1.3.5 Zoning
It can be observed that in groundwater systems, much more so than in surface water systems, current and potential transboundary effects vary enormously

within the aerial extent of the delineated system. As a result, it is often only necessary to control the transboundary interactions of a minor part of a large transboundary aquifer. This is due to the inherent inertia of groundwater systems and the usual fragmentation of groundwater flow into separate flow domains. Transboundary aquifer management should focus only on those parts of the aquifer systems that are likely to cause or receive transboundary impacts within a reasonable time frame. With this in mind it may be helpful to divide the aquifer into a number of zones. The zoning methodology should take into account the hydraulic characteristics of the aquifers (e.g. contrasting karstic versus weathered bedrock aquifers), flow direction and the type of transboundary interaction expected amongst others.

4.1.3.6 Data Harmonization and Information Management

The success of the characterization of any aquifer relies heavily on the availability and quality of related data. For internationally shared aquifers, however, the harmonization of data across the border plays an equally important role; if two data sets cannot be mutually compared (and further processed) they are not of much use. Besides, these data need to be made accessible internationally, which brings up the issue of information management.

Essentially, data harmonization and information management are technical activities related to the harmonization of formats, classifications, terminologies, reference systems and reference levels, software and hardware specifics and so on. Yet they are very much determined by the political, organizational, legal, cultural and economical situation and agenda. Progress, or lack of progress, made in harmonization and common information management is often a reflection of political willingness to cooperate, but also of other differences, such as technical and organizational. Finally, the complexity of data is also a major factor: the harmonization of hydrogeological maps is, for example, far more complex than one of groundwater levels.

As with the previously described assessment steps, data harmonization and information management are carried out at various levels, the level being largely defined by current data availability and the ambitions of involved countries. Many countries exchange and harmonize data *ad hoc*, for the purpose and duration of mostly common projects in border regions. This small-scale collaboration usually works, however inefficiently and without structural contribution to common information management.

In several GEF groundwater projects, relatively simple databases have been developed to accommodate data of basic groundwater variables. Usually, a Geographical Information System (GIS) is employed for visualization of the maps and spatial variability of groundwater variables. No cases have been reported of harmonization going beyond items such as reference levels and measurement scales. Equally, developed databases or systems are (according to the available information) neither web-based, nor real-time (i.e. automatically updated from the field). Databases available on the IGRAC (Figure 4.1.2) and The UNESCO Chair INWEB (International Network of Water-Environment Centres for the Balkans) [5] portals

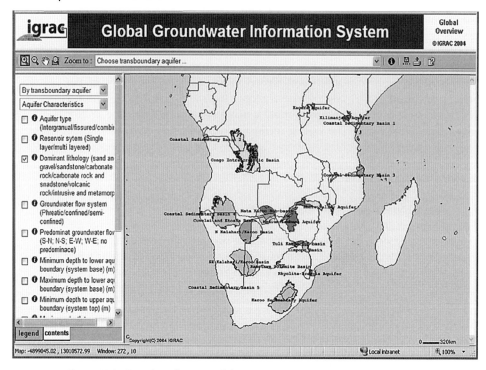

Figure 4.1.2 Transboundary view of the GGIS (IGRAC).

contain meta-information on transboundary aquifers. Water Information System Europe (WISE) accommodates delineated 'groundwater bodies' and the observations (with rather low density and frequency) of a selected set of groundwater variables. Ideally, all the transboundary data should be harmonized and made available on-line and in real-time. Such a sophisticated information system has been developed for instance for the border region of Germany and the Netherlands [6]. The system provides access to circa 20 000 piezometers and a semantic on-fly translation for lithological and hydrological units across the border.

4.1.4
Final Remarks

Assessment of internationally shared groundwaters is a challenging endeavour due to several factors such as groundwater invisibility, usually slow change, differing approaches to aquifer assessment, lack of information, political will and so on. There is a clear need for a generally accepted TBA assessment methodology. In this chapter the possible basic steps for such a methodology have been briefly discussed. Delineation and description steps are about collection and interpretation of hydro-geological information and subsequent classification, while diagnostic analysis and zoning are meant to provide the stakeholders with information necessary for decision-making. Finally, data harmonization and information management steps are more demanding at an international level than at a national level due to differences in language, classifications, terminologies, formats, reference systems and so on.

These concepts need to be further elaborated. This will be carried out under the ISARM umbrella to ensure the broad involvement of all partners and subsequently a general acceptance of the methodology.

References

1 UNESCO (2001) ISARM Framework Document (www.isarm.net) (accessed 15 October 2010).
2 IGRAC (2009) Transboundary Aquifer Map of the World, a special edition for the 5th World Water Forum Istanbul.
3 UNESCO/IGRAC (2009) ISARM Web Portal), www.isarm.net (accessed 15 October 2010).
4 http://www.isarm.net/publications/297 (accessed 15 October 2010).
5 http://www.inweb.gr (accessed 15 October 2010).
6 Kukuric, N. and Belien, W. (2006) Distributed information services for cross-border water management. Presented at the 7th International Conference on Hydroinformatics, 4–7 Sep 2006, Nice, France.

Further Reading

Van der Gun, J. (2006) Transboundary aquifer resources management: a topic in development. Third International Symposium on Transboundary Waters, 30 May–2 Jun 2006, University of Castilla–La Mancha/UNESCO, Ciudad Real. Spain.

4.2
Challenges in Transboundary Karst Water Resources Management – Sharing Data and Information

Ognjen Bonacci

4.2.1
Introduction

To obtain harmonious, reliable and sustainable development it is necessary to take the complex, interactive, technical, social, economic, environmental and cultural aspects of water resources management into account in decision-making. In the case of transboundary karst water resources management this is an especially hard task. Sustainable development should consider the global pressure on karst water resources. It should safeguard the quantity and quality of vulnerable karst water resources with its valuable habitats, home to various endangered and often endemic surface and, especially, underground species. The necessity of identifying a karst water system as a whole is becoming more significant, because karst terrains have recently become more densely populated, which has resulted in greater demands for water. One billion people in about 40 countries live in areas characterized by karst formation. Common environmental problems in these fragile landscapes are desertification, fast and massive groundwater pollution, the collapse of land, flooding and droughts.

Puri and Aureli [1] consider that transboundary aquifers are as important a component of global water resource systems as transboundary rivers. They warn that their recognition in international water policy and legislation is very limited. Existing international conventions and agreements barely address aquifers and their resources. This poses a special problem for transboundary karst water resources management.

The goal of this chapter is to point out the specific characteristics of karst water circulation that create difficulties in karst water resources management, and at the same time represent the main and real trigger for possible conflict when they are internationally shared. The karst system shows extreme heterogeneity and variability of geologic, morphologic, hydrogeologic, hydrologic, hydraulic, ecological and other parameters in time and space. A system this complex needs an interdisciplinary approach. It is highly important to understand the interaction of groundwater and surface water in karst and their influence on surface and underground biological processes [2].

When considering problems of transboundary water resources management the fact that there is a long history of water-related violence and conflicts should not be disregarded. The chance of terrorists targeting water is real. Gleick [3] considers that there is a long history of such destructive attacks. Water infrastructure can be destroyed or water can be contaminated. This latter threat is very real in vulnerable karst terrain.

4.2.2
Specific Characteristics of Karst

Karst terrains are typified by a wide range of closed and a few open surface depressions, a well-developed underground drainage system and strong interaction between the circulation of surface water and groundwater. Karstification is a continuous process governed by natural and man-made interventions [4].

Conditions for water circulation and storage in karstified medium are strongly dependent on the consideration of space and time scales, especially in the deep and morphologically complex vadose zone. This zone and underlying phreatic karst aquifer are a two-component system in which the major part of storage is in the form of true groundwater in narrow fissures and matrix, where diffuse or laminar flow prevails. On the other hand, the majority of the water is transmitted through the karst underground by quick or turbulent flow in solutionally enlarged conduits.

One of the root causes of problems in karst waters management is the fact that is it either impossible or very difficult to define the catchment areas and catchment boundaries, as well as their changes in time and space. The determination of these boundaries is the starting point in most hydrological analyses and one of the essential data that serve as a basis for water resources management. The differences between the topographic and hydrologic catchments in karst terrain are usually so large that data about the topographic catchment are useless in water management practice. Very often the position of the karst catchment boundaries depends upon the groundwater levels, which strongly and sharply change in time and space. In some situations at very high groundwater levels (caused by heavy rainfall and during karst flash floods) generally fossil and inactive underground karst conduits are activated, causing the redistribution of the catchment areas, that is overflow from one catchment to another [5].

Anthropogenic intervention, especially constructions of dams and reservoirs and inter-basin water transfers through long tunnels and pipelines, can introduce instantaneous and distinct changes in catchment areas and boundaries. In karst terrains those processes are very often uncontrolled, and result in hazardous consequences [6]. Natural and anthropogenic changes in karst regions frequently cause redistribution in karst catchments, which strongly, suddenly and dangerously affect processes of water circulation at the local and regional scale. The benefit resulting in one area is frequently less significant than the damage caused in another area. Occurrence of landslides, floods, collapses of dolines, regional water redistribution and the drying up of karst springs and open watercourses can be caused. In the case of transboundary shared karst surface water and groundwater these events can be a trigger for serious international conflicts.

Karst water resources, because of their unique hydrologic and ecologic characteristics, are extremely susceptible to contamination. The surface and especially subterranean environment in karst provide a range of habitats with very different chemical and biological processes. Karst ecosystems are sensitive to environmental changes. Karst species are often highly endangered because of the extreme vulnerability of karst terrains and because populations are so small. The importance of maintaining biological diversity goes far beyond the mere protection of endangered species and a beautiful landscape. It is necessary to obtain a thorough understanding of how aquatic and terrestrial ecosystems function and interact with each other in very complex and vulnerable karst systems, which at the same time are also extremely dynamic in time and space [2]. This means that it is imperative that the ecological aspects of transboundary karst water resources management are treated much more carefully than would be the case in other types of terrain.

4.2.3
Three Cases of Karst Transboundary Water Resources Management

Water crises concerning karst surface water and groundwater management are becoming increasingly serious all over the world. Three examples of internationally shared karst water resources are given.

The Ohrid and Prespa lakes are located in the Balkan Peninsula. The Ohrid Lake is shared by the FYR of Macedonia and Albania, and the Prespa lakes by the FYR of Macedonia, Albania and Greece [7]. From the hydrological and hydrogeological points of view the lakes are extremely interesting water bodies, but they have yet to be fully investigated. The main reasons for the lack of investigation are the very complex runoff processes in the karst underground of their catchments and the fact that the lakes' basin is shared among three countries. The lakes are the largest tectonic lakes in Europe. The Ohrid Lake is a resource of tremendous local and international significance and because of its biodiversity and unique cultural heritage was declared a UNESCO World Heritage Site in 1980. The Prespa lakes do not have surface outflow and its waters outflow through a karst underground massif into the Ohrid Lake. Therefore, from the hydrological as well as the transboundary water management aspect, the lakes cannot be analysed separately. The lakes are influenced by permanent and strong natural changes that are poorly monitored, but also by uncoordinated, uncontrolled and mainly dangerous anthropogenic activities.

For the sustainable development and protection of such ecologically and economically valuable karst transboundary water resources, it is very important to establish prerequisites for the definition of a reliable water balance. This will provide conditions for integrated and sustainable management of transboundary water resources and avoid potential conflicts related to water use. This is the only way to prevent an ecological catastrophe, which is a real threat mainly to the Prespa lakes, where the water level decreased 7.8 m from 1963 to 1995.

The Cetina River in southern Croatia and western Bosnia-Herzegovina has been regulated for hydroelectric power generation since the beginning of the twentieth century. Its catchment area is estimated to cover about 4000 km², although the precise hydrological catchment area and boundaries are not known. From the catchment area of the Cetina River located in Bosnia-Herzegovina the water flows to the Croatian part of its catchment only through karst underground connections. The analyses show that the operation of five hydroelectric power plants have caused changes in the hydrological regime throughout the greater part of the Cetina River course situated in Croatia, while about two-thirds of its catchment is in Bosnia-Herzegovina [8, 9]. The hydrotechnical system currently in existence was built in former Yugoslavia. The Cetina River catchment is a typical example of an unclear relationship between water in an open karst stream flow, many abundant karst springs and karst groundwater in a transboundary shared karst catchment. However, even though more systematic work on the management of the transboundary karst water resources of the Cetina River catchment between Croatia and Bosnia-Herzegovina has not yet started, it is positive that both sides left the discussions for a more opportune moment.

The Trebišnjica River catchment is internationally shared between Croatia and Bosnia-Herzegovina. In former Yugoslavia a complex hydrotechnical system was constructed in its catchment area (Figure 4.2.1). This area is part of deep and bare Dinaric karst [4, 6, 8, 10]. There are strong karstic underground connections between the Trebišnjica River and the Ombla karst spring, which supply the town of Dubrovnik with high quality water, as well as many other karst springs around the Adriatic Sea coast in Croatia. Croatia plans to build the hydro electric power plant at Ombla, which will exclusively use groundwater from the Ombla spring aquifer, while Bosnia-Herzegovina wishes to transfer water through a tunnel

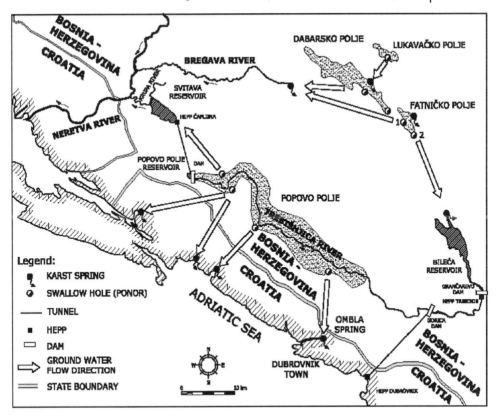

Figure 4.2.1 Map of the Trebišnjica River catchment with designated state boundary between Croatia and Bosnia-Herzegovina, reservoirs and hydroelectric power plants of the hydrotechnical system.

from the Fatničko polje to the Bileća reservoir [11]. Figure 4.2.1 shows the position of three poljes (Lukavačko, Dabarsko and Fatničko) with their altitudes, location of permanent and temporary karst springs and ponors, and groundwater flow directions based on groundwater tracing measurements. The poljes are hydrogeologically connected [10, 11]. Very interestingly, water swallowed up in the ponor zone 1 in Figure 4.2.1 reappears at the Bregava river karst spring, while the water sinking in ponor zone 2 in Figure 4.2.1 reappears at the Trebišnjica river karst spring. The 'straight-line' distance between these two springs is more than 65 km, while the distance between the two ponor zones in the Fatničko polje is only about 2 km. Consequences of the Croatian and Bosnian and Herzegovinian projects will change the hydrological and hydrogeological regime in the complex and little understood transboundary karst aquifer system of the Trebišnjica River. These changes could create serious water management problems between the two neighbouring countries sharing the Trebišnjica River catchment.

4.2.4
Conclusion

Because of the fact that the appearance, storage and circulation of water in karstified areas is significantly different from other more homogenous and isotropic terrains, transboundary karst water management should develop original methods and approaches based on the continuous monitoring of different climatological, hydrological and ecological parameters, amongst others. A first and essential step is to define the aquifer characteristics, the catchment areas and the parameters of their water budget. This should be based on the detailed and continuous monitoring of many different climatological, hydrological, geophysical, ecological and chemical parameters, amongst others. In close co-operation with other geosciences, hydrologists should publicly report the scientific issues that affect the interests of society.

The development of transboundary karst water resources involves complex issues. It requires careful consideration of costs and benefits, environmental issues, a long planning period and negotiation between stakeholders before any decision can be made. It is a multidisciplinary problem consisting of the following components: hydrology, hydrogeology, water quality, water resources planning, ecology, policy, culture, economy, administration and law.

The principles and lessons learned from conflict resolution should form an essential component of transboundary karst water resources management. Mostert [12] proposes a model of conflict resolution in water management consisting of the following four parts: (i) an inventory of the possible source of conflicts; (ii) analyses of the 'basic mechanisms' for addressing the sources; (iii) an overview of practical conflict resolution methods and procedures; and (iv) a discussion of several contextual factors affecting conflict resolution.

It is of paramount importance to ensure a stable exchange of information and to create institutions and space where public and open discussions among all partners can be conducted. These discussions should be based on assessments given by reliable and objective experts [13]. Nobody should imply that politicians and lawyers can be the main judges of disputes connected with transboundary karst water resources management. Karst water related problems will be better and more efficiently solved if professional and scientific principles are fully recognized and not affected or influenced by daily politics [8]. Cooperation between hydrologists and hydrogeologists from countries involved in the problem is the paramount prerequisite for understanding the dynamics of water resources in complex karst systems, and to avoid international conflict. There is no reason to imagine that conflicts over transboundary water resources all over the world will disappear in the future. On the contrary, the likelihood of ongoing interstate conflicts over water resources is substantial [14]. The main existing problem is that managers wish to implement projects quickly and cheaply while experts would like greater attention to be paid to the principles of sustainability and ecological compatibility. There is no doubt that new approaches should be found and used in transboundary karst water resources management to maximize the benefits both to society and to the environment.

References

1 Puri, S. and Aureli, A. (2005) Transboundary aquifers: a global program to assess, evaluate and develop policy. *Ground Water*, **43** (5), 661–668.

2 Bonacci, O., Pipan, T. and Culver, D. (2009) A framework for karst ecohydrology. *Environmental Geology*, **56** (5), 891–900.

3 Gleick, P.H. (2006) Water and terrorism. *Water Policy*, **8**, 481–503.

4 Bonacci, O. (1987) *Karst Hydrology with Special References to Dinaric Karst*, Springer Verlag, Berlin.

5 Bonacci, O., Ljubenkov, I. and Roje-Bonacci, T. (2006) Karst flash floods: an example from Dinaric karst (Croatia). *Natural Hazards and Earth System Sciences*, **6**, 195–203.

6 Bonacci, O. (2004) Hazards caused by natural and anthropogenic changes of catchment area in karst. *Natural Hazards and Earth System Sciences*, **4**, 655–661.

7 Popovska, C. and Bonacci, O. (2007) Basic data on the hydrology of Lakes Ohrid and Prespa. *Hydrological Processes*, **21**, 658–664.

8 Bonacci, O. (2003) Conflict and/or co-operation in transboundary karst groundwater resources management, IHP-VI Technical Documents in Hydrology, PC-CP Series no. 31, pp. 88–98.

9 Bonacci, O. and Roje-Bonacci, T. (2003) The influence of hydrotechnical development on the flow regime of the karstic river Cetina. *Hydrological Processes*, **17**, 1–15.

10 Milanović, P.T., (2004) *Water Resources Engineering in Karst*, CRC Press, Boca Raton.

11 Makropoulos, C., Koutsoyiannis, D., Stanić, M., Djordjević, S., Prodanović, D., Dašić, T., Prohaska, S., Maksimović, Č. and Wheater, H. (2008) A multi-model approach to the simulation of large scale karst flows. *Journal of Hydrology*, **348**, 412–424.

12 Mostert, E. (1998) Conflict and/or co-operation in the management of international freshwater resources: a global review, IHP-VI Technical documents in Hydrology, PC-CP series no. 19, http://unesdoc.unesco.org/images/0013/001333/133305e.pdf (accessed 01 October 2010).

13 Bonacci, O. (2000) The role of international co-operation in a more efficient, sustainable development of water resources management in the Danube basin. *European Water Management*, **3** (2), 26–34.

14 Sherk, G.W. (1994) Resolving interstate water conflicts in the eastern United States: the re-emergence of the federal-interstate compact. *Water Resources Bulletin*, **30** (3), 397–408.

4.3
The Importance of Modelling as a Tool for Assessing Transboundary Groundwaters

Irina Polshkova

4.3.1
Introduction

Research and forecasting of hydrogeological processes in transboundary aquifers of neighbouring countries demands an exact quantitative estimation, especially in cases when anthropogenic loads on groundwater are increasing. Such problems may be approached by creating regional mathematical models with borders that coincide with the natural borders of the hydrodynamic flow of the groundwater.

These borders are probably located on adjacent territories. Accurate monitoring data on groundwater from the territory of neighbouring countries is a prerequisite for creating such a model, therefore one of the primary goals is to get agreement on the granting of all monitoring data for modelling at interstate level. Information on existing or planned sources of possible groundwater pollution is of critical importance. A thorough analysis of the hydrogeological situation in the zone near the boundaries is necessary, where experts from both countries should collaborate. Only with such an analysis can it be shown which of the bordering countries overpumps the transboundary groundwater resources and inflicts damage on its neighbour.

4.3.2
Methodology and Instrument

The basic equation describing a geofiltration process is reduced to one with simple physical sense – the sum of flow rates at each point i of aquifer n is equal to 0 in natural conditions or time difference in capacity in broken conditions:

$$\sum_i Q_{xi}^n + \sum_i Q_{yi}^n + \sum_i Q_{zi}^{n-1} + \sum_i Q_{zi}^{n+1} + \sum_i Q_{IIi}^n$$
$$+ \sum_i Q_{IIIi}^n + \sum_i Q_{wi}^n = \sum_i Q_{ci}^n \qquad (4.3.1)$$

where:

$$Q_{xi}^n = \frac{\partial}{\partial x}\left(T_{xi}^n \frac{\partial H_i^n}{\partial x}\right)$$

is the plane flow along axis x (m per day), Q_{yi}^n is the plane flow along axis y, Q_{zi}^{n-1} and Q_{zi}^{n+1} are vertical flows between neighbouring aquifers, Q_{wi}^n is an infiltration, Q_{IIi}^n is an intensity of ground water extraction, $Q_{IIIi}^n = (H_{si}^n - H_i^n)G_{IIIi}^n$ is water exchange with surface water, H_{si}^n is surface water level, G_{IIIi}^n is conductivity of river-bed deposits, $Q_{ci}^n = S_i^n(\partial H_i^n/\partial t)$ is the change in the capacity for a transient filtration regime, H_i^n is the required function of the water pressure head at point i of aquifer n. The first four members of this equation describe a configuration and the hydrodynamic parameters of an aquifer system, the other three reflect external sources of disturbance.

This approach shows an important peculiarity of mathematical modelling, which lies in the fact that new information on regional groundwater flow conditions can be obtained, owing to the possibility of separate components of the total water balance calculation in accordance with Equation (4.3.1). These possibilities mean that the development pressure influence on underground hydrosphere can be studied and predicted.

The following issues can be solved by modelling:

- estimation and prediction of the degree of an admissible level of exploitation of groundwater;

- estimation and prediction of an admissible level of damage to an underground component of river runoff, which is a result of long-term groundwater extraction in comparison with natural conditions;
- forecast of the contaminated areas in groundwater flow from possible pollution sources, and tracking of their relative dynamics.

To produce optimal schemes of joint groundwater use in accordance with suggested criteria as a result of modelling the following data are output from the model database:

1) **Assessment of groundwater depletion:**
   - maps of hydroisopiese and maps of water head position relative to the top of the aquifer in check time steps, which allow a quantitative assessment of direction, velocity and the time of depression cones spreading towards the administrative boundaries;
   - graph of groundwater levels lowering in inner points of hydrodynamic flow values corresponding to the administrative boundaries;
   - graph of the full amount of plane groundwater flow in the case of a depression cone spreading on the territory of the neighbouring state.

2) **Assessment of surface water depletion** : [1]:
   - graph of value variations in water exchange between surface and groundwaters compared to natural conditions;
   - values of damage to the groundwater component of river runoff for all surface water sources.

3) **Assessment of groundwater contamination:**
   - maps of time and areas of contaminated groundwater distribution from the surface sources of pollution, which allow the sanitary zones of exploited or planned well-fields to be determined;
   - determining probable pollution source locations if contaminants in groundwater samples taken from aquifers are detected.

In several cases, a major factor defining the character and rate of migrant movement in groundwater flow is convectional mass transport, which develops according to the geofiltration process. In comparison with other methods, the modelling of groundwater pollution in accordance with the piston replacement scheme requires the least additional information. At each point of the modelled area speeds of migrant movement (VX, VY, VZ), according to the groundwater flow gradient, are calculated. The time of pollution front movement to the given node point is the minimum possible time of advancement of the migrant getting out from the initial pollution area. This time is calculated by the analysis of all transition times between adjacent node points in ten directions along each possible route of the migrant movement according to a speed vector direction. Calculated flow times in days are used as input in the model database and appear as output in the form of an isochrone map. This map depends on the chosen mode and can be a map of front pollution location or a map of sanitary protection zones of water intakes. Such

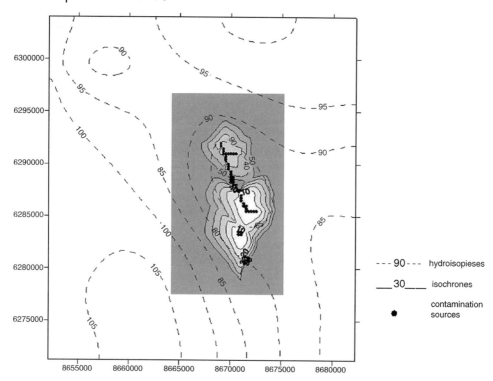

Figure 4.3.1 Map of time and areas spreading from contamination sources.

approximation is enough for a rough estimation of groundwater deterioration rates.

For a mathematical model to be created and to function properly, a cartographical database is necessary. Initial information as well as model results should be given in formats compatible with standard GIS-packages. In Figure 4.3.1 the isochrone map is given as an example of the scheme described.

4.3.3
Numerical Models

Nowadays transboundary problems have become especially acute for bordering territories of former Soviet republics. However, all the studies carried out focused mostly on hydrological constructions. The first quantitative estimation of the mutual hydrodynamics influence in the process of simultaneous exploitation of aquifers of zones near the border was made for the zone near the border between Russia and Estonia [2].

In this paper the methodology described was approved for the calculation of transboundary groundwater flows between Russia and Kazakhstan and Russia and Ukraine.

The modelling area of the zone near the border between Russia and Kazakhstan is a part of the west Siberian artesian basin and is 800 × 550 km in plane. The three-layer hydrogeological schematization including the first from the surface atlym aquifer and the two aquifers of chalk deposits was accepted for mathematical modelling with a dimension of 170 × 140 node points. The aquifer's borders are the natural borders of the aquifer outcrop on the earth's surface. A calculation scheme of non-rectangular (quadrangular) blocks was chosen for modelling purposes. Node points have exact geographical coordinates. Modelling is performed in two steps. First, a steady-state model is used to reproduce the natural conditions of groundwater flow existing and obtain suitable initial heads for transient model. The latter is performed in a second step and allows the estimation of water depletion because of the exploitation of chalk aquifers for 50 years. The correlation of model results and observed data was satisfactory.

In Figure 4.3.2 the model results for natural geofiltration and a variant of the broken mode are represented. Thus the sum groundwater extraction in Pavlodar city (Kazakhstan Republic) from two aquifers with a value of 530 000 m³ per day creates a depression cone that extends over to Russian territory. There is a gradual lowering of the groundwater level along the administrative boundary to an extent

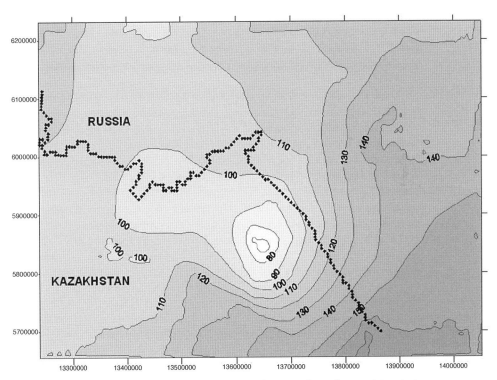

Figure 4.3.2 Map of model water heads for natural and broken groundwater flow conditions in the Russia–Kazakhstan border area.

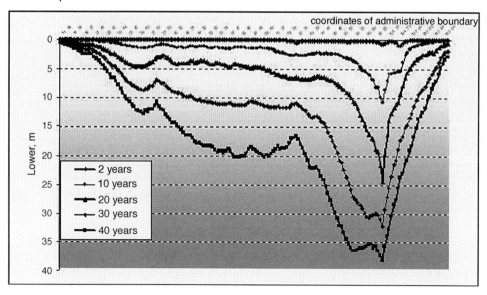

Figure 4.3.3 Lowering of the points along the administrative boundary for some time steps.

of 300 km, which increases to up to 40 m for a 40-year period of water intakes exploitation (Figure 4.3.3). The quantity of plane groundwater flow, which is caught by the depression cone along the border, increases from 60 000 up to 135 000 m^3 per day.

The Russian–Ukraine model is extended for 248 × 276 km on the territory of the Belgorod and Harkov regions. It reproduces the general groundwater flow for a 1 km step in plane and four aquifers in vertical. Natural conditions were modelled for 1970 and broken ones for the years 1970–2009. The results for the most exploited aquifer are presented in Figure 4.3.4. The maximum lowering of the groundwater head in depression cone areas is 55 m for Belgorod city and 80 m for Harkov city in the years 1980–1982.

From the groundwater head maps it is possible to conclude that the flow character does not change. This is confirmed by quantitative estimation of plane flows across the administrative boundary on the model. It increases by approximately 7–8000 m per day near Harkov city. An assessment of surface water depletion was made for the year 2009 in comparison with the year 1970. As can be seen from Figure 4.3.5, the recharge of groundwater from surface water increased from 300 000 up to 380 000 m^3 per day and the release of groundwater decreased from 3 200 000 to 2 600 000 m^3 per day.

Reducing water extraction until the middle of the 90th year would almost restore the natural condition of groundwater flow near Harkov city. However, an excess of water extraction can cause a conjoint depression cone on the transboundary groundwater flow between Russia and Ukraine and considerable depletion of the aquifer (Figure 4.3.6).

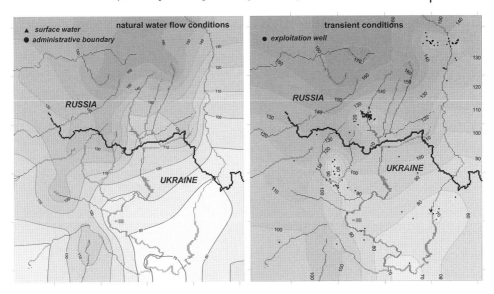

Figure 4.3.4 Maps of model water heads for natural and broken groundwater flow conditions in the Russia–Ukraine border area.

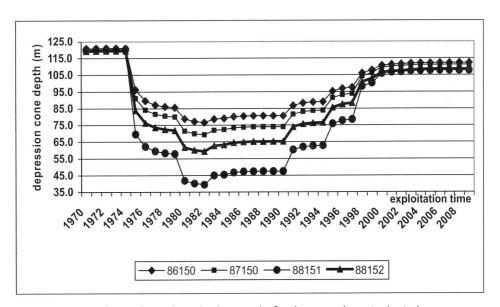

Figure 4.3.5 Time change of groundwater level as a result of working groundwater intakes in the Harkov city area.

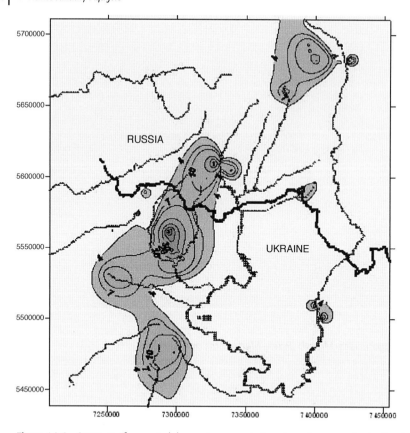

Figure 4.3.6 Contours of generated depression cones in transboundary groundwater flow in the Russian–Ukraine area from 1980 to 1982.

4.3.4
Conclusions

Mathematical models, as an instrument for hydrogeological forecasting, should form part of a total interior monitoring subsystem. Using such an approach, both geological institutions and governmental authorities can be provided with a real instrument for assessing present day and predicted conditions of the underground hydrosphere, which will allow the anthropogenic pressure load to be efficiently regulated. All technogenic loads on groundwater should be preliminarily estimated with the help of models to avert possible damage taking place.

Acknowledgements

This study was made under the framework of projects carried out using a grant from the Russian Fund of Fundamental Investigations. The author is very grateful to the leader of the Russian project Professor I. Zektser and the leader of the Ukrainian project Professor V. Shestopalov and also to Dr U. Rudenko for the factual material they provided.

References

1 Polshkova, I. and Reznik, D. (2009) Otsenka uscherba podzemnoi sostavliajuschei rechnogo ctoka v uslovijah regionalnogo vodootbora metodom matematicheskogo modelirovanija. (The estimation of damage to an underground component of river runoff in conditions of regional water extraction by mathematical modelling.) *Razvedka i Ohrana Nedr.*, 1 (40–45), 45.

2 Mironova, A., Mol'skii, E. and Rumynin, V. (2006) Transboundary problems associated with groundwater exploitation near the Russian-Estonian state boundary. *Water Resources and the Regime of Water Bodies*, 33 (4), 423–432.

4.4
Hydrogeological Characterization of the Yrenda–Toba–Tarijeño Transboundary Aquifer System, South America

Ofelia Tujchneider, Marcela Perez, Marta Paris and Mónica D'Elia

4.4.1
Introduction

The aquifer system located beneath the alluvial fan of the Pilcomayo river between Ibibobo and Misión La Paz–Pozo Hondo (Argentina, Bolivia and Paraguay, South America), covers an area of about $300\,000\,\mathrm{km}^2$, a fact that should be corroborated since its limits still need to be verified. It underlies the territory of the Tarija, Chuquisaca and Potosí departments in Bolivia, the provinces of Salta and Formosa in Argentina and the southwest sector of the Boquerón department in Paraguay (Figure 4.4.1). More than $500\,000$ inhabitants could benefit from the exploitation of this aquifer system, which until the 1990s had not been recognized as a transboundary aquifer by the sharing countries.

The groundwater system has been under study for a long time. Nevertheless, investigations have been carried out at different scales. They mainly define aspects such as regional geology and hydrogeology [1, 2]. Only a few sectors of the system have been studied in detail, for example, by means of isotopic hydrology to better identify the hydrochemical characteristics of ground waters [3].

In Bolivia the aquifer has been defined as multilayer, with interconnected and interbeded water levels. According to its behaviour it is possible to distinguish unconfined, confined and semi-confined layers. The water has an acceptable level of salinity, with electric conductivity of about $1500\,\mu\mathrm{S\,cm}^{-1}$. The flow rate of the existing wells does not generally exceed $3\,\mathrm{l\,s}^{-1}$.

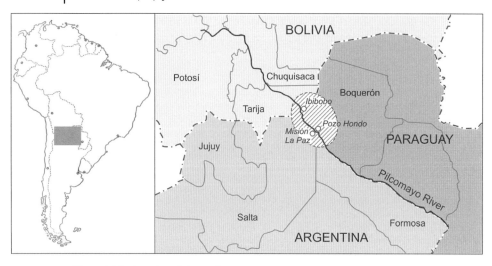

Figure 4.4.1 Location of the study area.

In Paraguay, the aquifer is included into the great region called Paraguayan Chaco (below 21° south latitude). Hydrogeologically, the confined and semi-confined layers consist of silts from the Tertiary and Quaternary Periods. In addition, fine and medium sands, interbeded with layers of loamy material, and plaster and carbonatic concretions constitute the main lithological characteristics of the geo-logic formations. Towards the west, the aquifer system is generally found at a depth lower than 50 m, decreasing to the east. In the Guaraní language (a native language used by most Paraguayan people) YRendá means: Y = place; Rendá = water, 'the place of the water'.

In Argentina the water quality, as well as the productive groundwater levels, have a great spatial variability, due to the great heterogeneity of the water-bearing geological formations. Fresh water levels interspersed with brackish layers are not yet properly identified and dimensioned. The aquifer system is also defined as multilayer, constituted by a succession of unconfined, confined and semi-confined layers. The main recharge of the groundwater reservoirs is allochthonous, probably coming from the west border which collects the runoff from the mountain area.

In some sectors of the area intensive exploitations that are not efficiently controlled by the local states have already been identified. This situation has generated a significant lowering of the groundwater levels. There is also a salinization hazard for the fresh groundwater aquifers due to constructive deficiencies of the pumping wells. In addition, because of the climatic characteristics of the area and the shortage of surface water, groundwater is of strategic interest for the socio-economic develop-ment of the region.

The three countries that share the aquifer have excellent international relations and the precedent of being part of the El Plata basin. For this reason, since the last century the countries have signed various treaties and agreements to guarantee cooperation and understanding of the shared administration of the surface water resources.

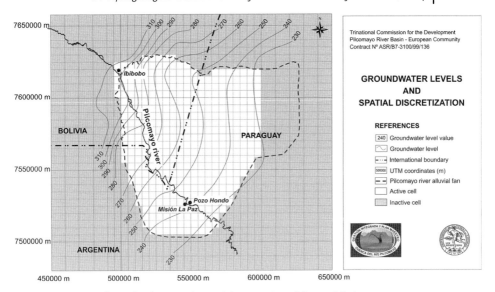

Figure 4.4.2 Groundwater level map and spatial discretization of the modelled area.

However, with regards to groundwater resources, up to now there are no agreements, conventions, treaties or any other legally valid tool.

In 2007, the National Universities of Salta and El Litoral (Argentina), seeking the integration of the available information, elaborated the preliminary groundwater conceptual model. The stratigraphic sequence of hydrogeological interest (assigned to the Tertiary and Quaternary geological formations) could be determined by interpreting the available data collected in the area. The sedimentary deposits ascribed to the Quaternary, from aeolian and fluvial origin, have variable thickness. They overlie discordantly with the so-called Upper sub-Andean Tertiary formation, which is characterized by abundant coarse-grained clastic sedimentary rock, sandstones and sandy shales [4, 5]. From the groundwater level map (Figure 4.4.2), hydraulic gradients, parameters and velocities were estimated.

4.4.2
Groundwater Model Design and Main Results

Because of the scarce basic information available in comparison to the size of the area and the purposes of the study, the implementation of the numerical model required a careful selection of data.

Visual MODFLOW [6] was the computer code selected for modelling the aquifer system. Spatial discretization in the horizontal plane is based on a regular grid of 5 × 5 km cells. The groundwater conceptual model was defined using the concept of hydrostratigraphic units (geologic units of similar hydrogeologic properties). Consequently, for the vertical discretization of the numerical model a two-layer

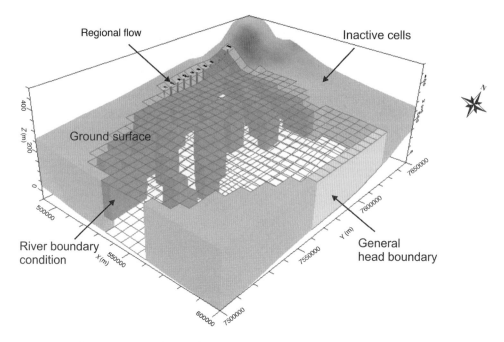

Figure 4.4.3 Boundary conditions.

aquifer was considered. There were also considered active and inactive cells (Figure 4.4.2).

The hydraulic properties of the aquifer system were estimated from pumping tests, as well as from the grain-size of the sediments. The river stages were obtained from previous hydrological studies and surface and groundwater use was adapted in response to the climatic conditions, in terms of the amount of rainfall and the reservoir storage.

Figure 4.4.3 shows the boundary conditions assumed and Figure 4.4.4 illustrates the inflow/outflow zones defined for analysing the groundwater–surface water interaction.

The model was calibrated in a steady state. During this process conductance values for the river boundary condition as well as discharge rates were adjusted. A rather good agreement between the observed and simulated steady-state groundwater levels was obtained, which is manifested in the normalized root mean square calibration error (normalized RMS = 5.42%).

The steady-state groundwater flow simulation shows that within the model domain the total flow rate is $20\,300\,m^3$ per day. The west border accounts for an inflow of $6000\,m^3$ per day, and $304\,m^3$ per day leaves the system by the evapotranspiration process (Figure 4.4.5). In Figure 4.4.5, the bi-directional relation between the Pilcomayo River and the aquifer system is shown, as well as other surface water bodies and the aquifer, a fact that had not previously been taken into account.

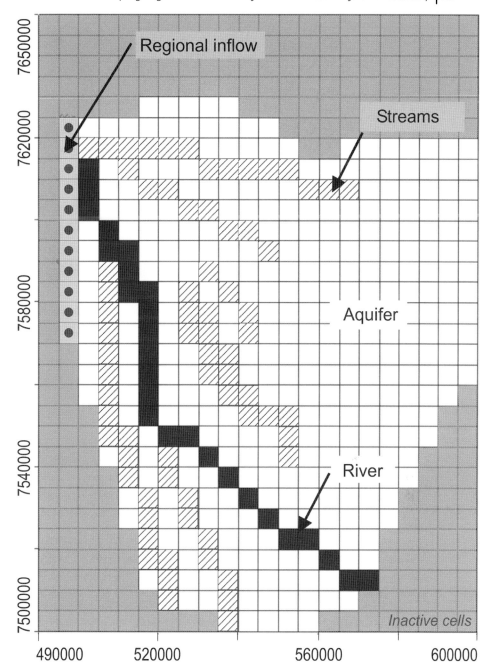

Figure 4.4.4 In–out zones.

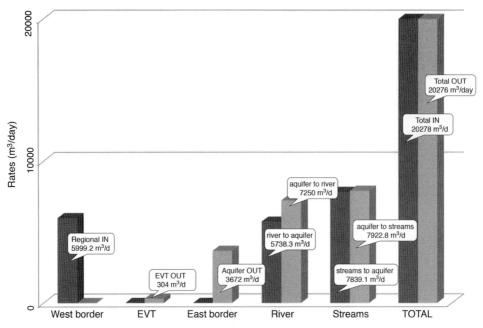

Figure 4.4.5 Hydrogeological budget.

4.4.3
Conclusion

The groundwater dynamic in the alluvial fan of the Pilcomayo River is complex and shows inflow and outflow relationships between the surface water bodies and the aquifer system. This is in contradistinction to previous concepts that considered the river and the streams as bodies losing water (they lose water to groundwater by outflow through their beds). This preconception comes from not considering the global circulation of groundwater.

Generally, the recharge to an aquifer is the water coming from any direction that effectively reaches the saturated zone, constituting an increase in the groundwater reservoir. Taking into account this concept it is understood that the relationships found reveal a situation that should be investigated in more detail in order to develop an integrated management model for the water resources of the Pilcomayo River basin.

Considering that the integrated management of water resources is based on the many different interdependent uses of the water resource, new hydrogeological studies should be undertaken to improve knowledge in a multi-disciplinary effort to support or verify the conceptual model.

It is recommended that the understanding and evolution of the aquifer system associated to the Pilcomayo River basin be deepened, taking into account especially the following tasks: measure the present groundwater levels, estimate the discharge into the streams and river to quantify the subsurface flow, determine tritium, deuterium and ^{18}O, both in rainwater as well as surface and groundwater, carry out *in situ* hydraulic tests for the later regionalization of the values, homogenize the existing and generated hydrochemical data, and articulate all the thematic results. These evaluations would be very useful in order to adequately manage the water resources so that decisions taken can be to the benefit of stakeholders.

Acknowledgements

The authors would like to thank the EU Project Integrated Management and Master Plan for the Pilcomayo River Basin for the financial support provided and the National University of El Litoral for allowing the necessary activities to carry out this study.

References

1 Spandre, R. (2005) Proyecto Gestión Integrada y Plan Maestro de la Cuenca del Río Pilcomayo. Convenio UE No. ASR/B7-3100/99/136. Resultados de la Segunda Misión del Experto Europeo en Hidrogeología Roberto Spandre (EU Project ASR/B7-3100/99/136: Integrated Management and Master Plan of the Pilcomayo River Basin. Second Mission Results, Roberto Spandre, European Expert in Hydrogeology).

2 Stransky, D. (2005) Proyecto Gestión Integrada y Plan Maestro de la Cuenca del Río Pilcomayo. Convenio UE No. ASR/B7-3100/99/136. Informe Final Componente: Estudios Especiales: Estudio hidrogeológico de la porción apical del abanico aluvial del río Pilcomayo entre Ibibobo y Misión La Paz-Pozo Hondo (Argentina, Bolivia y Paraguay) (EU Project ASR/B7-3100/99/136: Integrated Management and Master Plan of the Pilcomayo River Basin. Hydrogeological studies of the apical portion of the Pilcomayo River alluvial fan. Final Report).

3 Pasig, R. (2005) Origen y dinámica del agua subterránea en el noroeste del Chaco sudamericano (Origin and dynamics of groundwater in the northwestern of South American Chaco). PhD dissertation. Bayrischen Julios-Maximilians-Universitat Würzburg, Germany.

4 Tujchneider, O., Perez, M., Paris, M. and D'Elia, M. (2007) Caracterización Hidrogeológica del Abanico Aluvial del Río Pilcomayo, entre Ibibobo (Bolivia) y Misión La Paz (Argentina), Taller de Acuíferos Transfronterizos, 81–90. V Congreso Argentino de Hidrogeología. Octubre 16-19, 2007. Paraná, Argentina. Tujchneider & Pasig (compiladores). Editor responsable: Grupo Argentino de la Asociación Internacional de Hidrogeólogos (Hydrogeologic characterization of the Pilcomayo River alluvial fan, between Ibibobo (Bolivia) and Misión La Paz (Argentina). Workshop on Transboundary Aquifers).

5 Baudino, G., Pitzzú, G., Bercheñi, G., Fuertes, A. and Manjarrés, C. (2006) Estudio Hidrogeológico de la porción apical del abanico aluvial del Río Pilcomayo entre Ibibobo y Misión La Paz – Pozo Hondo (Argentina, Bolivia, Paraguay), Segunda Etapa: Correlación hidroestratigráfica. Informe Final. Convenio Proyecto Pilcomayo – INASLA, Universidad Nacional de Salta. Proyecto de Gestión Integrada y Plan Maestro de la Cuenca del Río Pilcomayo (Hydrogeologic studies in the apical portion of the Pilcomayo River alluvial fan, Ibibobo, Misión La Paz – Pozo Hondo (Argentina, Bolivia, Paraguay), Second phase: hydrostratigraphic correlation. Final Report).

6 Waterloo Hydrogeologic, Inc. (2005) Visual MODFLOW Professional Edition. Users' Manual.

4.5

The State of Understanding on Groundwater Recharge for the Sustainable Management of Transboundary Aquifers in the Lake Chad Basin

Benjamin Ngounou Ngatcha and Jacques Mudry

4.5.1
Introduction

In the Lake Chad basin (Figure 4.5.1) the use of groundwater as supply has increased dramatically, but the long-term sustainable yield of these resources is not known. It is anticipated that the decrease in freshwater sources will intensify stress on groundwater resources [1]. To ensure water supply to the growing population under changes in human activities, changes in climate variability, growing demands and the degradation of the catchment area an investigation into groundwater recharge will be one of the main concerns for water managers in the coming decades. Such information will be of great use to all those concerned with issues of water resources management in arid and semi-arid areas. This study is part of a larger effort to improve the knowledge and understanding of groundwater resources and their proper management in the Lake Chad basin, a transboundary aquifer where groundwater is a vital resource that plays an increasingly strategic role in sustainable development. This role will be of greater importance in years to come, as water scarcity and increased climatic variability become major global concerns [2, 3].

4.5.2
Hydrogeological Contexts

The Lake Chad basin is a sedimentary basin formed in the Mesozoic era. According to the geological classification, two aquifers shared by several countries may be distinguished, the Quaternary and the Continental Terminal aquifers. The Quaternary aquifers (Figure 4.5.2) consist of sandy, deltaic and lacustrine deposits. This aquifer is unconfined and locally known as the shallow aquifer. It is in direct hydraulic contact with the Logone–Chari River. The Quaternary aquifers are a key water resource that, if exploited in a sustainable way, could alleviate the high levels of poverty in the Lake Chad basin [4]. The Continental Terminal (CT) aquifer, including the Pliocen and Oligo-Miocen [5, 6], consists of sandstone and argillaceous sands. Key characteristics of the CT are that it has generally been considered to be a classic example of an artesian aquifer system.

4.5.3
Methods for Groundwater Recharge Investigation in the Lake Chad Basin

Over the last 40 years, much attention has been given to improving groundwater recharge investigation in the Lake Chad basin [7–19]. Table 4.5.1 shows the main

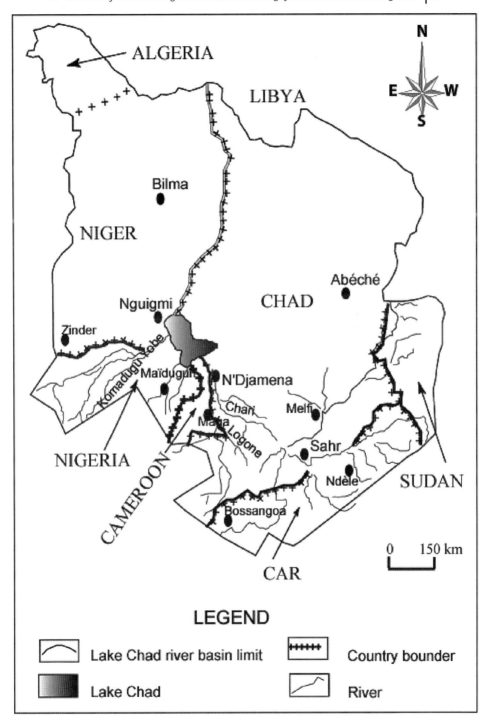

Figure 4.5.1 Lake Chad watershed.

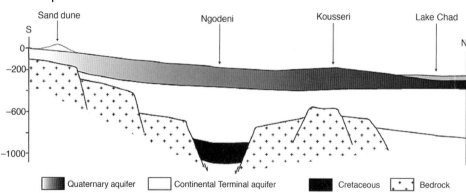

Figure 4.5.2 Geological and hydrogeological formations in the Lake Chad basin.

methods of investigation. The more classical methods of groundwater investigation like hydrogeology and hydrogeochemistry are complemented with the methods of isotope hydrology and mathematical modelling.

4.5.4
Groundwater Recharge in the Lake Chad Basin

There are three sources of recharge in the Lake Chad basin (Figure 4.5.3). UNESCO carried out the first systematic study of groundwater recharge, using hydrograph records and groundwater dating methods with the isotopes ^{18}O, ^{2}H, ^{3}H and ^{14}C in parts of the Chad basin [7]. Fluctuations of groundwater levels are direct reflections of the conditions that affect the aquifer, such as geology and climate. Near the river, water table fluctuations correspond closely with the dynamic water level of the river. Stable isotopes in rainwater for the region have average concentrations of $-4‰$ $\delta^{18}O$ and $-20‰$ $\delta^{2}H$. Surface water samples from rivers and Lake Chad fall on the evaporation line of this average value (Figure 4.5.4).

The $\delta^{18}O$ and $\delta^{2}H$ values of groundwater were close to each other. There was evidence of enrichment of heavy isotopes in groundwater due to evaporation. Plotting $\delta^{18}O$ and $\delta^{2}H$, one obtains a linear relationship that follows the equation $\delta^{2}H = 6.3$ $\delta^{18}O + 4.2$, which is typical for water that was subject to kinetic evaporation and therefore exhibits a slope significantly different from that of the global meteoric water line (GMWL: $\delta^{2}H = 8\delta^{18}O + 10$). Measured tritium activities indicate that a large variation exists in groundwater age throughout the Quaternary aquifer. Tritium results in groundwater samples gave evidence of ongoing recharge. The shallow aquifers contain more than 50% ^{14}C and the deep aquifers less than about 4% ^{14}C. Recharge rates range from 14–49 mm yr^{-1}, on the basis of unsaturated zone Cl values and rainfall chemistry. Nitrate is generally the contaminant of most concern in the Chad basin [20].

Table 4.5.1 Methods for groundwater recharge investigation in the Lake Chad basin (methods tested are in black and methods not tested are in grey).

Methods	Water zone				Area of application in the Lake Chad basin
	Rainfall	Surface water	Unsaturated zone	Saturated zone	
Thornthwaite method	■				Cameroon
Hydrogram of separation		■			Komadugu Yobe
Water balance					Hydrological basin
Water table fluctuation			■	▨ / ■	Cameroon, Chad, Niger, Nigeria
Lysimeter and Soil Physics					Cameroon
Mathematical modelling		▨ / ■		■	Quaternary aquifer
Stable isotopes (^{18}O, ^{2}H)			▨		Cameroon, Chad, Niger, Nigeria
Tritium			■		Cameroon, Chad, Niger, Nigeria
Carbon-14	■				Cameroon, Chad, Niger
Chlore mass balance	■				Nigeria
Chlore chemistry					Nigeria

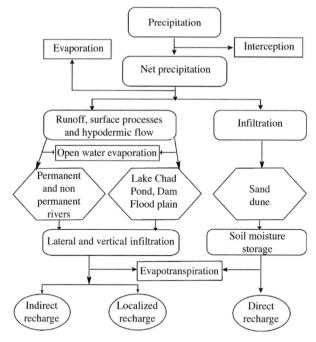

Figure 4.5.3 Sources of recharge in the Lake Chad basin.

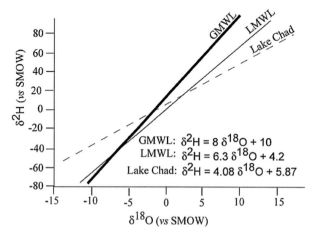

GMWL: $\delta^2H = 8\ \delta^{18}O + 10$
LMWL: $\delta^2H = 6.3\ \delta^{18}O + 4.2$
Lake Chad: $\delta^2H = 4.08\ \delta^{18}O + 5.87$

Figure 4.5.4 Deuterium and oxygen-18 content compared to the global meteoric line (GMWL); LMWL = local meteoric water line; SMOW = standard mean ocean water.

4.5.5
Conclusions and Further Research

A critical component to managing water resources is in understanding the source of the groundwater that is extracted from a well. This chapter contributes to improving the existing knowledge on Lake Chad shared aquifer systems. Despite the lack of knowledge on basic hydrogeological parameters of the aquifers involved and pie-zometric data, environmental isotopes enabled the identification of different types of recharge. The mechanism of the recharge for wells along Lake Chad is not uniform. As for any transboundary resource, management of these aquifers could prove a daunting challenge, as it requires collaboration amongst the different countries involved. To improve the supply guarantee for a given water demand, the joint management of shared groundwater will be a strategic component in the overall management of the Chad basin. In this context, further research in the framework of the RIPIECSA (Recherche Interdisciplinaire et Participative sur les Interactions entre les Ecosystèmes, le Climat et les Sociétés d'Afrique de l'Ouest) project would (i) allow an analysis of the new realities of groundwater use, both quantitatively and qualitatively, in the Lake Chad basin in current settings, (ii) afford a better under-standing of the interdependency between groundwater and surface water (Lake Chad, local ponds) from a hydrodynamic and hydrogeochemical point of view and (iii) develop an uncontested database (for analysis and interpretation) and monitor-ing programmes. The implementation of these actions is ongoing, and the results will provide a starting point for the achievement of effective international water management.

Acknowledgements

We thank the French Ministry of Foreign Affairs (RIPIECSA Project) for its financial and logistic support. Thanks go to the UNESCO Chair INWEB at the Aristotle University of Thessaloniki for making possible our participation at the IV Interna-tional Symposium on Transboundary Waters Management held in Thessaloniki, Greece, 15–18th October 2008.

References

1 Rakhmatullaev, Sh., Huneau, F., Kazbekov, J., Le Coustumer, Ph., Jumanov, J., El Oifi, B., Motelica-Heino, M. and Hrkal, Z. (2010) Groundwater resources use and management in the Amu Darya River Basin (Central Asia). *Environment Earth Sciences*, **59**, 1183–1193.

2 Puri, S. and Aureli, A. (2005) Transboundary aquifers: A global program to assess, evaluate, and develop policy. *Ground Water*, **43** (5), 661–668.

3 Zaisheng, H., Jayakumar, R., Ke, L., Hao, W. and Rui, C. (2008) Review on transboundary aquifers in People's Republic of China with case study of Heilongjiang-Amur River Basin. *Environmental Geology*, **54**, 1411–1422.

4 Ngounou Ngatcha, B., Mudry, J. and Leduc, C. (2008) Water resources

management in the Lake Chad basin: Diagnosis and action plan. *IAH Selected Papers on Hydrogeology*, **13**, 65–84.

5 Kilian, C. (1931) Des principaux complexes continentaux du Sahara. *Comptes Rendus Société Géologique France*, **9**, 109–111.

6 Ngounou Ngatcha, B., Mudry, J. and Leduc, C. (2006) A propos des aquifères du Continental terminal dans le bassin du lac Tchad. 21st Colloquium on African Geology, Geoscience for Poverty Relief, Maputo, Mozambique, 3th–5th July 2006, Abstract book, p. 374.

7 UNESCO (1969) Synthèse hydrologique du bassin du lac Tchad. Projet UNESCO/Fonds spécial, 1966–1969, Rapport technique présentant les principaux résultats des opérations, 217 pp.

8 Fontes, J.-Ch., Gonfiantini, R. and Roche, M.A. (1970) Deutérium et oxygène-18 dans les eaux du lac Tchad, *Isotope Hydrology. Proceeding Symposium Vienna, 1970*, IAEA, Vienna, 387 pp.

9 Ketchemen, B. (1992) Etude hydrogéologique du Grand Yaéré (Extrême Nord du Cameroun). Synthèse hydrogéologique et étude de la recharge par les isotopes de l'environnement. Thèse de Doctorat 3ème cycle, Université Cheikh Anta Diop Dakar, Sénégal.

10 Njitchoua, R. and Ngounou Ngatcha, B. (1997) Hydrogeochemistry and environmental isotopic investigations of the North Diamaré plain, Extreme-North of Cameroon. *Journal of African Earth Sciences*, **25** (2), 307–316.

11 Edmunds, W.M., Fellman, E, Baba Goni, I., McNeill, G. and Harkness, D.D. (1998) Groundwater, paleoclimate and paleorecharge in the SW Chad basin, Borno State, Nigeria, in *Isotope Techniques in the Study of Past and Current Environmental Changes in the Hydrosphere and the Atmosphere*, IAEA, Vienna, pp. 693–707.

12 Goes, B.J.M. (1999) Estimate of shallow groundwater recharge in the Hadeija-Nguru Wetlands, semi-arid northeastern Nigeria. *Journal of Hydrogeology*, **7**, 305–316.

13 Djoret, D. and Travi, Y. (2001) Groundwater vulnerability and recharge or paleorecharge in the southeastern Chad basin, Chari Baguirmi aquifer, in Isotope Techniques in Water Resource Investigations in Arid and Semi-Arid Regions, IAEA-TECDOC-1207, pp. 33–40.

14 Leduc, C., Sabljak, S., Taupin, J.D., Marlin, C. and Favreau, G. (2000) Estimation de la recharge de la nappe quaternaire dans le Nord-Ouest du bassin du Lac Tchad (Niger oriental) à partir de mesures isotopiques. *Comptes Rendus Académie des Sciences Paris*, **330**, 355–361.

15 Ngounou Ngatcha, B., Mudry, J., Wakponou, A., Ekodeck, G.E., Njitchoua, R. and Sarrot-Reynauld, J. (2001) Le cordon sableux Limani-Yagoua (Nord Cameroun) et son rôle hydraulique. *Journal of African Earth Sciences*, **32** (4), 889–898.

16 Gaultier, G. (2004) Recharge et paléorecharge d'une nappe libre en milieu sahélien (Niger Oriental: approches géochimique et hydrodynamique. Thèse Doctorat, Université Paris Sud, Orsay, 179 pp.

17 Goni, I.B. (2006) Tracing stable isotope values from meteoric water to groundwater in the southwestern part of the Chad basin. *Journal of Hydrogeology*, **14** (5), 742–752.

18 Ngounou Ngatcha, B., Mudry, J., Aranyossy, J.F., Naah, E. and Sarrot-Reynauld, J. (2007) Apport de la géologie, de l'hydrogéologie et des isotopes de l'environnement à la connaissance des "nappes en creux" du Grand Yaéré (Nord Cameroun). *Journal of Water Science/Revue des Sciences de l'eau*, **20** (1), 29–43.

19 Ngounou Ngatcha, B., Mudry, J. and Sarrot-Reynauld, J. (2007) Groundwater recharge from rainfall in the southern border of Lake Chad in Cameroon. *World Applied Sciences Journal*, **2** (2), 125–131.

20 Ngounou Ngatcha, B. and Djoret, D. (2010) Nitrate pollution in groundwater in two selected areas from Cameroon and Chad in the Lake Chad basin. *Water Policy*, **12**, 722–733.

4.6

Development, Management and Impact of Climate Change on Transboundary Aquifers of Indus Basin

Devinder Kumar Chadha

4.6.1
Introduction

The transboundary aquifers of the Indian continent pertain to the Indus and Ganges basins and these belong to different geological periods. Table 4.6.1 gives the transboundary countries, type of the aquifers present, their regional continuity and areal extent. The transboundary aquifers of the Indus Basin are shared with Pakistan only as the basin areas with Afghanistan and China are mountainous and have no defined aquifer system The aquifer systems with Pakistan range in age from Quaternary to Proterozoic periods but it is the Quaternary to Recent alluvial aquifers that are of great significance as they are continuously being developed to meet the water demand from various sectors (Figure 4.6.1).

4.6.2
Indus Basin

The Indus basin has a geographical area of 1.16 million km^2 of which 0.32 million km^2 (i.e. 28%) is in India and more than 50% lies in Pakistan. In the Indian part of the Indus Basin, the high mountain ranges cover 85% (i.e. 0.27 million km^2) and only 15% is covered by the foothills and plain areas. The rainfall varies from 2400 mm in the northeast to 200 mm in the southwest and the average monsoon runoff is estimated as 58 640 million cubic metres (MCMs). The Indus plains have a close network of canal systems; the net area irrigated is 5.80 million hectares and the net area sown is 9.59 million hectares. The groundwater potential for irrigation in India is estimated as 5428 MCM per year.

The sharing of the Indus river water between India and Pakistan is based on the Indus Water Treaty but, as groundwater development was much less during the 1950s, provisions for groundwater were neither considered necessary nor envisaged. Now, however, groundwater development has increased greatly, and there are more than 0.15 million structures in operation, which exploit 70% of the dynamic potential and also the deep aquifer zones, the cumulative pumpage of which has not been quantified. Similarly in Pakistan, groundwater from shallow and deep aquifers is being exploited, resulting in a continuous lowering of the water table and also the piezometric head. Keeping in view future water demand and the continuous development of groundwater to mitigate the demand, it has become imperative to study in detail the transboundary aquifers in the vicinity of international borders and to share the information gathered to achieve controlled development and better management practices.

Table 4.6.1 Transboundary Aquifers of India.

Serial number	Age of aquifers	Name of transboundary aquifer systems	Countries sharing the aquifer system	Type of aquifer	Extent
1.	Upper Tertiary-Quaternary	Upper Siwalik and alluvial aquifers (Bhabar-Tarai)	Nepal, India	Porous sedimentary: unconfined, confined	Extensive and continuous
2.	Quaternary (Pleistocene-Recent)	Alluvial/deltaic aquifer	India, Bangladesh	Porous: unconfined, confined	Extensive and continuous
3.	Quaternary (Pleistocene-Recent)	Alluvial aquifer (older alluvium)	Bhutan, Indian	Porous: unconfined, confined	Extensive and continuous
4.	Quaternary-Recent; Mesozoic-Tertiary (?)	Aeolian, alluvial and sandstone	India, Pakistan	Porous: confined, Artesian	Porous: extensive/discontinuous
5.	Upper Tertiary-Quaternary	Alluvium, Tipam sandstone (Shiwalik)	India, Bangladesh	Porous confined, Artesian	Locally extensive and disposed in synclinal valleys as intermontane 'hollows'
6.	Mountain Aquifer (Proterozoic)	Crystalline Rock Aquifers (granite, phyllite, quartzite, gneiss, limestone, etc.)	India, China, Pakistan, Nepal	Fissured: non-aquiferous zones	Local and discontinuous
7.	Tertiary-Quaternary	sandstone, siltstone, claystone, alluvium	India, Myanmar	Porous	Discontinuous
8.	Proterozoic	Granite-gneiss	India, Bangladesh	Fissured and joined aquifer	Local and discontinuous

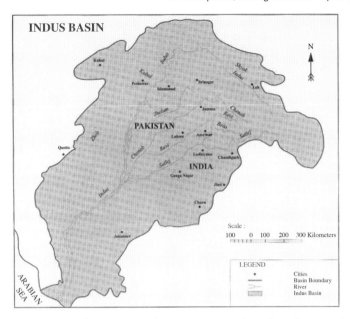

Figure 4.6.1 Indus Basin map showing its extent and transboundary countries.

4.6.3
Groundwater Potential and Development

The geological formations range in age from pre-Cambrian to Quaternary deposits (recent) and thus constitute different aquifer systems with different potentialities. The groundwater potential of the mountain areas, which cover about 85% of the basin, has not been estimated, but in the 15% of the foot hills and plain areas the dynamic (annual replenishment) and in-storage of fresh groundwater potential is 26.5 billion cubic metres (BCMs) and 1338 BCM respectively (Table 4.6.2) [1]. The saline aquifers cover an area of about 25 100 km^2; by and large they remain unutilized but contribute towards water logging and soil salinity.

4.6.4
Aquifer System

The Indus plain area encompasses a multi-aquifer system consisting of different grades of sand, gravel and pebbles. The Kandi zone, a thick boulder bed formation occurring between the Siwalik rocks (Upper Tertiary Age) and recent alluvial sediments, acts as a recharging zone for the deep aquifers in both India and Pakistan. The synthesis of test drillings and lithological logs of tube wells shows that the Quaternary aquifer system has a variable thickness of aquifer zones separated by intercalated impervious clay horizons. The whole aquifer system up to the explored depth of 650 m below ground level can be classified as one unified aquifer system with locally confined/semi-confined layers. The cumulative thickness of pervious

Table 4.6.2 Groundwater assessment and development status for the Indus Basin, India.

Serial number	State/ district	Geographical area falling in Indus basin (km²)	Annual gross recharge (MCM)	Net draft (MCM)	Area covered (km²)	Potential (MCM)	Water logged area (km²)
			Fresh groundwater		Saline groundwater		
1	Jammu Kashmir	117 683	4425.6	40.3	Nil	Nil	Nil
2	Himachal Pradesh	47 436	247.72	49.3	Nil	Nil	321
3	Punjab	50 362	18 192.27	16 101.95	2980	4162	5525
4	Haryana	14 679	3645.83	2067.85	3750	5430	4350
5	Rajasthan	14 624	603.96	210.09	18 347	1158	1171
6	Chandigarh (UT)	114	0.023	0.018	Nil	Nil	Nil

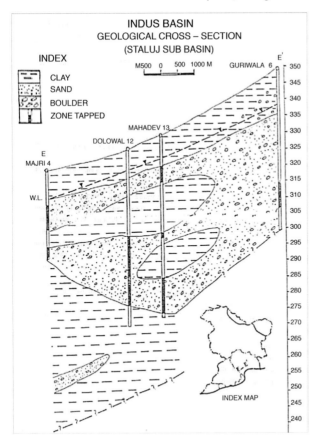

Figure 4.6.2 Lithological cross section of Satluj Sub-basin. (After the Central Ground Water Board, India).

zones varies from 70 to 85% of the depth explored in upper Bari Doab and therefore the yield from the tube wells is very high (about 1000 000 l per day). The sub-surface lithological sections described below show the disposition of pervious and impervious zones, which are laterally and vertically extensive and also transboundary. A brief description of the sub-surface lithology of two of the sub-basins in the plain area is given below.

4.6.5
Satluj Sub-basin

The extensive exploration carried out in this sub-basin shows the marked lateral and vertical variation in different alluvial sediments. This section (Figure 4.6.2) indicates that the uppermost layer is clayey and is about 10 m thick. Below this clay layer, a single aquifer consisting of sand, gravel, pebbles and boulders exists in the

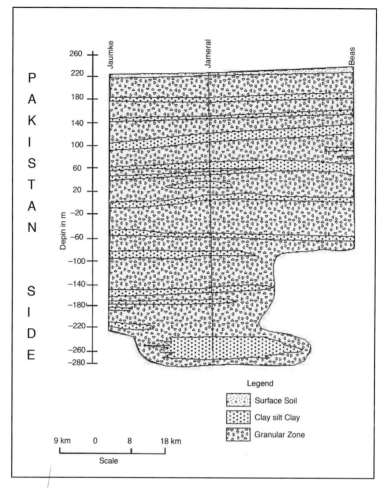

Figure 4.6.3 Sub-surface lithological Upper Bari Doab (Beas-Ravi sub-basin).

north-eastern part and extends down to a depth of 50 m. In the south-western part, the same aquifer bifurcates into two, and is separated by clay lenses. The aquifer is of maximum thickness in the north-eastern part and this thickness decreases towards the south-west area. The cumulative thickness of the impervious zone is more than that of the pervious zones.

4.6.6
Beas Sub-basin

The lithological section extends between Jaunke and Beas (Figure 4.6.3) and reveals the presence of several granular layers 7–9 m thick, intercalated with impervious clay horizons down to a depth of 300–500 m below ground level.

It may be observed that the different layers terminate at short distances, except for two regionally extensive clay layers. The detailed description of sub-surface lithology and the presence of granular zones at different depths highlight the fact that the particular aquifer zones may not necessarily continue on a trans-boundary basis as the lithology may change from sand layer (pervious) to silty clay (impervious) at the same depth. The extensive depth of the quaternary aquifer system is to be considered by the transboundary countries as one unified aquifer system for planning its sustainable development.

4.6.7
Groundwater Management

Since groundwater development is a necessity to meet irrigation demand, which uses 90% of the resource, the resultant decline in water levels is being restored through the practices of Managed Aquifer Recharge (MAR) and regulatory measures. The saline groundwater resources have only been developed on a limited scale as conjunctive use with surface canal water. It is estimated that feasible groundwater storage based on the surface water availability and sub-surface storage is 29 800 MCM. A non-committed run-off of about 2100 MCM has been planned for MAR for creating sub-surface storages to check the declining trends of water levels and reduce energy costs. Several recharge structures have already been constructed.

4.6.8
Impact of Climate Change

Initial studies on the impact of climate change on water availability in India covering a part of the Indus basin show an increase in the water stressed conditions in the arid areas falling in the south-west, whereas the melting of glaciers in the Himalayas, at the origin of the Indus River and other major tributaries, will increase flooding. This anomalous situation will impact food security, as the lower Indus basin is the food basket for both India and Pakistan. It is thus evident that the Indus basin is vulnerable to climate change and this will adversely affect the socio-economic conditions of the region.

4.6.9
Problems Related to Transboundary Aquifers

India has a long border (about 2900 km) with Pakistan and the transboundary aquifers are of different geological ages with variable groundwater quality and hydro geological conditions. In the Indus basin, most of the precipitation is on the Indian side, the groundwater flow direction is towards Pakistan, both sides have a close network of canal systems for irrigation, whose condition is water logged, and there is inland salinity in shallow aquifers. Moreover the quantum jump in groundwater development in downstream Pakistan has disturbed the hydro-dynamics and general hydraulic gradient, inducing salinity in groundwater. The emergence of these problems can be attributed to poorly-developed institutional mechanisms and regulatory laws. It is proposed that any future study of the transboundary aquifers should not be limited to freshwater aquifers but should also include saline aquifers, as these are a continuation of the fresh aquifer zone but have different lithology. Moreover India has embarked upon the implementation of large MAR schemes that are mandatory for industries and urban developers/infrastructure, and has established an authority to regulate the development of groundwater resources. However, such actions are not envisaged at a transboundary level. The laws on transboundary aquifers give broad guidelines for resolving problems related to transboundary aquifers [2]. The likely conflict issues can be resolved through mutual discussion of the UN articles, such as sharing of water and environmental studies.

Presently, there is no regional agreement between the transboundary countries for sharing hydrogeological data or cooperation on any identified conflict issues. However, with the increased demand for water for food security and domestic requirements and the development of deep fresh aquifers, new problems will emerge that will require regional agreement on transboundary aquifer management using the UN articles.

4.6.10
Conclusions

The Indus basin serves as a food basket for both India and Pakistan and plays an important role in mitigating poverty alleviation and social economic development. Therefore, any suggestion for an agreement to cover the entire basin may not be viable or acceptable to the countries concerned. Any studies performed should be extended to a maximum of 5 km from the international boundary, so as to have a definite picture of the continuity of the different aquifer systems and their development needs. It is suggested that if any breakthrough is to be achieved for bringing the transboundary countries to a common platform it is important to show proven case studies from other parts of the world that illustrate the benefits to be obtained from the joint management of transboundary aquifers.

References

1 Central Groundwater Balance (2005) Dynamic Groundwater Resources of India.

2 United Nations, General Assembly (2009) The Law of Transboundary Aquifers.

4.7

Natural Background Levels for Groundwater in the Upper Rhine Valley

Frank Wendland, Georg Berthold, Adriane Blum, Hans-Gerhard Fritsche,
Ralf Kunkel and Rüdiger Wolter

4.7.1

Introduction

The Commission's proposal of Groundwater Directive COM(2003)550 developed under article 17 of the Water Framework Directive (WFD) sets out criteria for the assessment of the chemical status of groundwater, which is based on existing Community quality standards (nitrates, pesticides and biocides) and on the requirement for Member States to identify pollutants and threshold values (TVs) that are representative of groundwater bodies found as being at risk, in accordance with the analysis of pressures and impacts carried out under the WFD. In this context, natural background levels (NBLs) for groundwater are required as a reference to quantitatively evaluate whether groundwater is significantly modified by anthropogenic influences [1].

In the light of the above, in the framework of the EU-Specific Targeted Research Project BRIDGE (Background CRiteria for the IDentification of Groundwater thrEsholds) a scientifically based and generally applicable approach to derive natural background levels for the groundwater and groundwater threshold values was derived. The applicability and validity of this approach is checked in 14 case study areas at the level of aquifer typologies throughout Europe [2], including the Upper Rhine Valley as a transboundary French–German–Suisse case study.

4.7.2

General Applicable Approach for Deriving NBLs and TVs

Groundwater quality in aquifers taking part in the active water cycle (surface-near aquifers) is particularly influenced by anthropogenic inputs, for example, from agriculture and atmosphere [3]. Whereas some of these inputs (e.g. pesticides) are a direct indicator of human impacts, most inorganic contents occurring in the groundwater originate both from natural and anthropogenic sources [4]. This makes it difficult to decide whether an observed groundwater concentration pattern in a certain area is influenced by pollution intakes or still represents an (almost) natural state.

An evaluation of existing approaches for NBL assessment [5–9] has shown that geostatistical procedures are very useful in this kind of research. Preselection methods are, for example, appropriate for deriving natural background levels on the level of aquifer typologies as defined by Wendland *et al.* [2] on the basis of petrographic characteristics.

The basic idea of preselection methods is that there is a correlation between the concentration of certain indicator substances and the presence of anthropogenic influences. To preselect the groundwater samples the following criteria were chosen based on the experiences in several French and German studies on NBL assessment [5–8], where they proved to be appropriate:

- exclusion of groundwater samples displaying purely anthropogenic substances;
- exclusion of samples displaying concentrations of indicator substances exceeding:

$-NO_3 > 10 \, mg \, l^{-1}$ in oxidized aquifers ($O_2 > 2 \, mg \, l^{-1}$ and $Fe(II) < 0.2 \, mg \, l^{-1}$)

or

$-NH_4 > 0.5 \, mg \, l^{-1}$ in reduced aquifers ($O_2 < 2 \, mg \, l^{-1}$ and $Fe(II) > 0.2 \, mg \, l^{-1}$).

The NBLs are defined subsequently for all groundwater parameters as the 90th percentiles (P_{90}) of the concentration distributions from the remaining samples. For substances that are purely synthetic with no natural sources (e.g. TCE), NBLs are set to zero.

The TVs are established with reference to the NBLs and a chosen reference standard (REF). The latter may either be a drinking water standard (DWS) value, an environmental quality standard (EQS) or ecotoxicological value (EToxV). The calculation of threshold values is presented for three cases:

- **Case 1**: NBL $\leq$ REF: if the NBL of a certain substance is below the REF value the TV is set as the concentration in the middle between the NBL and the REF values: TV = (REF + NBL)/2.
- **Case 2**: NBL < one-third of REF: when the NBL is considerably below the REF value, the TV is limited to twice the NBL: TV = 2NBL.
- **Case 3**: NBL $\geq$ REF: when the NBL is larger than the REF value, the TV is set equal to REF itself: TV = NBL.

The TV for purely synthetic substances with no natural sources (e.g. TCE) is defined as the detection limit.

4.7.3
Application to the Case Study Area Upper Rhine Valley

The case study area of the Upper Rhine Valley is a transboundary river basin located between France and Germany with smaller parts in Switzerland. It belongs as a whole to the aquifer typology 'fluviatile deposits of major streams' (Figure 4.7.1) and represents one of the biggest rift structures in Europe, filled up with alternating layers of slit, clay, sand and gravel during the Pliocene and Quaternary period. The total area comprises $9.290 \, km^2$. Owing to the high population density, various anthropogenic impacts on groundwater quality (intensive agriculture, industry and water withdrawal) are present.

In recent years many joint German–French–Swiss projects dealing with the implementation of the EU–WFD have been carried out. In this framework the

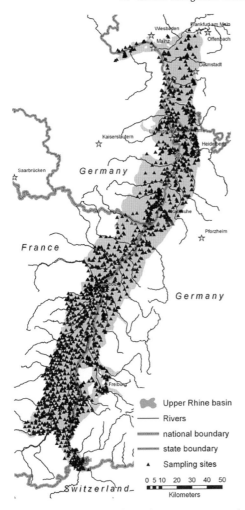

Figure 4.7.1 Location of sampling sites in Upper Rhine Valley.

Bureau de Recherches Géologiques et Minières (BRGM) (France) developed a joint groundwater quality database, fed by different German, French and Swiss State authorities (BRGM Alsace, LUA Baden-Wurtemberg, LUA Rheinland-Pfalz, HLUG Hessen, Bâle Ville, Bâle Campagne). Access to this database was guaranteed because of the participation of BRGM (France) and HLUG (Germany) in the BRIDGE project. Owing to this fact, the database could be used for evaluating NBL and threshold values and it consists of almost 1700 groundwater samples for the years 2002 and 2003 from French, German and Swiss monitoring networks. For each of the monitoring stations one sample containing the solution contents for integral and chemical environment parameters, characteristic major and minor parameters and WFD pollutants was available. However, no information on heavy metals was included in the database.

Table 4.7.1 Natural background levels (P_{90}), drinking water standard values as reference standard (REF) and groundwater threshold values (TV1) for selected parameters.

Parameter	Unit	P_{90}	REF	TV1	Case
B	$mg\,l^{-1}$	0.1	1	0.2	2
Cl	$mg\,l^{-1}$	84	250	167	1
Fe(II)	$mg\,l^{-1}$	3.6	0.2	3.6	3
K	$mg\,l^{-1}$	7.2	10	8.6	1
Mg	$mg\,l^{-1}$	25	50	37	1
Mn(II)	$mg\,l^{-1}$	0.82	0.05	0.8	3
Na	$mg\,l^{-1}$	41	200	83	2
SO$_4$	$mg\,l^{-1}$	173	250	211	1
LF	$\mu S\,cm^{-1}$	951	2500	1726	1
As	$\mu g\,l^{-1}$	4	10	7	1
NH$_4$	$mg\,l^{-1}$	0.39	0.5	0.45	1
NO$_2$	$mg\,l^{-1}$	0.04	0.5	0.08	2
NO$_3$	$mg\,l^{-1}$	8.2	50	16.4	2
PO$_4$	$mg\,l^{-1}$	0.17	6.7	0.34	2

Preselection according to nitrate ($NO_3 < 10\ mg\,l^{-1}$) has led to the exclusion of 1094 samples (64%). Evaluation of the redox status of groundwater has shown that 188 samples indicate reduced aquifer conditions and 35 of those samples were excluded from the NBL derivation because of their high NH$_4$-contents ($NH_4 > 0.5\ mg\,l^{-1}$). In the end 594 groundwater samples regarded as being appropriate for the NBL derivation in the Upper Rhine Valley remained.

Table 4.7.1 shows the derived NBL and TV for 14 substances, for which NBLs have been defined and REF values were available. As the last column in the table indicates, case 1 is typical for TV derivation for most of the investigated parameters (Cl, K, Mg, SO$_4$, LF, As, NH$_4$), that is, the TV was set to half of the difference between NBL and the REF. For B, Na, NO$_2$, NO$_3$ and PO$_4$, however, parameters for which the NBL is significantly below the REF value, case 2 was applied. Only for two parameters, Mn (II) and Fe(II), was the NBL above the REF value, so that the NBL was regarded to be equal to the TV (case 3).

4.7.4
Conclusion and Discussion

After data preselection 594 samples were used to derive the NBLs and TVs. However, it is essential to compare these values to the situation assessed on the basis of the total data set containing about 1700 groundwater samples. An analysis shows that about 90–95% of the 1700 available samples display concentrations below the derived threshold values. In these cases a good status of groundwater can be postulated and no further measures to improve groundwater quality are required.

However, 5–10% of the samples exceed the TV. For these cases the reasons for this exceedence need to be checked. An assessment should be made of the impacts of natural influences, like salinization, redox-conditions and hydrodynamics, which may lead to natural (geogenic) anomalies, as well as those of anthropogenic influences like diffuse and point source pollution. If the reasons for exceeding the TV are due to natural influences, the 'good status of groundwater' can still be achieved. Only in cases where the reasons are due to anthropogenic influences can the 'good status of groundwater' not be achieved, and measures to improve the status have to be implemented.

Acknowledgements

The groundwater monitoring data sets provided by French, German and Swiss water authorities are greatly appreciated. The derivation of natural background levels and threshold values for groundwater bodies in the Upper Rhine Valley has been compiled in the framework of the EU-Specific Targeted Research Project 'BRIDGE' (Background cRiteria for the IDentification of Groundwater thrEsholds) – Contract No 006538 (SSPI). The views expressed in the chapter are purely those of the author (s) and may not in any circumstances be regarded as stating an official position of the European Commission.

References

1 Nieto, P., Custodio, E. and Manzano, M. (2005) Baseline groundwater quality: a European approach. *Environmental Science & Policy*, 8, 399–409.

2 Wendland, F., Blum, A., Coetsiers, M., Gorova, R., Griffioen, J., Grima, J., Hinsby, K., Kunkel, R., Marandi, A., Melo, T., Panagopoulos, A., Pauwels, H., Ruisi, M., Traversa, P., Vermooten, J.S.A. and Walraevens, K. (2008) European aquifer typology: a practical framework for an overview of major groundwater composition at European scale. *Environmental Geology*, 55 (1), 77–85.

3 Campbell, N., D'Arcy, B., Frost, A., Novotny, V. and Sansom, A. (2004) *Diffuse Pollution: An Introduction to the Problems and Solutions*, IWA Publishing, London.

4 Plant, J., Smith, D., Smith, B. and Williams, L. (2001) Environmental geochemistry at the global scale. *Applied Geochemistry*, 16, 1291–1308.

5 Kunkel, R., Voigt, H.-J. Wendland, F. and Hannappel, S. (2004) Die natürliche, ubiquitär überprägte

Grundwasserbeschaffenheit in Deutschland, Schriften des Forschungszentrums Jülich, Reihe Umwelt/Environment, Band 47, Forschungszentrum Jülich GmbH, Jülich, Germany, 204 S.

6 Chery, L.(coord.) (2006) Qualité naturelle des eaux souterraines. – Méthode de caractéri-sation des états référence des aquifères (Francais), Rapp. BRGM editions – Orlèans, 238 pp.

7 HLUG (1998) Grundwasserbe schaffenheit in Hessen, Auswertung von Grund- und Rohwasseranalysen bis 1997, ed: Hessische Landesanstalt für Umwelt, Umweltplanung, Arbeits- und Umweltschutz, Heft 250, 102 pp.

8 Wendland, F., Hannappel, S., Kunkel, R., Schenk, R., Voigt, H.-J. and Wolter, R. (2005) A procedure to define natural groundwater conditions of groundwater bodies in Germany. *Water Science and Technology*, 51 (3–4), 249–257.

9 Christensen, T.H., Bjerg, P.L., Banwart, S.A., Jakobsen, R., Heron, G. and

Albertsen, H.-J. (2000) Characterisation of redox conditions in groundwater contaminant plumes. *Journal of Contaminant Hydrology,* **45**, 165–241.

Further Reading

Salminen, R. and Tarvainen, T. (1997) The problem of defining geochemical baselines: A study of selected elements and geological materials in Finland. *Journal of Geochemical Exploration,* **60**, 91–98.

4.8
Hydrogeological Study of Somes-Szamos Transboundary Alluvial Aquifer

Radu Drobot, Peter Szucs, Serge Brouyere, Marin-Nelu Minciuna, László Lenart and Alain Dassargues

4.8.1
Introduction

The development of transboundary groundwater resources has generated, and will continue to generate, acrimony amongst states, nations and provinces. However, groundwater resources can also promote peace and accommodation, as jurisdictions and decision-makers, who share a common groundwater resource, realise that cooperation is the only way to ensure effective management and thus resource protection and sustainability.

The Somes-Szamos aquifer, which extends on both sides of the Romanian–Hungarian border, supplies drinking water to a population of about 395 000 inhabitants in Romania and 50 000 inhabitants in Hungary (Figure 4.8.1). The research project named SQUASH (Somes/Szamos Quantitative/Qualitative Study of the Hydrogeology) carried out by a group of research teams formed by Belgian, Romanian and Hungarian partners, in the scope of and supported by the NATO Science for Peace Programme, was intended to develop common tools and guidelines for local end-users (water supply companies and regulatory authorities from Romania and Hungary) to be able to manage groundwater quantity and quality. The project focused on improving the previous understanding of the groundwater conditions, including flow and pollutant transport across many scales, using data acquisition techniques and computer simulation models.

The main steps/tasks of the project were:

- collecting all existing data,
- building and development of a common database,
- new measurements campaigns,
- regional modelling,
- groundwater quality studies in pilot zones.

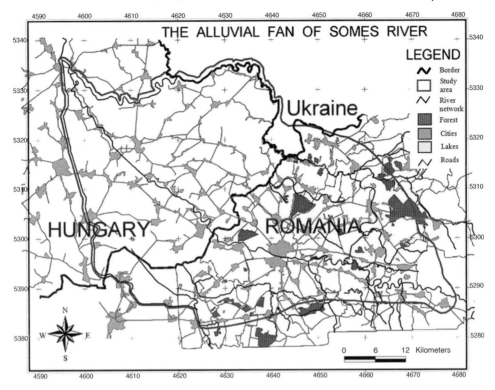

Figure 4.8.1 Location of Somes-Szamos aquifer in Romania and Hungary.

4.8.2
Transboundary Project Activities to Achieve Sustainable Groundwater Management

After adopting a common database structure based on the database developed by the Hydrogeology Group of the University of Liège [1, 2], each team was in charge of feeding the database with all the data collected. The database, linked to a geographical information system (GIS), insures perfect compatibility for data exchanges.

On the basis of analysis of the available data, new campaigns of measurements were carried out focusing on the following aspects: piezometric levels, pumping tests for hydrodynamic parameters, groundwater quality campaign (sampling + analysis), and tracer tests for assessment of solute transport parameters (locally). Priority was given to measurements in areas with low density of observation wells, in order to ideally prepare all the needed data to allow reliable groundwater modelling.

The lithology revealed a meander sedimentological system, whose layers developed from 30 m to a maximum depth of 130 m on the Romanian territory and 190 m on the western limit of the Hungarian territory. These layers represent the main aquifer due to their extent, hydrogeological parameters and abstracted discharges (Figure 4.8.2).

Cross-Section II-II'

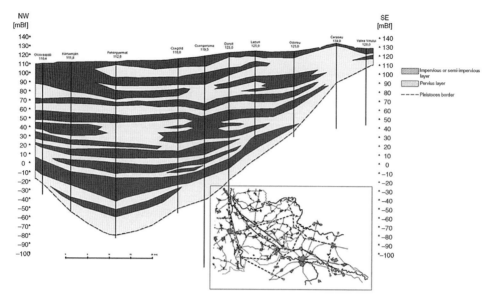

Figure 4.8.2 Common concept: a new geological cross-section.

One of the most important steps in the mathematical modelling was the choice of the conceptual model of the aquifer. By keeping the essential features of the system, a reasonable compromise between the complexity of the multi-layered aquifer and the available reliable data concerning the actual structure and hydrogeological parameters was proposed. Belgian, Romanian and Hungarian teams agreed on a conceptual model consisting of two aquifer systems with an aquitard between them. At the same time, a field campaign showed on the Romanian side differences of hydraulic head between the so-called 'shallow' and 'deep' aquifer systems in the range of 1 to 5 m, thus justifying the selected simplified model structure.

The regional groundwater mathematical model based on all the available data (preprocessed using GIS/database) represents one of the most important outputs of the project [3]. The groundwater model (GWM) and the database coupled to GIS integrate all the available data and the new knowledge about the studied aquifer system, allowing a better evaluation of groundwater resources and their sustainable management.

Groundwater flow and contaminant transport simulations were performed using MODFLOW and MT3D. Calibration of the hydrogeological parameters was achieved in steady-state (Figure 4.8.3) and in transient conditions.

The zonation of the hydraulic conductivities obtained at the end of the steady state calibration was kept unchanged during the transient calibration. The history of groundwater levels, pumping rates, recharge values and time dependent boundary conditions for stress periods of one month were used as additional data. The transient

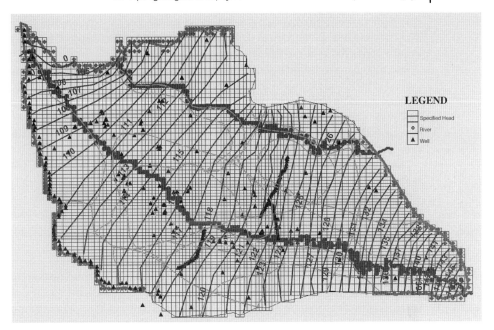

Figure 4.8.3 Calibrated hydraulic head map for the shallow aquifer in the case of the steady-state regional groundwater model (October, 2001).

state simulation for the calibration of the specific yield, in respective of the specific storage coefficients, was undertaken for the interval January 2001–April 2002. Following the calibration, a transient simulation for the interval May 2002–December 2002 represented the base for model validation.

After a sensitivity analysis and a validation procedure, the mathematical model was used to simulate different scenarios for the future [3]. These scenarios were decided in consultation with the end-users and decision-makers (changes in pumping rate, irrigation, drainage, etc.). Of these scenarios, the most critical that was tested consisted of doubling the present pumping rates for all production wells. A steady state simulation showed that the resulting drawdowns in the first layer are less than 0.5 m, while in the third layer, which represents the main aquifer for water supply, the maximum drawdowns are 1.5 m. Still, the tested increase of the pumping rates is not likely in the foreseen future. According to these results the aquifer does not seem to be at risk from a quantitative point of view.

Locally, some groundwater quality problems were observed or anticipated in the basin. Waste disposal sites, due to the lack of appropriate technical design, are often major environmental issues throughout Hungary as well as Romania. Owing to the lack of a centralized waste management concept in past decades, thousands of legal and illegal waste disposal sites were created. Basically every settlement had at least one disposal site, usually without any design or lining concept. Thanks to nation-wide programmes the establishment of regional waste disposal sites is now on a dynamic track; however, the remediation of former waste disposal sites and their environmental effect remains a problem to be solved. For this reason, the waste disposal areas near Fehérgyarmat (Hungary) and Satu-Mare (Romania) were chosen as pilot zones for local modelling purposes.

The modular three-dimensional transport model referred to as MT3DMS was applied to build and run two local solute transport models in the investigated region. To build a reliable transport model, some additional and accurate information about the hydrodispersive properties was required from the different hydraulic units of the investigated aquifer. Detailed multi-tracer test investigations were carried out in both countries to obtain hydrodispersive parameters.

Based on available geological data and results from chemical analyses of the samples taken during the last field campaign, local groundwater transport models were developed. For each site, the boundary conditions were deduced from the regional flow model. The flow conditions were re-calibrated at the local-scale, while the transport model was calibrated based on the measured maps of the concentration. As required for protection zones delineation, solute transport times were calculated for the main pumping wells.

With the available data at this stage, the model showed that the municipal waste disposal in Hungary did not actually represent a direct hazard for the water quality and the ecosystem of the Szamos-Someş River; the simulated contamination plumes do not reach the well-head protection zones of the local producing water wells (Figure 4.8.4).

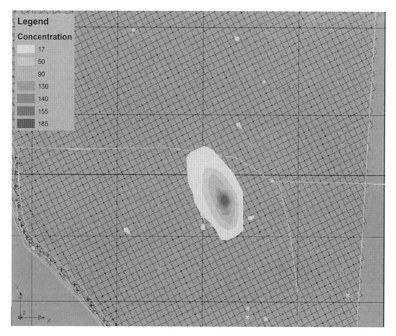

Figure 4.8.4 Computed Cl$^-$ concentration distribution after 50 years (MOC, Method of characteristics numerical solution) at Fehérgyarmat in Hungary.

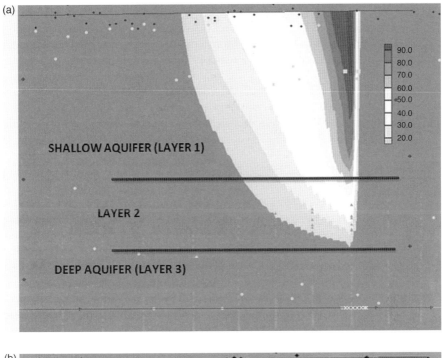

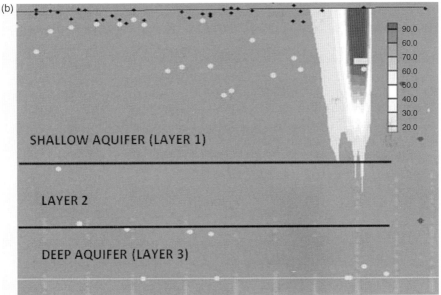

Figure 4.8.5 Comparison of the plume extension at Satu-Mare landfill: (a) no measures taken; (b) two pumping wells inside the landfill.

Since the phenomena of sorption and decay were neglected, the computed results can be considered as the worst case scenario.

On the contrary, according to the simulations at the municipal solid waste landfill of Satu-Mare (Romanian side of the aquifer) after 20 years the whole thickness of the aquitard is contaminated and the plume penetrates into the third layer (Figure 4.8.5a). Different solutions for preventing the plume extension were imagined and simulated. Thus, two pumping wells abstracting a discharge of $5\,l\,s^{-1}$ and located inside the landfill were designed to intercept the leakage and prevent the pollution of the aquitard (Figure 4.8.5b).

4.8.3
Conclusion

One of the main objectives of the groundwater modelling work was to make possible the sustainable management of the regional groundwater resources of Somes-Szamos alluvial aquifer.

The NATO Science for Peace Programme provided the participants with a perfect framework to carry out a healthy and fruitful scientific and technical collaboration. On the basis of the obtained modelling results, the project developed guidelines to provide both water supply companies and regulatory authorities the means to manage the groundwater quantity and quality simultaneously. A common monitoring and an excellent exchange of data is taking place between the water authorities on both sides of the Hungarian-Romanian border.

4.8.4
Acknowledgments

The partners of the SQUASH Project gratefully acknowledge the NATO Science for Peace Programme for support of this research.

Special thanks go to Dr Atle Dagestad from NGU (Geological Institute of Norway), the NATO observer, for his assistance and advice in critical phases of the project development.

References

1 Gogu, R.C., Carabin, G., Hallet, V., Peters, V. and Dassargues, A. (2001) GIS-based hydrogeological database and groundwater modelling. *Journal of Hydrogeology*, **9**, 555–569.
2 Orban, P., Popescu, I.C., Ruthy, I. and Brouyere, S. (2004) Database and general modeling concepts for groundwater modeling in the SQUASH project. *Hidrotehnica*, **49** (9–10), 51–57.
3 Lénárt, L., Madarász, T., Mikó, L., Szabó, A., Szűcs, P., Juhászné, M.V., Karsai, M., Bretotean, M., Drobot, R., Filip, A., Jianu, M., Minciuna, M., Brouyere, S., Dassargues, A. and Popescu, C. (2003) Complex hydrogeological study of the alluvial transboundary aquifer of Somes/Szamos (Romania–Hungary). Presented at the XI. World Water Congress. Madrid (Spain).

4.9

Towards Sustainable Management of Transboundary Hungarian–Serbian Aquifer

Zoran Stevanović, Péter Kozák, Milojko Lazić, János Szanyi, Dušan Polomčić, Balázs Kovács, József Török, Saša Milanović, Bojan Hajdin and Petar Papic

4.9.1
Introduction

During the last decade there have been many activities aiming to improve water management amongst neighbouring countries [1, 2]. Many such projects have been initiated worldwide but, with the exception of the European Union (EU) territory, only very few have been successfully implemented or even started.

The project Sustainable Development of Hungarian–Serbian Transboundary Aquifer (SUDEHSTRA) was carried out from June 2007 till August 2008. It was one of the cross-border cooperation programmes funded by the European Union (ERDF/INTERREG IIIA/Community Initiative) whose objective and tasks were fully in line with the EU Water Framework Directive (WFD; EC 2000/60) and EU Groundwater Daughter Directive targets.

The benefits of this cross-border project are supposed to range from the national level to the very local one:

1) strengthening bilateral technical cooperation and enabling the exchange of information between the neighbouring countries;
2) improving national, regional and local water practice;
3) creating an ambience to facilitate the realization of the targets of the EU Water Framework Directives;
4) building local technical capacities;
5) increasing public awareness of the importance of water issues and of sustainable use of water and its protection from pollution;
6) developing tools for appropriate groundwater management and monitoring at all levels.

4.9.2
Study Area

The aquifer system under study is located between the Danube and Tisa (Tisza) rivers and extends to the vicinity of Kiskunfelegyhaza on the Hungarian side (north) and to Vrbas in Serbia (south). The main groundwater consumers are the cities and industries of Szeged, Kiskunhalas, Baja, Tompa, Hódmezővásárhely (Csongrád and Bács-Kiskun Counties in Hungary) and Subotica, Sombor, Bačka Topola, Vrbas, Kula (in total 16 municipalities in Serbia, Vojvodina province, Bačka region). The study area is populated by over 800 000 inhabitants, about 40% of whom are on the Hungarian side of the border.

The Pannonian basin (or the Great Hungarian Basin) represents a geographical and geological entity that spreads over the territory of several countries. The central portion is in Hungary while the marginal parts belong to Romania, Slovakia, Slovenia, Croatia and Serbia (Figure 4.9.1).

The Pannonian basin is a typical lowland, whose height above sea level is mainly 70–120 m. The flat relief features various geological structures, such as loess plateau, loess terraces and alluvial plains, and small depressions in between prevail.

The region has a moderate continental climate with an average annual air temperature of 11 °C and rainfall of 575 mm per year. The whole hydrographic network belongs to the river basin of the Black Sea. It is assumed that about 99% of surface waters are water flows of a transitory nature, while only about 1% comes from rainfall within the region. The major rivers are the Danube and the Tisa. At the Hungarian–Serbian border the average annual river flow of the Danube is around 2500 m^3 s^{-1}, while for the Tisa the average is 700 m^3 s^{-1}.

Figure 4.9.1 Location map of study area.

4.9.3
Groundwater Distribution and Use

Systematic geological research in this region began at the end of the nineteenth century. The first data about the geological structure date came from the drilling of artesian wells. Later on, systematic searches for oil reserves resulted in numerous studies and maps reconstructing paleogeographic conditions and improving the knowledge of geological structures [3–5]:

- Pre-Cambrium and Paleozooic metamorphic schists are the oldest rocks and represent a deep fundament for younger sedimentary rocks, mainly of Tertiary age. Mesozoic sediments mostly of Triassic and Cretaceous ages are also widely distributed in the basin's basement.
- The thickness of the Tertiary (Badenian, Sarmatian, Pannonian and Pliocene) sediments is variable in different parts of the terrain, from a few tens to over two thousand metres.
- Quaternary (Pleistocene and Holocene) sediments are dominant throughout the study area and cover the older sediments. Their thickness varies from a dozen to 200 m.

In the vertical section, water-bearing layers with intergranular porosity interfinger with impermeable strata. Sandy-gravel sediments represent major aquifer systems. Their thickness varies from less than 10 m up to 50 m, but is generally of the order of 10 to 20 m. The transmissivity coefficient ranges from 10^{-5} to $10^{-4}\,\mathrm{m}^2\,\mathrm{s}^{-1}$. The semi-permeable or fully-impermeable sandy clays often disconnect aquifer layers, which has resulted in the artesian character of the latter.

Groundwater (GW) resources are vital for the economy and society, as well as for the development of both countries. Within the framework of the International Commission for the Protection of the Danube River (ICPDR) activities and the Roof Report for 2004 [6], the transboundary aquifer system of Hungary–Serbia was preliminary separated into two parts: one large GW body in Serbia named CS_DU 10, and five in Hungary (P.1. and P.2. groups). The total area is assumed to cover around 27 000 km².

The region completely satisfies the demand for drinking water from the ground. There are more than 300 sources and centralized waterworks of different sizes, from the very small demand of small villages to the very large ones of supplying water to more than 100 000 consumers. There are over 1200 operational deep wells, belonging either to those waterworks or to small industrial enterprises and individual farms. However, data concerning their pumping rates and drawdown are not always collected, particularly on the Serbian side of the border.

Most of the wells for water supply are drilled to a depth of 50–150 m, but some reach 250 m or even more (600 m in Szeged). The well capacity ranges from 5 to $25\,\mathrm{l\,s}^{-1}$, depending on geology, applied drilling techniques, construction materials, well development and other factors [7]. In the past, many wells were over-pumped and forced beyond their optimal capacities, which caused them to deteriorate quickly. Therefore, many were replaced or revitalized. There are some indications of local drawdown; many wells which were previously artesian are characterized today by a static groundwater table lower than 10 m below the surface.

The conceptual hydrogeological model includes five main aquifer layers to a depth of some 2500 m (Figure 4.9.2). The first two (Quaternary and Upper Pliocene), which are most prominent and characterized by the presence of fresh groundwater, are utilized mostly for drinking water supply and for irrigation. The deeper layers are also used in the water supply of some cities (e.g. Szeged) or for geothermal or balneotherapeutical purposes: thermal water is used for recreation and medical purposes in several spas in both countries, while geothermal energy is more efficiently used in Hungary. The number of wells that tap deeper aquifer layers with thermal waters is over 100 on the Hungarian side, while in Serbia there are some 15–20 such wells.

In the Hungarian part of the study area the estimated exploitation is $2 \, m^3 \, s^{-1}$, 60% of which is used for municipal water supply and irrigation, and the rest is employed for industrial purposes. In Serbia it is assumed that the current exploitation of the transboundary aquifer is around $2.8 \, m^3 \, s^{-1}$, of which $1.2 \, m^3 \, s^{-1}$ are used by individual industry and farms. Groundwater is also used for irrigation purposes, but to a lesser extent because of rain-fed agriculture and the surface waters from the rivers and channels. Tapping of shallow aquifer layers is prevalent in Serbia, whereas the deeper aquifer layers are more exploited in Hungary.

4.9.4
Proposed Measures for Sustainable Utilization of the Aquifer Systems

On the Hungarian side of the model domain a MODFLOW based model was built [8] to investigate the hydrodynamic regime of the shallow aquifers.

The preliminary SUDEHSTRA hydrodynamical model was conceived and built as a multilayer model with ten layers (five water-bearings and five semi-permeables between them). Hydraulic parameters are approximated on the basis of provided documentation as representative values for the whole layer. Owing to insufficient data of the groundwater regime this initial model covered only steady state flow conditions.

The obtained results are useful for preliminary evaluation. They show insignificant regional depletion of groundwater for the simulated extraction rates (similar to the existing ones or with a slight increase in Hungary), and positive effects for simulated artificial recharge. The former is most likely due to the fact that the main pressure on groundwater extraction is shared between different water-bearing layers. Whereas in Serbia the main tapping is from the second and third aquifer layers, for the water supply of the main Hungarian consumer Szeged and for geothermal energy, the deeper aquifer layers (fourth and fifth) are intensively used.

Some of the concrete proposals of the SUDEHSTRA project include [9]:

- centralize the waterworks at municipal and regional levels (incorporate small village waterworks into municipal ones);
- keep existing sources functional as an integral part of future water supply systems;

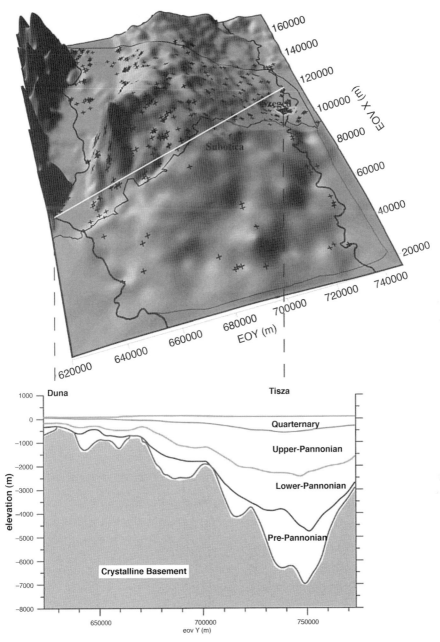

Figure 4.9.2 Three-dimensional model and general cross section through the study area between Danube (Duna, in the west) and Tisa River (Tisza in the east). Note: 'Pannonian' layers are considered to be the local Hungarian classification; in fact, the Upper Pannonian is of Pliocene Age.

- establish systematic monitoring of extracted water quantities and groundwater table fluctuations;
- reduce the pollution of water and soil to the maximal extent;
- introduce complex water treatment, particularly where problems with organic components or increased concentration of Fe, Mn, As is recognized;
- delineate the sanitary protection zones of water sources in accordance with criteria, taking into consideration primarily the local hydrogeological conditions and residence time for the pollutant transport (as recently reviewed legislation in the two countries required);
- introduce river bank filtration along major rivers such as the Danube and the Tisa;
- initiate a survey for artificial recharge, where natural conditions would be assumed to be promising for such interventions.

In line with last two proposals the alternative water sources are also analysed. Alluvial deposits along the Danube have the largest potential for water supply of the population and industry, but Tisa alluvium should not be neglected. The Danube alluvium has been studied as one of the most promising sources for the regional water supply of northern Serbia. In the vicinity of Apatin, 40–50 m thick and very permeable alluvial deposits, gravel and sands (average transmissivity $T = 2 \times 10^{-2}$ $m^2 s^{-1}$) enable the tapping of some 0.25–$0.3 \, m^3 s^{-1}$ per 1 km distance along the riverbank [10].

4.9.5
Conclusion

For this project numerous activities were undertaken, such as a local waterworks enquiry, field survey and measurements, the establishment of a GW monitoring network, the creation of a GW initial database, common workshops and seminars for local capacity building, and water-saving promotional activities. As one of the results of the project, an initial conceptual hydrogeological and hydrodynamical model was created and tested.

The results of the conducted survey enabled the proposal of many concrete measures and activities for better management and sustainable development of the groundwater as a very important and even sole resource in large parts of this region. The proposals that can be considered most important are: (i) opening of new sources for regional water supply based on alluvial groundwater and (ii) centralizing waterworks at municipality levels so that water will be adequately treated and protected from pollution.

Acknowledgements

The authors of this chapter gratefully acknowledge the financial and technical assistance of ERDF and the EU Delegation to the Republic of Serbia (previously the European Agency for Reconstruction) provided to the project SUDEHSTRA.

References

1 UN/ECE (2000) *Guidelines on Monitoring and Assessment of Transboundary Groundwaters. Work Programme 1996–1999*, UN/ECE, Lelystad, Netherlands. ISBN 9036953154.

2 Puri, S., Appelgren, B., Arnold, G., Aureli, A., Burchi, S., Burke, J., Margat, J. and Pallas, P. (2001) Internationally Shared (Transboundary) Aquifer Resources Management: Their Significance and Sustainable Management. A Framework Document, in IHP-VI, IHP Series on Groundwater No. 1, UNESCO, Paris. Available at http://unesdoc.unesco.org/images/0012/001243/124386e.pdf

3 Vadasz, E. (1960) *Magyaroszag Földtana (Geology of Hungary)*, Akadémiai Kladó, Budapest, 532 pp.

4 Milosavljević, S., Vasiljević, M. and Vilovski, S. (1997) Hydrogeological explorations in Vojvodina, in *100 Years of Hydrogeology in Yugoslavia*, vol. 1 (ed. Z. Stevanovic), FMG, Belgrade, pp. 117–146.

5 Almasi, I. (2000) Petroleum hydrogeology of the Great Hungarian Plain, East Pannonian basin. PhD thesis, University of Alberta, Alberta.

6 Djurić, D., Josipović, J., Zogović, D., Komatina, M., Stevanović, Z., Djokić, I. and Lukić, V. (2005) ICPDR Roof Report for 2004, Institute WM "Jaroslav Cerni", Document of Ministry of Agriculture Forest and Water, Belgrade.

7 Soro, A., Dimkić, M. and Josipović, J. (1997) Hydrogeological investigations related to water supply in Vojvodina, in *100 Years of Hydrogeology in Yugoslavia*, vol. 2 (ed. Z. Stevanović), FMG, Belgrade, pp. 101–112.

8 Völgyesi, I. (2006) A Homokhátság felszínalatti vízháztartása. Vízpótlási és visszatartási leheto ségek, MHT XXIV. Országos Vándgyu lés Kiadványa. Pécs.

9 Stevanović, Z., Lazić, M., Polomčić, D., Milanović, S., Hajdin, B., Papić, P., Sorajić, S. and Kljajić, Z. (2008) Sustainable Development of Hungarian–Serbian Transboundary Aquifer (SUDEHSTRA), Final Report, Fac. Min. Geol. University of Belgrade, Belgrade.

10 Institute of Water Management "Jaroslav Černi" (2004) *Alternative Solutions in Water Supply of Vojvodina, Hydrodynamical Study*, Ministry of Science Serbia, Belgrade (unpublished).

4.10
Transboundary Groundwater Resources Extending over Slovenian Territory

Petra Meglič and Joerg Prestor

4.10.1
Introduction

Despite its small area, Slovenia has a very and complex hydrogeological structure. Therefore different hydrogeological boundaries and characteristics for groundwater bodies and transboundary groundwater bodies were used for their delineation.

4.10.2
Transboundary Groundwater Resources

Slovenian territory ($20.273 \, \text{km}^2$), as seen on the world map of groundwater resources $1 : 50\,000\,000$, is an area with a complex hydrogeological structure [1]. Only the western and north-eastern border sections cross major groundwater basins. The north-eastern part belongs to a bigger major groundwater basin (Panonian basin) with high groundwater recharge of between 100 and 300 mm per year [1]. The western edge only touches a major groundwater basin (river Po) with a very high groundwater recharge of more than 300 mm per year [1]. Slovenian territory extends from the north-west from Italy and Austria over the Alps to the Dinarides mountains. These mountains then pass into Croatian territory, and further to the south-east onto the Balkan Peninsula. Indeed, most Slovenian territory has a complex hydrogeological structure, with a very high level of groundwater recharge of more than 300 mm per year [1]. Very important water resources are exploited from the Dinaric karst and fissured aquifers extending over the Slovene–Croatian state border. The major "Panonian" groundwater basin has been developed over a wide area between Vienna, Budapest and Belgrade, and has some very important drinking and thermal water resources, which are actually exploited by Slovenia, Austria and Hungary. The sustainable cross border management of these resources will certainly be crucial for their future use. The major groundwater basin of the river Po and the neighbouring Adriatic rivers plain has been developed mainly on Italian territory below the Alps and has some very important groundwater resources for the highly populated Italian territory and represents very important resources for future use in the Slovene coastal and border area, which is highly developed.

Figure 4.10.1 shows all the larger and important aquifer systems in Slovenia that cross the state border (marked with circled numbers), as well as the groundwater bodies in Slovenia and the surface river basin divide between the Adriatic Sea and the Danube river. The important aquifers that cross the state border to the neighbouring countries of Croatia, Italy, Austria and Hungary are shown in Figure 4.10.1 are [3, 4]: (1) limestone and dolomite aquifers in Karavanke (extending over the state border to Austria); (2) alluvial gravel aquifers of Drava and Mura (extending over the state border to Austria); (3) sand and clay aquifers on Goričko (extending over the state border to Austria and Hungary); (4) alluvial gravel aquifers of Drava and Mura (extending over the state border to Hungary and Croatia); (5) local carbonate and sand aquifers in the Sotla river basin (extending over the state border to Croatia); (6) alluvial gravel aquifer of Sava (extending over the state border to Croatia); (7) carbonate fissured and karst aquifers in the Kolpa river basin (extending over the state border to Croatia); (8) very karstified carbonate aquifers in the river basin of the Adriatic Sea between Kvarner and Trieste bay (extending over the state border to Italy and Croatia); (9) very karstified, generally limestone aquifers in the river basin of the Adriatic Sea and Soča–Brestovica (extending over the state border to Italy); (10) alluvial gravel aquifers of the Vipava and Soča River basin (extending over the state border to Italy); (11) fissured, predominantly dolomite and limestone aquifers of the west part of Soča and the Sava Dolinka watershed (extending over the state border to Italy).

Every aquifer is further divided and characterized according to the existing knowledge on aquifer systems, where direct cross border groundwater flow is recognized. There are also aquifer systems where direct cross border groundwater flow has not yet been recognized or has been recognized but is of low importance. It is

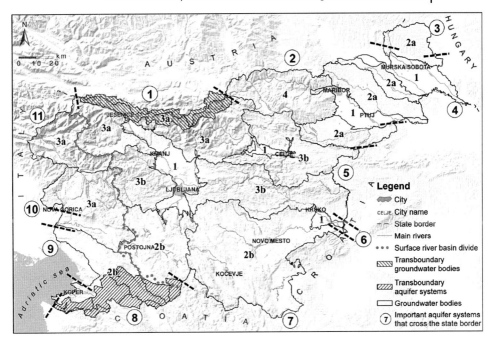

Figure 4.10.1 Map with transboundary aquifer systems and groundwater bodies in Slovenia marked. The result of the first delineation [2] of groundwater bodies, which are marked on the map with a number and a number + letter as follows: (1) aquifer systems in alluvial sediments (intergranular porosity); (2) aquifer systems in sedimentary rocks and unconsolidated, dominantly non-alluvial sediments, (a) dominantly intergranular porosity and (b) dominantly karst porosity; (3) aquifer systems in a complex hydraulic system adapted to intensely folded mountain zones, (a) dominantly karst porosity and (b) dominantly fissured porosity; (4) aquifer systems in basement geological strata. The encircled numbers are defined in the text.

expected that the database on characterized aquifer systems will grow quickly in the next few years.

Groundwater bodies (GWBs) ground water body (GWB) and transboundary groundwater bodies (TGWBs) transboundary groundwater body (TGWB) in Slovenia were delineated and characterized according to the WFD (Water Framework Directive) and Slovenian legal regulations, namely, 'Rules on methods for determining water bodies of groundwater' [5] and 'Rules of determining water bodies of groundwater' [6]. The delineation of individual groundwater body and groups of groundwater bodies was performed according to the regulations and was based on porosity and lithology boundaries, productivity and extent boundaries, catchments basin boundaries, flow lines, interstream boundaries, junctions with large affluent, recovering and potential use boundaries (water protection areas) and tracer experiments results. The methodology used for delineation was that of the Bureau de Recherches Géologiques et Minières (BRGM) [7], where groundwater bodies are classified as (i) aquifer systems in alluvial sediments, (ii) aquifer systems in sedimentary rocks and unconsolidated dominantly non-alluvial sediments, (iii) aquifer systems in hydraulic complex system adapted to intensely folded mountain zones, (iv) aquifer systems in basement geological strata and (v) aquifer systems in low permeable strata and local and limited water resources. The latter was not defined in Slovenia. A more detailed methodology is described in Prestor *et al.* [8]. Each groundwater body is then further divided into one to four important aquifers, which are laterally or vertically subsequent.

The cartographic base for delineation of GWBs and TGWBs in Slovenia is the latest hydrogeological map (1 : 250 000), which is based on an international standard legend adopted after the IAH (International Association of Hydrogeologists) [9].

There are three main tectonic units in Slovenia that also play an important role in the groundwater bodies' delimitation: (i) the southern and eastern Alps, which are dominantly karst and also fissured aquifers, with a complex hydraulic system, intensely folded mountain areas including Periadriatic igneous rocks as basement rocks and also volcanoclastic rocks and fissured aquifers; (ii) the external and internal Dinarides, which are dominantly karst aquifers, but with a minor part of fissured aquifers and also very low permeability layers (e.g. flysch) with local and limited small aquifers; (iii) the Tertiary and Quaternary sediments of the Panonian basin, which are dominantly alluvial Quaternary gravel, sandy gravel aquifers with a strong connection to surface streams and non-alluvial sediments, dominantly sandy and silty gravel aquifers with low or no connection to surface streams and often of confined or semi-confined type.

4.10.3
Conclusions

Until now, the only transboundary groundwater body that has been delineated by both neighbouring countries is the Karavanke GWB, which extends over the state border to Austria. From both the Slovenian and Austrian side, the major factors in determining the boundaries for this shared GWB were its geological and hydrogeological characteristics. On the Austrian side of the border this GWB was delineated as a 'group of groundwater bodies with prevailingly karst aquifers', whereas on the Slovenian side of the border it was delineated as a group of groundwater bodies in 'a complex hydraulic system adapted to an intensely folded mountain area made up of different aquifers and aquifer systems'. During the preparation stage of the water management plan for this shared GWB, consideration was given to the fact that in some parts a cross-boundary groundwater flow is actually recognized while in other parts there is no such flow. Only those areas where a cross-boundary flow is detected need common management.

A second and larger transboundary project focused on the area between Slovenia and Croatia, on the Dinaric karst on the Istria peninsula. The main objective of the project was the common hydrogeological characterization of the cross-border groundwater flows and the development of transboundary water protection areas.

A third very important transboundary project co-financed by the Interreg programme characterized the transboundary thermal water resources between Slovenia and Austria [10]. The mutual and common development of geographical information systems (GISs) and hydrogeological data was performed, including the compilation and improvement of expert knowledge about transboundary structures and thermodynamics, extending also further to the Panning basin on Hungarian territory. In future a trilateral transboundary aquifer system can be expected.

The Geological Survey of Slovenia also participated in the project eWater. One of the principal objectives of the project, and its 'Multilingual cross-border access to groundwater databases' was to increase the cross-border availability, accessibility and re-usability of spatial data on quality, location and use of subsurface waters. The objective of the project was to develop a WEB GIS portal for participating countries' hydrogeological data, accessible on www.ewater.eu. This envisaged cross-border portal is targeted at the EC itself, at national and river basin water authorities, water suppliers, added-value data service providers, insurance companies, planning and controlling organizations, as well as the general public. The participating countries are Slovenia, Austria, the Czech Republic, Denmark, France, Hungary, Italy, Lithuania, the Netherlands and Slovakia.

References

1 (a) BGR & UNESCO (2008) Groundwater Resources of the World 1 : 50 000 000, Maps, chief editors W. Struckmeier and A. Richts (b) The Groundwater Resources Map of the World at the scale of 1:50.000.000. BGR, UNESCO, Paris at http://www.whymap.org (accessed 18 October 2010).

2 Tavzes, R., Vodopivec, N., Matoz, H., Turk, I., Planinšič-Kolar, V. and Nared, N. (2004) Izvajanje vodne direktive na Vodnem območju Donave. Republika Slovenija. Ministrstvo za okolje in proctor, National Report on WFD Implementation in Slovenia, 183 pp., 2 Annexes.

3 Lapanje, A., Urbanc, J., Prestor, J., Mali, N., Meglič, P., Janža, M. and Krivic, J. (2007) Definiranje potencialnih prekomejnih vodonosnikov – Italija (Defining potential transboundary aquifers – Italy), Report in the archive of Geological Survey of Slovenia, Geological Survey of Slovenia, Ljubljana.

4 Prestor, J., Lapanje, A., Urbanc, J., Mali, N., Meglič, P., Janža, M. and Krivic, J. (2007) Definiranje potencialnih prekomejnih

vodonosnikov – Hrvaška (Črnomorsko povodje) (Defining potential transboundary aquifers – Croatia), Report in archive of Geological Survey of Slovenia), Geological Survey of Slovenia, Ljubljana.

5 Government of the Republic of Slovenia (2003) Pravilnik o metodologiji za določanje vodnih teles podzemnih voda - Rules on methods for determining water bodies of groundwater. Official Gazette of Republic of Slovenia (Uradni list Republike Slovenije) at http://www. uradni-list.si/, Ur. l. RS, no. 65/2003 (accessed 18 October 2010).

6 Government of the Republic of Slovenia (2005) Pravilnik o določitvi vodnih teles podzemnih voda - Rules of determining water bodies of groundwater. Official Gazette of Republic of Slovenia (Uradni list Republike Slovenije) at http://www. uradni-list.si/, Ur. l. RS, no. 63/2005 (accessed 18 October 2010).

7 BRGM (2003) Mise en oeuvre de la DCE identification et délimitation des masses d'eaux souterraines. Guide

méthodologique, Document public, BRGM/RP-52266-FR, janvier 2003.

8 Prestor, J., Urbanc, J., Meglič, P., Lapanje, A., Rajver, D., Hribernik, K., Šinigoj, J., Strojan, M. and Bizjak, M. (2004) Nacionalna baza hidrogeoloških podatkov za opredelitev vodnih teles podzemne vode Republike Slovenije 2004, Zvezek VI – Metodologija (National base of hydrogeological data for delineation of groundwater bodies in Slovenia 2004, Fascicle VI - Methodology), Report in the archive of the Geological Survey of Slovenia, Geological Survey of Slovenia, Ljubljana, 62 pp.

9 Struckmeier, W.F. and Margat, J. (1995) Hydrogeological Maps – A Guide and a Standard Legend, IAH International Contribution to Hydrogeology 17, Heise (Hannover).

10 Bäk, R., Budkovič, T., Domberger, G., Götzl, G., Hribernik, K., Kumelj, Š., Lapanje, A., Letouzé, G., Liparski, P., Poltnig, W. and Rajver, D. (2007) Geotermalni viri severne in severovzhodne Slovenije (Geothermal resources of northern and north-eastern Slovenia), Geological Survey of Slovenia, Ljubljana, 124 pp., 15 Appendices.

5
Transboundary Lakes and Rivers

5.1
Do We Have Comparable Hydrological Data for Transboundary Cooperation?

Zsuzsanna Buzás

5.1.1
Introduction

In 2008 the WMO RA VI Working Group on Hydrology [1] made a survey on the use of national and international standards for hydrological observation and data processing for the main quantity parameters of surface and groundwater.

5.1.2
Institutional Background

The World Meteorological Organization (WMO) is a specialized agency of the United Nations [2]. As weather, climate and the water cycle know no national boundaries, international cooperation at a global scale is essential for the development of meteorology and operational hydrology as well as to reap the benefits from their application. The WMO provides the framework for such international cooperation [3].

The WMO promotes cooperation in the establishment of networks for making meteorological, climatological, hydrological and geophysical observations, as well as the exchange, processing and standardization of related data, and assists technology transfer, training and research. It also fosters collaboration between the National Meteorological and Hydrological Services (NMHS) and furthers the application of meteorology to public weather services, agriculture, aviation, shipping, the environment, water issues and the mitigation of the impacts of natural disasters. The WMO facilitates the free and unrestricted exchange of data and information, products and services in real- or near-real time on matters relating to the safety and security of society, economic welfare and the protection of the

Transboundary Water Resources Management: A Multidisciplinary Approach, First Edition.
Edited by Jacques Ganoulis, Alice Aureli and Jean Fried.
© 2011 Wiley-VCH Verlag GmbH & Co. KGaA. Published 2011 by Wiley-VCH Verlag GmbH & Co. KGaA.

environment. It contributes to policy formulation in these areas at national and international levels.

The Hydrology and Water Resources Programme (HWRP) is concerned with the assessment of the quantity and quality of water resources, both surface and groundwater, in order to meet the needs of society, to permit mitigation of water-related hazards and to maintain or enhance the condition of the global environment. It includes the standardization of various aspects of hydrological observations and the organized transfer of technologies for enabling National Hydrological Services (NHS) to provide the hydrological data and information required for the sustainable development of their countries. It provides advice to members on flood management policy and assists them in their effort to adopt Integrated Water Resources Management (IWRM) with an emphasis on practical applications.

The six World Meteorological Organization Regional Associations (RA I – Africa, RA II – Asia, RA III – South America, RA IV – North and Central America, RA V – South-west Pacific and RA VI – Europe) of the WMO have each established a Working Group on Hydrology (WGH) with terms of reference encompassing the range of topics covered by the HWRP. 'The Directive 2000/60 of the European Parliament and Council established a framework for community action in the field of water policy', well-known as the EU Water Framework Directive (WFD) entered into force on 23 October 2000 [4]. The 27 EU Member States (MS) are obliged to enforce this directive and it has to be applied for river basins entering the sea. Many non-EU neighbouring countries joined the implementation process. Article 8 of the WFD states that 'Member States shall ensure the establishment of programmes for monitoring of water status in order to establish a coherent and comprehensive overview of water status within each river basin district'. These programmes had to be operational by the end of 2006. This Directive also contributes to the implementation of the UNECE Convention on the Protection and Use of Transboundary Water Courses and International Lakes. It means that the EU MS implement the WFD on national, transboundary, river basin and EU levels.

All the EU countries belong to the WMO RA VI region and are usually represented by their respective hydrological services in the WGH.

5.1.3
Survey Based on a Questionnaire

In the Action Plan 2008–2011 of the RA VI Working Group on Hydrology there is a task

'to initiate joint effort with the International Organization for Standardization (ISO) and European Committee for Standardization (CEN) on international standards for hydrological observation and processing'.

The aim is the improvement of the comparable monitoring results. To survey the situation an overview of European practice related to standardization is needed.

For this reason a survey was made in the first quarter of 2008 based on a questionnaire that related to the surface and groundwater quantity parameters both for measurement and observation, and for data processing.

In this questionnaire the following parameters were used for surface water: water level, water temperature, discharge, suspended and bottom sediment; for groundwater the parameters were: water level and water temperature. Countries were requested to indicate which type of regulation–guidance was used by their NHS for field measurement and data processing, hydrological guidelines (these are technical guidance published internally by the NHS and have no binding value), national standards, ISO standards or CEN standards. The following 14 countries' NHS sent back the completed questionnaire: Croatia, Czech Republic, UK, Estonia, Finland, Germany, Hungary, Iceland, Latvia, FYR of Macedonia, Poland, Slovakia, Spain and Sweden.

Based on the results of this questionnaire it can be concluded that most of these countries basically use hydrological and national standards both for field measurement and data processing. The use of the international standards is the highest when measuring discharge (ten countries) and surface water level (six countries), but, even for these parameters, for data processing only five countries indicated using ISO for discharge and three countries for water level. For surface water temperature only one country used ISO standards, for groundwater level and temperature again only one country indicated ISO for measurement and no country indicated using ISO for data processing. Suspended sediment was measured only by eight countries and the bottom sediment only by four countries. Only one country used ISO for field measurement of suspended sediment, and no country for data processing. No country used ISO for bottom sediment.

The WMO hydrological technical regulations, guides and manuals declared the aim 'to ensure adequate uniformity and standardization in practices and procedures'. However, beside the general rules and procedures these documents also refer to ISO standards.

The ISO hydrological frame is the Hydrometry TC (Technical Commission) 113, which has the areas and items listed in Table 5.1.1.

Table 5.1.1 Hydrometry TC 113.

Area	Number of items
Velocity area methods	23
Flow measurement structures	17
Terminology and symbols	3
Instruments, equipment and data management	13
Sediment transport	10
Groundwater	5

One question in the RA VI questionnaire related to the participation either in ISO or CEN activities. Only one country answered yes. This means that the national hydrological services do not send representatives to these committees. The ISO, established in 1947, is the world's largest standards developing organization. It has two technical committees dealing mainly with hydrometry related tasks, namely the TC 113 Hydrometry [5], which has five subcommittees (velocity area methods; flow measurement structures; instruments, equipment and data management; sediment transport; and ground water), and the TC 147 [6], which covers water quality related tasks.

The CEN has signed the Vienna Agreement with the ISO through which common European and international standards can be developed in parallel. The CEN/TC 318 Hydrometry [7] has its own business plan. The RA VI is represented by an observer at this technical committee, where the other members are representatives mainly from European hydrological services. Since the committee's first meeting in 1994 six standards have been published, the majority of which were European adoptions of international ISO standards. Regarding their business plan 'the priority of the committee is to develop standards for making hydrometric measurements across the water cycle. This will become increasingly important as Governments and water resource authorities develop and maintain the river basin management plans required by the Water Framework Directive.'

5.1.4
Conclusions

Rivers and other water bodies do not respect national boundaries. Consequently, the management of river basins and groundwater resources cannot be achieved at either the catchment or national level. Cooperation between organizations at both the sub- and supranational level is imperative. These bodies need standards to ensure that they are all producing hydrometric data based upon agreed procedures. Accurate measurements across the whole of the hydrological cycle are of great importance to facilitate the efficient management of water resources, as required by the European Union WFD.

Correct hydrometric measurements, data processing and assessment will be even more important in the future for predicting changes in weather patterns affecting the whole world. If we are to have comparable hydrological data produced by the NHS for international data exchange on bilateral, catchment, European and international levels, the importance of the joint efforts of the ISO and the CEN cannot be stressed enough. There is an ongoing process for future cooperation between the ISO and the WMO. Further information on the WMO and ISO partnership on international standardization may be found in Annex 1.

Thanks to CEN TC/318 Hydrometry, the RA VI also has its own platform to fulfil the requests of the NHSs. For effective and fruitful work to be achieved, it is important that all services send representatives and give priority to the work plan. There is also a need that these standards to be prepared by hydrologists, who are able to give their practical views on them.

It is also necessary for RA VI to create a priority list of parameters based on the requests of the NHSs. Standards for measurements and data processing, for surface and groundwater, and for water quality and quantity should be prepared for these parameters. Taking into account that not all services are able to send representatives and finance their participation, it should be ensured that they at least receive information about the ongoing work and are offered the opportunity not only to submit their priority list but also to express their opinion on draft standards.

Annex 1: WMO and ISO Strengthen Partnership on International Standardization

Geneva, 16 September 2008 – WMO/ISO: The World Meteorological Organization (WMO) and the International Organization for Standardization (ISO) have agreed to increase their cooperation in the development of international standards related to meteorological and hydrological data, products and services. ISO Secretary-General Alan Bryden and WMO Secretary-General Michel Jarraud today signed an agreement on Working arrangements in Geneva, Switzerland, to formalise the partnership.

The WMO has liaison status with nearly 30 of ISOs technical committees developing standards with relevance to hydrometry, air quality, water quality, soil quality, geographic information, solar energy, petroleum and gas industry, information technologies, marine, quantities and units.

The Working arrangements between the WMO and ISO aim to strengthen the development of International Standards and to avoid duplication of work on standards related to meteorological, climatological, hydrological, marine and related environmental data, products and services.

Procedures are now in place for the accelerated adoption by ISO of WMO documents as ISO standards. WMO and ISO will develop, approve and publish common standards based on WMO technical regulations, manuals and guides. Upon signing the arrangements, Michel Jarraud, Secretary-General of WMO, stated that the new procedures 'would clarify the authority of WMO documents and enhance their international recognition and dissemination. This will be of particular importance to the activities of National Hydrological and Meteorological Services in addressing standard issues'.

Alan Bryden, Secretary General of ISO, underlined that the agreement was an illustration of the increasing collaboration between the UN System and ISO, as well as of the contribution of international standards to responding to the challenges of climate change.

ISO and WMO have been working in close cooperation since the granting of consultative status to ISO by the WMO Executive Council at its fifth session in 1954.

ISO has recognized WMO as an international standardization body through ISO Council Resolution 43/2007 approved in December 2007.

References

1 WMO RA-VI (2008) http://www.wmo.int/pages/members/region6_en.html (accessed 04 October 2010).

2 World Meteorological Organization (2010) WMO in brief. Available at http://www.wmo.int/pages/about/index_en.html (accessed 04 October 2010).

3 World Meteorological Organization (2010) WMO commission for hydrology. Available at http://www.wmo.int/pages/prog/hwrp/chy/chy_index.html (accessed 04 October 2010).

4 European Parliament, Council (2010) Directive 2000/60/EC of 23 October 2000

establishing a framework for Community action in field of water policy. Available at http://eur-lex.europa.eu/LexUriServ/LexUriServ.do?uri=CELEX:32000L0060:EN:NOT (accessed 04 October 2010).

5 International Organization for Standardization (2010) TC 113 – Hydrometry. Available at http://www.iso.org/iso/iso_catalogue/catalogue_tc/catalogue_tc_browse.htm?commid=51678 (accessed 04 October 2010).

6 International Organization for Standardization (2010) TC 147 – Water Quality. Available at http://www.iso.org/iso/iso_catalogue/catalogue_tc/catalogue_tc_browse.htm?commid=52834 (accessed 04 October 2010).

7 European Committee for Standardization (2010) CEN/TC 318 Hydrometry – published standards. Available at http://www.cen.eu/cen/Sectors/TechnicalCommitteesWorkshops/CENTechnicalCommittees/Pages/Standards.aspx?param=6299&title=CEN/TC%20318 (accessed 04 October 2010).

5.2
Limnological and Palaeolimnological Research on Lake Maggiore as a Contribution to Transboundary Cooperation Between Italy and Switzerland

Rosario Mosello, Roberto Bertoni and Piero Guilizzoni

5.2.1
Introduction

Lake Maggiore (Figure 5.2.1), the second largest (212 km^2; volume of 37 km^3) and deepest (370 m) Italian lake, is one of the best studied lakes in Europe. The 6600 km^2

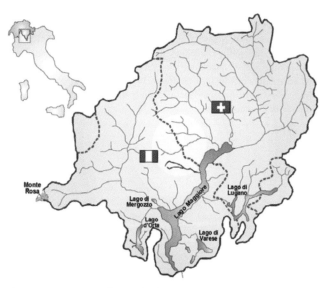

Figure 5.2.1 Drainage basin of Lake Maggiore.

of the watershed are shared in approximately equal parts by Italy (Regions Piedmont and Lombardy) and Switzerland (Canton Ticino). As much as 50% of this area lies at over 1283 m above sea-level, with 1.1% of it composed of glaciers. It includes several other lakes, the most important being lakes Lugano and Orta, the first again shared between Italy and Switzerland. The socio-economic system of this region depends strongly on the presence of the lake, whose waters are used for many purposes, including tourism, recreational-environmental activities, public and private navigation, professional and sport fishery, hydroelectric production, irrigation and to a growing extent drinking water. The inflowing waters are widely used in artificial reservoirs for the production of hydroelectric power, both in Italy and Switzerland, to a total of 600 million m^3. The lake also plays an important role in the context of the Po river plain, the main agricultural and urban area in Italy, both as a source of water for irrigation and for flood control. Since the 20th a dense web of navigable canals, originating from the sub-lacustrine Ticino River, has been present over a wide area of the plain. These channels were later used to irrigate valuable crops such as rice and forage. To rationalize these applications, a weir with mobile sluices was constructed on the lake outflow in 1943.

However, due to the morphology of the catchment and its location in a high precipitation climatic zone, Lake Maggiore is a water body with a high hydro-geological risk. Furthermore, the human activities of the area's 600 000 inhabitants, equally distributed between Italy and Switzerland, resulted in a progressive deterioration of the quality of the waters in the decades following the Second World War, primarily due to eutrophication [1].

Other early pollution problems were related to the mercury [2, 3] used in industry, for example, for the hydrolysis of sodium chloride or to produce chlorine for industrial purposes, including the synthesis of DDT [4].

To handle the many problems rising from the management of the lake and its watershed, in 1882 an International Commission for Fisheries in Italian Swiss waters was created in a cooperative venture between Italy and Switzerland. Almost a hundred years later, in 1972, a second body (International Commission for the Protection of Italian Swiss Waters, CIPAIS [5] was established with the aim of studying the increasing eutrophication of the water, locating the main sources of algal nutrients and proposing possible remediation actions. Since 1978 these studies have been organized in five-year programmes, covering the main aspects of basic limnology, climatology, hydrology and chemical loads of the tributaries, physical, chemical and biological characteristics of the lake water, the latter including topics such as phyto- and zooplankton, bacteria and macrophytes. The same programmes were carried out at the same time on lakes Maggiore and Lugano, the first by CNR-ISE, the second by the Cantonal Agency for Water and Soil Protection.

Hydrochemical and sediment analyses were performed by CNR ISE [6], Guilizzoni *et al.* [7], Marchetto *et al.* [1], Bertoni *et al.* [8] and Galassi *et al.* [9].

This chapter summarizes studies of the ecosystem-scale responses of Lake Maggiore to changes in human impact (e.g. nutrient sewage discharges) and recent warming.

5.2.2
Results and Discussion

Limnological and palaeolimnological studies enabled the documentation of the deterioration of the trophic conditions of the lake in the 1970s and the identification of phosphorus as the main cause of eutrophication (data summarized in many review papers [1, 7, 10]). Studies on the tributaries highlighted the areas of the watershed making the greatest contribution to the P load, and allowed identification of the main activities responsible for the P discharge (Figure 5.2.2). The evidence generated by these studies convinced the authorities, through the agency of the CIPAIS, to adopt measures reducing P in detergents, which they did in the first half of the 1980s in Switzerland and Italy. Furthermore, it was indicated to add the tertiary treatment to sewage plants, in part already under construction for sanitary uses, for a more effective P reduction. Sewage was treated using different strategies in Canton Ticino and in the two Italian regions, Piedmont and Lombardy. Continuous monitoring of the tributaries documented the positive effects of the treatment in Ticino and Piedmont, while the situation, although improved, remained unsatisfactory in Lombardy. Lake Maggiore's volume and resilience processes meant that it reacted slowly to the decreased P load. However, total P concentrations, which reached their highest values of 30 µg P l^{-1} over the whole water volume at the end of 1997, started decreasing in the 1980s, reaching their present level of 10–12 µg P l^{-1} – corresponding to oligo-mesotrophic conditions – in 1995 (Figure 5.2.3). The biological components also responded to the changed hydrochemistry, with a strong reduction in chlorophyll *a* (Figure 5.2.4) and total organic carbon concentration (Figure 5.2.5). It should be pointed out that, at present, conditions in Lake Maggiore are better than those of all the other deep southern alpine lakes (Garda, Como, Iseo and Lugano), which together constitute the main Italian lake district. While phosphorus has been successfully controlled, the same has not been true of nitrogen, and levels

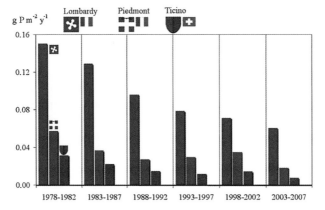

Figure 5.2.2 Areal contribution (5 years average) of the Italian and Swiss administrative regions to total phosphorus load of Lake Maggiore.

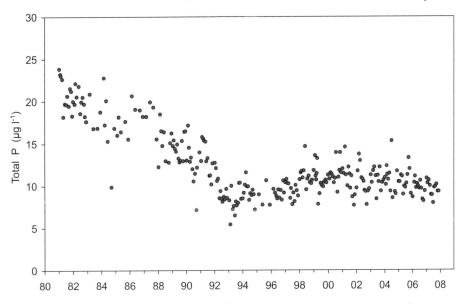

Figure 5.2.3 Total phosphorus concentration (0–370 m water column weighted average) in Lake Maggiore in the period 1980–2007.

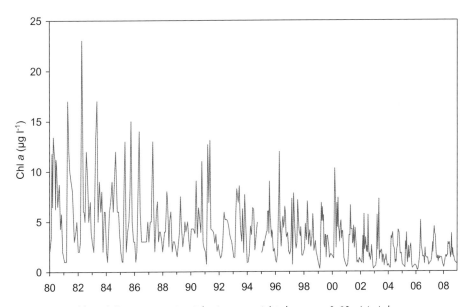

Figure 5.2.4 Chlorophyll *a* concentration (photic zone weighted average: 0–20 m) in Lake Maggiore in the period 1980–2008.

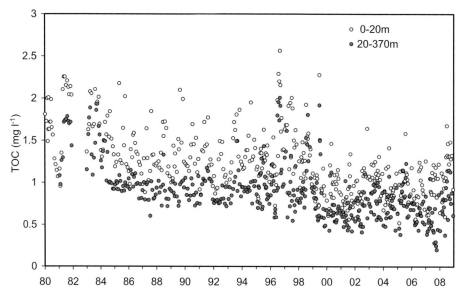

Figure 5.2.5 Total organic carbon (TOC) concentration in epilimnion (weighted average 0–20 m) and hypolimnion (weighted average 20–370 m) of Lake Maggiore in the period 1980–2008.

are at a historic high. This is typical of many lakes in this region, probably because it is relatively easy to control phosphorus and difficult to control nitrogen, especially when it derives from non-point sources like agriculture and atmospheric deposition.

In addition to the aspects directly related to the eutrophication processes, these studies yielded extremely valuable information on the physics of the lake, in particular as regards heat accumulation in the hypolimnium [11]. This may be related to global warming and is affecting the mixing dynamics of the lake, with consequences on the distribution along the water column of oxygen and nutrients. Concomitant studies on the acid deposition affecting the southern part of the watershed, which is added to the chemical loads transported from the different tributaries, pointed out the different origins of P and N. Results showed that P derived mainly from human activities, with a very low contribution from watershed weathering, while the main source of N was, and still is, atmospheric deposition, followed by human activities [12]. The interaction of atmospheric deposition with the watershed vegetation resulted in N retention.

From the mid-1990s research activities were extended to other types of pollution affecting the lake water, such as DDT, other persistent organic pollutants (e.g. PCB) and trace metals. In fact, in 1993–1995 high concentrations of pp'-DDT were found in some fish species by the Laboratorio Cantonale of Lugano. Accordingly, fishing (both professional and recreational) of the main commercial species was banned by the Italian Government. The ban lasted until 2000, since when it has been partially removed (at present the fishing of only one species, *Alosa fallax*, is forbidden).

Curiously, the laws governing the edibility of fish are different in Switzerland and Italy – Italian law is much more restrictive – so that there are different approaches on how to tackle this environmental and social problem.

The Italian Government was therefore asked by the Italian Commission for the Protection of Italian–Swiss Waters to investigate the origin of DDT contamination in Lake Maggiore. Specific projects were undertaken on several ecosystem compartments such as water [9], molluscs [13], aquatic birds, atmospheric precipitations and lake and tributary sediments as a whole. At present a new research programme, which is also included in a European project, has been set up to study how climatic changes (temperature and precipitations) may determine the redistribution of toxic pollutants in ecosystems and to identify and quantify the main processes and parameters controlling the mobilization of metals and POPs.

Despite these negative aspects, the limnological conditions of Lake Maggiore as a whole have improved significantly since studies began in the 1970s. This is true mainly for eutrophication, whereas there are persistent problems with the above-mentioned POPs (though their concentration trends are decreasing) and with the relatively high metal concentrations in lake sediments [4, 7, 14, 15]. The studies on sediment cores show that the concentrations of these pollutants in the sediment were much higher during the industrial and economic boom in northern Italy in the 1960s.

Nevertheless, the lake is facing other problems, which demand new scientific research: these include the effects of climatic changes on the biological and physical structure of the lake [8, 16].

The increase in water temperature, already well documented, is likely to extend the 'trophogenic layer' and increase the occurrence of cyanobacteria blooms [17]. The effects of warming would be similar to those of eutrophication, with a cascading effect over the whole trophic chain [18, 19]. Another aspect is that of the winter conditions, which are generally milder than in the past. These improved meteorological conditions reduce the depth of the spring water overturn, decreasing the frequency of a full homogenization of the water column, which may result in oxygen depletion in the deepest water. Parallel variations in the water chemistry, with likely effects on the biology, are to be expected because of the increased weathering and leaching of the watershed soils.

5.2.3
Conclusion

Most of the studies performed on Lake Maggiore have been possible thanks to financial support from the CIPAIS, as well as from the National Research Council and the European Union; and have resulted in Lake Maggiore and its watershed being the best documented and best known natural ecosystem in Italy, also from a socio-economic standpoint. The very positive impact of the cooperative venture supported by the CIPAIS is even more impressive if one considers that Switzerland is not a member of the EU, unlike all the other neighbouring countries. In addition, the fact that Swiss environmental laws are different from those of the EU (although the attention to environmental problems is the same) merely increases the value of the political and scientific agreement between Italy and Switzerland. The role of coordinator has always been and continues to be performed by the CIPAIS, in a dynamic that is essential considering the emergence of a range of new environmental problems.

These studies make it clear that human impact (eutrophication) on the trophic evolution of Lake Maggiore, and meteorological changes in terms of increasing trends of the lake water temperature during the last ca. 55 years, have had a profound effect on the biological component of this freshwater ecosystem.

Acknowledgements

Most of these studies were supported by the International Commission for the Protection of the Italian Swiss Common Waters.

References

1 Marchetto, A., Lami, A., Musazzi, S., Masaferro, J., Langone, L. and Guilizzoni, P. (2004) Lake Maggiore (N. Italy) trophic history: fossil diatoms, plant pigments, chironomids and comparison with long-term limnological data. *Quaternary International*, **113**, 97–110.

2 Damiani, V. and Thomas, R.L. (1974) Mercury in the sediments of the Pallanza Basin. *Nature*, **251**, 696–697.

3 Baudo, R., Cenci, R.M., Sena, F. and Dabergami, D. (2002) Chemical composition of lake Maggiore sediments. *Fresenius Environmental Bulletin*, **11** (9b), 675–680.

4 Guzzella, L., Patrolecco, L., Pagnotta, R., Langone, L. and Guilizzoni, P. (1997) DDT and other organochlorine compounds in the Lake Maggiore sediments: a recent point source of contamination. *Fresenius Environmental Bulletin*, **7**, 79–89.

5 Convenzione tra l'Italia e la Svizzera concernente la protezione delle acque Italo-Svizzere dall'inquinamento (1972) Trattati e Convenzioni, Roma, 20 Aprile 1972, 19 pp.

6 C.N.R.-I.S.E. (2007) Ricerche sull'evoluzione del Lago Maggiore. Aspetti limnologici. Programma quinquennale 2003-2007. Campagna 2007 e Rapporto quinquennale 2003-2007. Commissione Internazionale per la protezione delle acque italo svizzere (Ed.): 132 pp.

7 Salmaso, N. and Mosello, R. (2010) Limnological research in the deep southern subalpine lakes: synthesis directions and perspectives. Advances in Oceanography and Limnology, **1**, 29–66.

8 Bertoni, R., Piscia, R. and Callieri, C. (2004) Horizontal heterogeneity of seston, organic carbon and picoplankton in the photic zone of Lago Maggiore, Northern Italy. *Journal of Limnology*, **63**, 244–249.

9 Galassi, S., Volta, P., Calderoni, A. and Guzzella, L. (2006) Cycling pp'DDT and pp'DDE at a watershed scale. The case of Lago Maggiore (Italy). *Journal of Limnology*, **65**, 100–106.

10 Salmaso, N., Morabito, G., Garibaldi, L. and Mosello, R. (2007) Trophic development of the deep lakes south of the Alps: a comparative analysis. *Archiv für Hydrobiologie*, **170**, 177–196.

11 Ambrosetti, W. and Sala, N. (2006). Climatic memory. *Chaos and Complexity Letters*, **2**, 121–123.

12 Rogora, M., Mosello, R., Calderoni, A. and Barbieri, A. (2006) Nitrogen budget of a subalpine lake in northwestern Italy: the role of atmospheric input in the upward trend of nitrogen concentrations. *Verhandlungen International Verein Limnology*, **29**, 2027–2030.

13 Binelli, A. and Provini, A. (2003) DDT is still a problem in developed countries: the

heavy pollution of Lake Maggiore.
Chemosphere, **52**, 717–723.

14 Cenci, R., Baudo, R. and Muntau, H.
(1991) Mercury deposition history
of Pallanza Bay, Lake Maggiore,
Italy. *Environmental Technology*, **12**,
705–712.

15 Provini, A., Galassi, S., Guzzella, L. and
Valli, G. (1995) PCB profiles in sediments
of lakes Maggiore, Como and Garda (Italy).
Marine and Freshwater Research, **46**,
129–136.

16 Manca, M. and DeMott, W.R. (2009)
Response of the invertebrate predator
Bythotrephes to a climate-linked increase in
the duration of a refuge from fish
predation. *Limnology and Oceanography*,
54, 2506–2512.

17 Bertoni, R., Callieri, C., Caravati, E., Corno,
G., Contesini, M., Morabito, G., Panzani,
P. and Giardino, C. (2007) Cambiamenti

climatici e fioriture di cianobatteri
potenzialmente tossici nel Lago Maggiore,
in *Clima e Cambiamenti Climatici*
(eds B. Carli, G. Cavarretta, M. Colacino
and S. Fuzzi), Le attività di ricerca del CNR,
Consiglio Nazionale delle Ricerche, Rome,
pp. 613–618.

18 Manca, M., Torretta, B., Comoli, P.,
Amsinck, S. and Jeppesen, E. (2007)
Major changes in trophic dynamics in
large, deep sub-alpine Lake Maggiore
from 1940s to 2002: a high resolution
comparative palaeo-neolimnological study.
Freshwater Biology, **52**, 2256–2269.

19 Manca, M., Portogallo, M. and Brown,
M.E. (2007) Shifts in phenology of
Bythotrephes longimanus and its
modern success in Lake Maggiore as a
result of changes in climate and
trophy. *Journal of Plankton Research*,
29, 515–525.

5.3
Monitoring in Shared Waters: Developing a Transboundary Monitoring System for the Prespa Park

Miltos Gletsos and Christian Perennou

5.3.1
Introduction

Prespa is a transboundary basin in Southeast Europe, shared by Albania, Greece and the FYR of Macedonia. It consists of two lakes (Lake Greater Prespa, shared by all three countries, and Lake Lesser Prespa shared between Albania and Greece) surrounded by mountain ranges up to 2600 m above sea-level, and with a catchment area of approximately 1519 km^2 (figure adapted from Crivelli and Catsadorakis [1]). Prespa is a region of high biodiversity and endemism, where more than 1500 plant species, 260 bird species as well as nine endemic fish taxa have been recorded. The wetlands of Lake Lesser Prespa host rare and threatened avifauna, including the largest Dalmatian Pelican breeding colony in the world. Four national parks and a nature reserve are located wholly or partly within the Prespa basin (Figure 5.3.1), as well as two Ramsar sites (wetlands of international importance) [2]. The international significance of the area in terms of biodiversity and the need to manage the shared water bodies led in February 2000 to the establishment of the Prespa Park, by a joint declaration of the Prime Ministers of the three littoral states. The Prespa Park covers the two lakes and their surrounding catchment basin and is the first transboundary protected area in the Balkans [3]. Ten years after the Prespa Park Declaration, the

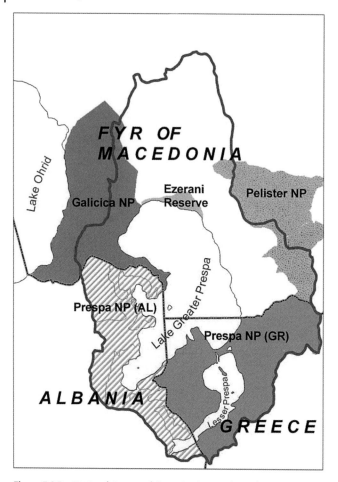

Figure 5.3.1 National Protected Areas in the transboundary Prespa Park basin (Map: SPP).

three ministers of environment of the littoral states and the European Commissioner for the Environment met in Prespa and reaffirmed their commitment by signing a legally binding agreement in February 2010.

Transboundary environmental monitoring, a prerequisite for informed decision-making for the management, protection and sustainable development of the trans-boundary basin, has been a priority for the Prespa Park stakeholders since the declaration of the park. The need for transboundary monitoring was explicitly mentioned in the Strategic Action Plan for the sustainable development of the Prespa Park – the first trilateral strategy document for the transboundary protected area [2]. This also led to the inclusion of monitoring in the priority activities of the Global Environment Facility/United Nations Development Programme (GEF/UNDP) project on Integrated Ecosystem Management, which is being funded mainly by GEF and has been implemented by UNDP since 2007 [4].

The Society for the Protection of Prespa (SPP), a local conservation NGO active in the Prespa Park process, secured co-funding from private funds for this activity and in 2007 it initiated a project for the development of a transboundary monitoring system (TMS) for Prespa. Technical expertise was provided by Tour du Valat, France, a research centre with long-term experience in the Prespa area and expertise in environmental monitoring. The TMS project is implemented by the SPP in full coordination with the GEF/UNDP Prespa project.

In 2007 the GEF/UNDP project established a Monitoring and Conservation Working Group (MCWG), consisting of experts from ministries, academia/research, national parks and the NGO sector from all three countries. The MCWG aims to satisfy the need to provide scientific and political guidance, ensure ownership of the TMS project, as well as tap expert resources from the three countries. The MCWG regularly convenes two to three times a year, with funding and support from the GEF/UNDP Prespa project.

5.3.2
A Workflow Process for Setting-Up an Environmental Monitoring System

The setting-up of a monitoring system with a transboundary perspective is particularly demanding. In general, a monitoring system by itself cannot solve all the problems or improve the management at which it is targeted. A simplified version of the typical workflow process for designing a monitoring system (adapted from Tomàs Vives [5]) is as follows:

Stage 1: defining the aims of monitoring;
Stage 2: defining protocols suited to the key monitoring aims;
Stage 3: actual monitoring;
Stage 4: data analysis, interpretation and reporting to managers/decision-makers (possibly with suggestions on needed actions);
Stage 5: decisions made on 'what to do', using monitoring data as well as other information available (e.g. socio-economic context, etc.);
Stage 6: action taken on the ground (possibly including extra monitoring).

It can be seen that only stages 1–4 (and partly 6) deal with the actual monitoring system. Stages 5 and 6 deal with decision-making and actual management; in these stages it may be decided for non-technical reasons whether or not to use some of the information gained through monitoring. Indeed, even the most wide-ranging and well designed monitoring programme cannot automatically guarantee that the ultimate aim of better-informed management will be achieved.

5.3.3
Methods Applied During the Preparatory Stage of the Prespa TMS

In the preparatory stage of the TMS project (October 2007–July 2008) Tour du Valat, SPP and national consultants from the three countries focused on ('Defining the aims of monitoring' (stage 1 of the above-mentioned process). They investigated in depth issues including the aims of the Prespa TMS; geographical scale; significant elements and values to be monitored; stakeholders; sources of funding; connection to EU Law and Policy; guidelines on indicators, institutions for implementing the TMS, national resources and trilateral administration, equipment, and training.

In recognition of the inherent difficulties of transboundary cooperation and the significant institutional differences between the three countries, a broad participatory approach was attempted. Project fact-finding missions resulted in consultations with over 30 public authorities, monitoring institutions, scientific and research institutions, as well as donors (UNDP) in eleven towns in all three countries. This gave a first indication of the dispersal of operations and responsibilities for monitoring, and of the complexity of the issue.

The project team also compiled a meta-database consisting of the environmental monitoring programmes active in Prespa, together with analytical description sheets for each monitoring programme or family of monitoring parameters.

5.3.4
Results Obtained

During the preparatory stage of the TMS, the meta-database of environmental monitoring programmes active in Prespa resulted in the identification of 27 institutions implementing monitoring in the three countries and covering more than 400 parameters (physicochemical, meteorological, hydrological, biodiversity, land-uses, demographics). One-off monitoring projects or surveys were not included in the meta-database, which, however, could be further amended or enriched when needed in an on-going process.

The availability of data monitored in the above programmes was varied. Some monitoring programmes publish data on the Internet or in printed reports, with varied regularity. Others make their data available upon request, free of charge or subject to a nominal fee. A few institutions charge high fees, or do not make their data available at all before their formal publication. Often only syntheses or processed data are made available to the public.

Concerning the aims of the TMS, the authors proposed five potential goals for the system, based on literature and review of existing or proposed monitoring systems for wetlands (and their watersheds):

1) **Routine Surveillance**: helps establish the baseline information for the site at transboundary level, including the usual range of variation for key elements selected.
2) **Adaptive Management**: assesses the impact of transboundary and national management activities undertaken at Prespa basin level for stopping and reversing its degradation, to help decision-makers and/or managers adapt/change their management practices.
3) **Knowledge-Oriented**: assesses the evolution of Prespa and the root causes of environmental problems, for example, according to the P-S-R model: pressure, state, response.

4) **Crisis/Emergency Management**: alarm system for competent authorities to be able to respond to emergencies.
5) **Policy-Evaluation Oriented**: Evaluation of public policies in operation in the catchment.

In the long term, the specific aims of any monitoring system for a natural area should be to facilitate the continual improvement of management practices, that is, 'Adaptive Management', and this was also incorporated in the relevant project document of the GEF/UNDP Prespa project [4]. Nevertheless, a pre-requisite for adaptive management is the existence of an effective management body/mechanism, which is able to translate rapidly into field action any conclusion resulting from a monitoring programme. The MCWG considered this as being under development and decided to propose alternative goals for the shorter term.

Therefore, the project team proposed 'Routine Surveillance' as the core short-term goal of the TMS. This consists of obtaining reliable data on important parameters over a period of a few years to establish the range of 'normal' or acceptable variations. Then the monitoring may be upgraded so as, for example, to ring an 'alarm bell' to the decision-maker when the indicator steps out of this range. Such basic surveillance is crucially missing from the Prespa Park area, at least at transboundary level. Some routine surveillance exists for some countries and some monitoring indicators (e.g. pelicans in Greek Prespa, human demography in Albanian Prespa, water quality in Greek Prespa and the part of Prespa in the FYR of Macedonia), but only at a national level. At the time of writing there was no jointly agreed and shared transboundary baseline in Prespa for any of the key environmental values and issues. The TMS will therefore play a crucial role in helping establish this common ground between all three countries.

Nevertheless, the MCWG recommended that if for some specific aspects the possibility for adaptive management appears in the short term, the system should be flexible enough to accommodate this. Furthermore, the three remaining options for the Aim of the TMS (i.e. 'Knowledge-Oriented', 'Crisis Management' and 'Policy-Evaluation') were not discarded totally, but instead were retained as possibilities in case the needs arise, the prerequisites mentioned above are met and the necessary budget is available.

Concerning the geographical range of the Prespa TMS, and following the expert proposal by the project team and the recommendations of the MCWG, it was decided that for most monitoring parameters or indicators this should focus exclusively on the watershed. For instance, streams within the Galicica National Park that flow into Lake Ohrid rather than Lake Greater Prespa (Figure 5.3.1) would normally be excluded from transboundary monitoring. However, the TMS will leave open the option to extend the geographical scope outside the existing known surface catchment – but within the National Parks' boundaries. This will only be the case for some selected monitoring parameters that need to be monitored beyond watershed borders in order to be meaningful, especially those linked to terrestrial ecosystems (e.g. alpine meadows, large carnivores, forests).

All outputs of the preparatory stage of the TMS project were reviewed and endorsed by the expert members of the MCWG. The MCWG meetings took place in the Prespa region between October 2007 and June 2008, with the aim of achieving trilateral consensus – a prerequisite for a transboundary process – as well as providing scientific and political guidance.

5.3.5
Preliminary Assessment of the Difficulties of Transboundary Cooperation

Notably, the fact-finding missions and the development of an inventory of the monitoring meta-data during the preparatory stage of the TMS showed noteworthy differences between the three littoral states. These differences pertain to the legal frameworks of the three states, the availability of resources and occasionally the technical methods employed by the national monitoring institutions.

Nevertheless, the process of the Transboundary Prespa Park, which has been going on since February 2000 with functioning political and technical institutions (such as the Prespa Park Coordination Committee established following the Prespa Park Declaration, and the MCWG), has increased mutual understanding and assisted in the creation of consensus. The use of international and European Union standards by the project team (e.g. reference to EU Directives, such as the 2000/60/EC Water Framework Directive) during the preparatory stage of the TMS has also attenuated the differences between the three countries and strengthened the common aim of designing a transboundary monitoring system for Prespa. Finally, the Prespa Park Agreement signed in 2010 states that the signatories will establish a common standard system of monitoring to observe, manage and control the environmental state of the lakes and their watershed [6].

5.3.6
Next Stage Towards the Development of a Monitoring System

The next stage of the TMS process involved the development of an Expert Study. Trilateral thematic expert groups consisting of experts from the three littoral countries, and international lead experts from France, Austria and the USA met in the Prespa area in 2009 and worked on seven monitoring themes: water resources; aquatic vegetation and habitats; forests and terrestrial habitats; fish and fisheries; birds and other biodiversity; socio-economics; land-use.

The Expert Study also recommended the national institutions responsible for the future monitoring system, whenever it will be deployed – in accordance with the national legislations, the conclusions of the trilateral thematic groups and the recommendations of and the supervision by the MCWG members. The final draft of the Expert Study [7] was presented to the MCWG in November 2009.

The development of the Expert Study is followed by the purchase and installation of equipment, supported by the GEF/UNDP Prespa Park Project, the SPP and the national monitoring institutions. In parallel, a pilot application for testing the monitoring parameters of the TMS, training, and networking is taking place at the time of writing and is expected to be finalized in July 2011. The evaluation of the pilot application and the final adjustment of the system by the technical/scientific consultants will follow and it will culminate in the endorsement by the supervising bodies and approval by the national authorities (end 2011).

5.3.7
Conclusions

During the TMS project scientists and decision-makers of the three littoral countries have come together for this common aim for the first time. It is likely that this achievement will encourage and enable the transfer of knowledge and skills to other Mediterranean or Balkan contexts, beyond the Prespa lakes region.

Following the development of the Expert Study of the TMS (2009) and the pilot application (2010), the major challenge will then be the concrete implementation by the three countries of the technical proposals, on the basis of their commitments made during this process. Crucial questions must be examined by the MCWG, such as local coordination and the choice of the national monitoring institutions to implement the monitoring system. The local partners will have to take the leadership of its implementation, and their supervising authorities at central level will have to commit institutional, technical and financial resources.

References

1 Crivelli, A.J. and Catsadorakis, G. (1997) *Lake Prespa, Northwestern Greece, A Unique Balkan Wetland*, Kluwer Academic Publishers, Dordrecht.

2 Society for the Protection of Prespa (SPP), WWF, Greece, Protection and Preservation of the Natural Environment in Albania (PPNEA), Macedonian Alliance for Prespa (MAP) (2005) Strategic Action Plan for the Sustainable Development of the Prespa Park. Executive Summary, Society for the Protection of Prespa, Aghios Germanos. Available at http://www.spp.gr/sap_executive_summary_edition_en.pdf (accessed 6 May 2011).

3 Christopoulou, I. and Roumeliotou, V. (2006) Uniting people through nature in Southeast Europe: the role (and limits) of nongovernmental organizations in the transboundary Prespa Park. *Southeast European and Black Sea Studies*, 6 (3), 335–354.

4 United Nations Development Programme (UNDP) (2005) Integrated Ecosystem Management in the Prespa Lakes Basin of Albania, FYR-Macedonia and Greece,

UNDP Full size Project Document, UNDP.

5 Tomàs Vives, P. (1996) *Monitoring Mediterranean Wetlands: A Methodological Guide*, MedWet Publication; Wetlands International, Slimbridge, UK and ICN, Lisbon.

6 (2010) Agreement on the Protection and Sustainable Development of the Prespa Park Area, 2 February 2010, Pyli, Greece. Available at. Available at http://www.ramsar.org/pdf/wwd/10/wwd2010_rpts_prespa_agreement.pdf (accessed 6 May 2011).

7 Perennou, C., Gletsos, M., Chauvelon, P., Crivelli, A., DeCoursey, M., Dokulil, M., Grillas, P., Grovel, R. and Sandoz, A. (2009) Development of a Transboundary Environmental Monitoring System (2007–2011). Society for the Protection of Prespa, Aghios Germanos, http://www.spp.gr/index.php?option=com_content&view=article&id=63&Itemid=68&lang=en (accessed 15 November 2010).

5.4

Integrated Remote Sensing and Geographical Information System Techniques for Improving Transboundary Water Management: The Case of Prespa Region

Marianthi Stefouli, Eleni Charou, Alexei Kouraev and Alkis Stamos

5.4.1
Introduction

Water resources management requires sufficient, long-term, frequent and reliable data, [1–3]. In the case of transboundary water resources management, additional problems can arise due to the different socio-economic conditions, and the different legal, regulatory frameworks and institutional settings of the countries that share the resources. Since water has no frontiers and does not recognize administrative boundaries or political levels of competence, methodologies and tools that overcome these difficulties are badly needed. The methodology presented here aims to provide information for transboundary water management by integrating remote sensing and geographic information system (GIS) techniques. The unique ensemble of the Macro Prespa, Micro Prespa and Ochrid Lakes that are shared by Albania, Greece and the FYR of Macedonia in the Balkan Peninsula (Figure 5.4.1) is used as the study area. The international ecological significance of this area, and especially the need for sustainable water management for the benefit of both the physical environment and the inhabitants, led to the establishment of the Prespa Park – the first transboundary protected area in the Balkans – with a Joint Declaration of the Prime Ministers of the three littoral countries, in February 2000.

5.4.2
Data and Methods

The methodology presented in this chapter provides information regarding (i) catchment characterization after the extraction of selected hydrological parameters, (ii)

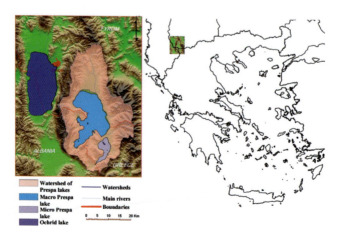

Figure 5.4.1 Pilot project area of Prespa/Ochrid Lakes.

surface water monitoring of lakes by estimating the surface area of lakes, the water level and the water quality parameters and (iii) geological support.

Topographic, geologic, land cover maps of scale 1 : 200 000 and 1 : 50 000 as well as Landsat (Global Land Cover Facility), Envisat MERIS and ERS satellite data were processed and integrated into the GIS. The outline of the watersheds, coastlines, the road/railway network, some major rivers, springs, major cities and land cover information published on the Internet (TRABOREMA FP6 project) were also included in the GIS. A digital terrain model (DTM) was obtained from SRTM-satellite data [Shuttle Radar Topography Mission (SRTM) of US NASA and German DLR].The TNTmips Integrated Image Processing/GIS software package and the BEAM VISAT toolbox were used for this study.

5.4.2.1 Catchment Characterization, Land Cover Mapping

Analysis of the DEM (digital elevation model) resulted in delineation of the hydrographic network. The following parameters are stored for each line of the drainage pattern in the GIS data base: the classification is according to Horton, Strahler and Shreve and their geometric properties (Figure 5.4.2a). Generalization of catchments was also obtained as each river network was described by the various stream ordering procedures. Summary results for the watershed of the Prespa Lakes are shown in Table 5.4.1. A key action that it is needed for qualitative improvement at the dataset level is experimentation with higher precision and resolution DEMs.

The catchments of the three lakes were described by the GIS based analysis of 'CORINE Land Cover Classification' (Figure 5.4.2b).

MERIS data was used to update the CORINE land cover map. The burnt areas identified on the MERIS data (Figure 5.4.b) are about 36 km^2 and include broad-leaved forest, mixed forest, natural grasslands and transitional woodland-shrub (according to level 3 CORINE nomenclature). Up-to-date information about land cover, land use, vegetation status and their changes over time (e.g. seasonally) can be obtained and is important for the understanding and modelling of hydrological processes such as infiltration, runoff rates, evapotranspiration and water needs.

5.4.2.2 Surface Water Monitoring

Landsat time series images were used for the identification of the surface area of the lakes. Measurements for the Macro Prespa Lake surface were as follows: November 1973 ~276.5 km^2, August 1988 ~273.7 km^2, August 2000 ~265.2 km^2, August 2008 ~257.2 km^2. The Macro Prespa Lake lost nearly 19.5 km^2 of its surface in the period from 1973 to 2008 and it is evident that this surface area is continuing to decline.

The size of the Ochrid Lake and orientation of ENVISAT track 657 (cycles 10 to 68, October 2002 to May 2008) make it possible to select a sufficient number of altimetry observations for calculating lake level (Figure 5.4.3). Similar observations cannot be made for the Prespa Lakes due to the limited intersection between satellite track and water surface. Temporal level variability estimates were made and these show a fluctuation of the lake water level up to 3 m. In general both *in situ* and altimetric observations are in very good agreement (Figure 5.4.3).

Multi-temporal MERIS Envisat data were used [4] to extract water quality (WQ) parameters related to total suspended matter (tsm) and chlorophyll concentrations (chl_conc). Results for May 2008 are shown in Figure 5.4.4.

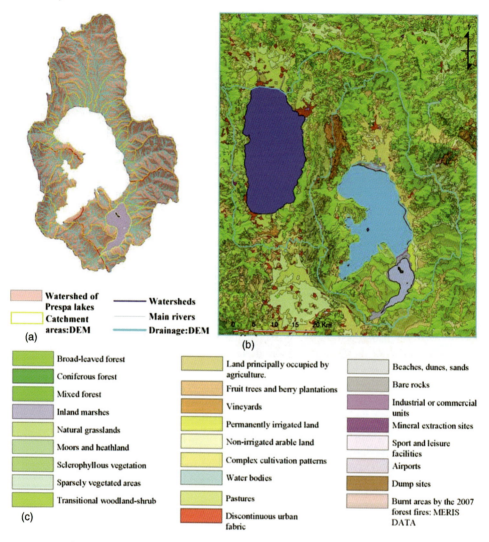

Figure 5.4.2 Catchments (a), land cover map (b), CORINE legend (c) of the region.

Table 5.4.1 Selected statistics for Prespa Lakes watershed.

Watershed	Stream order by Strahler	Number of streams	Total stream length	Average length	Drainage area: fill depressions
Prespa Lakes	1 to 5	3459	4715.8 km	102.9 m	407 km^2

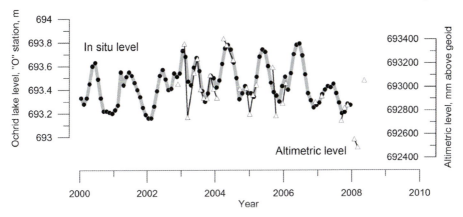

Figure 5.4.3 *In situ* (black dots) and altimetric (white triangles) water level time series of Ochrid Lake.

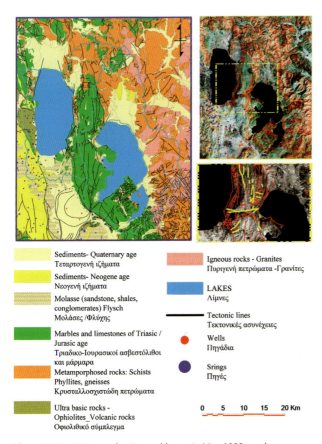

Sediments- Quaternary age Τεταρτογενή ιζήματα	Igneous rocks - Granites Πυριγενή πετρώματα -Γρανίτες
Sediments- Neogene age Νεογενή ιζήματα	LAKES Λίμνες
Molasse (sandstone, shales, conglomerates) Flysch Μολάσες /Φλύσης	Tectonic lines Τεκτονικές ασυνέχειες
Marbles and limestones of Triasic / Jurasic age Τριαδικο-Ιουρασικοί ασβεστόλιθοι και μάρμαρα	Wells Πηγάδια
Metamporphosed rocks: Schists Phyllites, gneisses Κρυσταλλοσχιστώδη πετρώματα	Srings Πηγές
Ultra basic rocks - Ophiolites_Volcanic rocks Οφιολιθικό σύμπλεγμα	0 5 10 15 20 Km

Figure 5.4.4 Water quality (tsm, chl_conc), May 2008 results.

The oligotrophic state of Ochrid Lake in contrast to the nearly eutrophic state of the Prespa Lakes can be interpreted in images in Figure 5.4.4. Sensor resolution may hinder the retrieval of such parameters over small water bodies, and adequate *in situ* measurements for calibration and validation purposes are needed.

5.4.2.3 Supporting Geology and Groundwater Surveys

Groundwaters cannot be observed directly by existing Earth observation (EO) satellites; however, location, orientation and length of lineaments can be derived from EO and used as input for studies of fractured aquifers (e.g. location of sites for water harvesting). Available geologic maps were scanned, geo-referenced and digitized for the whole region (Figure 5.4.5). The legend of all rock formations that is stored in the GIS database was homogenized and is bilingual (English/Greek) (Figure 5.4.5). Tectonic features and the location of springs were also integrated in the GIS database.

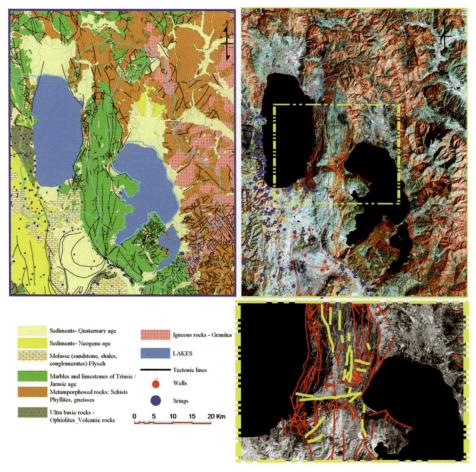

Figure 5.4.5 Geology and map update with interpreted lineaments.

Lineaments of the fractured limestones were mapped (Figure 5.4.5) using the Landsat 2000 visible/infrared spectral bands (30 m) fused with the panchromatic band (15 m). Mapped lineaments may act as conduits for the confirmed water outflow of Macro Prespa to Ochrid Lake, as the location of spring outflows are related to the location of the faulting and satellite derived lineaments. Catchments can be described according to their hydro-geological characteristics.

5.4.3
Conclusion

Earth observation techniques can be used effectively for monitoring selected parameters of transboundary lake basins. Several parameters were extracted related to the description of catchments, surface area, water level, hydrogeology and water quality characteristics of the lakes. Results obtained by applying the methodology to a transboundary area of 4769 km^2 were used to update the GIS database. The methodology proved to be cost-effective and to improve significantly the understanding of hydrological processes affecting lake basins in response to climate variability. Although the three states that share the lakes are not directly involved in this work they contributed to the measurements and data collection. Issues regarding accessibility of the data to all three countries as well as problems regarding the three states' maps, which are neither on the same scale nor provide the same thematic information and are in different languages, had to be faced and would need to be taken into consideration in the future. A preparatory programme should be set up for raising awareness on the use of Earth observation techniques for monitoring transboundary lake basins.

Acknowledgements

ESA Cat.-1 project no. 4864 is acknowledged for providing the MERIS data. Eleni Katsimpra supported the activities of the GIS system. Support was also provided by Micro Images, Inc. and Dr K. Shawse (NATO SfP Project No. 98116).

References

1 Schultz, G.A. and Engman, E.T. (eds) (2000) *Remote Sensing in Hydrology and Water Management*, Springer, Berlin.

2 Giri, C., Zhu, Z. and Reed, B. (2005) A comparative analysis of the global land cover 2000 and MODIS land cover data sets. *Remote Sensing of Environment*, **94** (1), 123–132.

3 Töyrä, J. and Pietroniro, A. (2005) Towards operational monitoring of a northern wetland using geomatics-based techniques. *Remote Sensing of Environment*, **97** (2), 174–191.

4 Candiani, G., Floricioiu, D., Giardino, C. and Rott, H. (2005) Monitoring water quality of the perialpine Italian Lake Garda through multi temporal MERIS data. Proceedings of MERIS-(A) ATSR Workshop, Frascati, Italy, 26–30 September 2005. Available on CD-ROM, ISBN 92-9092-908-1 ESA.

5.5
Transboundary Integrated Water Management of the Kobilje Stream Watershed

Mitja Brilly, Stanka Koren, Jožef Novak and Zsuzsanna Engi

5.5.1
Introduction

Water management is a very broad, sustainable and complex process, which is difficult to implement, even within national borders. The integrated approach of water management has many different aspects and efforts should be made to find solutions that are integrated in terms of space, time, hydrologic cycle, professional disciplines, administration, services, stakeholders and so on [1].

For a more transparent implementation the decision-making process should be divided into strategic, action-based and operational levels. The strategic decisions consist of the overall interests and their derivatives in the international governmental institutions. Actions can be performed at a lower level with a small number of actors sharing common interests; at the operational level just the direct involvement of stakeholders is needed.

Integration in space means basin wide analysis and the use of the watershed as the modelling unit. A master plan based on the water basin has been fundamental for water management from the beginning. We cannot analyse a solution for the downstream impact of upstream structures, for the allocation of financial resources from downstream urban areas to soil conservation and torrential control in an upstream rural part of the watershed, or for flooding rural upstream areas for the protection of a downstream town. A geographic information system (GIS) is the most useful tool for space integration in water management. This is also the reason why GIS was chosen as the basis for water information systems.

Integration in time is a well known but unpredictable problem in water management. An optimal solution today will not be an optimal one tomorrow. How can we integrate paleohydrological data with historical floods, climatic changes, and landscape and riverbed dynamics in time? Hydrologic measurements are only available for a short period of time, which is insufficient to allow hydrological processes with a long return period to be estimated. Paleohydrology and the study of historical floods can help extend the period of observation, but unpredictable climatic changes and landscape development will indicate whether extend information from the past can be extended into the future. Society also changes over time, more so than hydrology, and water management is changed by the economy, legislation and levels of population density.

Surface water flow and ground water flow, including atmospheric processes and erosion processes on land surface, are part of the integrated hydrologic cycle that are frequently simulated or analysed separately. Nowadays there is the additional problem of integrating climatic change prediction in the hydrological cycle, as well as that of down-scaling the modelling results or up-scaling the impact of variability of the water regime.

Hydrology, hydraulics, surveys, geology, sociology, economy, psychology, political science and so on are professional disciplines involved in the complex action of flood management. There is a lack of research that incorporates technical and human science; also, misunderstandings, different methodologies and concepts impede common research activities. Interdisciplinary research in the field of water management is urgently needed.

Structural and non-structural measures alone do not result in optimal water management. Optimal integrated water management needs the integrated policy of stakeholders, insurance, administration, community, province, state agencies and ministries. There is a lack of responsibility, willingness and trust to take common action. The problem is literacy and misunderstandings in communication between responsible services that fail to make timely decisions and take action.

When dealing with transboundary water management problems that are integrated in interstate relations, the result is often that no action at all is taken or that the solution is unilateral. Such kinds of situations generally lead to lose–lose or win–lose solutions. The importance of water as an irreplaceable substance essential for life and for different economical activities gives water a high ranking in relations between countries. Country borders also differ from watershed contour lines or, even worse, natural river borders and official border lines are very often in the middle of a stream.

5.5.2
The Kobilje Stream Floods

The headwater part of the Kobilje stream is in Slovenia, from which point the stream flows through Hungary, again passes into Slovene territory and finally flows into the Ledava River. The shared watershed is 296 km^2. There is a high flood risk in its downstream part, where several villages on both sides of the border are exposed to flood risk (Figure 5.5.1) [2].

The flood risk of the Kobilje stream has long been known and river training works have been developed since 1908. However, flood risk still exists. In 1998 and 2005 eight villages were flooded on both sides of the border (Figures 5.5.2 and 5.5.3). A joint study of flood risk conducted by engineers from both sides suggests that the best solution is to store floodwater in a dry detention pond on Hungarian territory (Figures 5.5.1, 5.5.2 and 5.5.4). The watershed area at the detention pond is 177.5 km^2. Table 5.5.1 presents the probability of discharges.

The dry detention pond covers 272 hectares and has a storage capacity of 2.84×10^6 m^3, and a maximum depth of water of 2.5 m. The flood peak will drop from 94 to 38 m^3 s^{-1}.

5.5.3
Project Development

The developed project is part of a long tradition of well-developed cooperation in the area of water management between Hungary and Slovenia. In 1994 a special

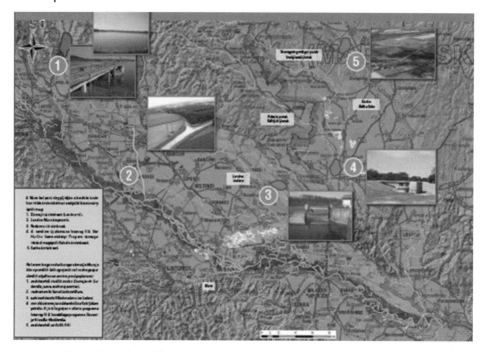

Figure 5.5.1 The Kobilje stream watershed with detention ponds 4 and 5 [2].

agreement between Hungary and Slovenia was signed to develop common objectives and tasks in water management. In October 2005 a joint agreement was signed to find technical solutions after 20 years of action development. In 2006 agreement on a project proposal was reached and both sides approved the project. The project finished in February 2008.

Figure 5.5.2 Map of the Kobilje stream with villages at risk and the detention pond [2].

Figure 5.5.3   Flooded village in the 2005 flood.

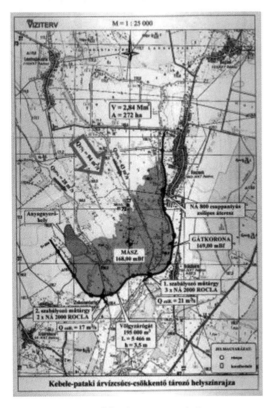

Figure 5.5.4   Map of the detention pond [2].

Table 5.5.1 Probability of discharges of the Kobilje stream at the detention pond.

Return period in years	Discharge in m^3 s^{-1}
100	94
50	81
30	67
10	56
5	45
2	30

Table 5.5.2 Structure of the cost [2].

Actions	Slovenia Value in Euros	Hungary Value in Euros	Sum Value in Euros	Ratio (%)
Preparation	0	8 458	8 458	0.3
Administrative cost	3 602	9 809	13 411	0.5
Management of the project	7 081	140 639	211 453	8.5
Information management	3 797	10 280	14 077	0.6
Preparation work	0	297 536	297 536	12.0
Construction	1 291 721	390 734	1 682 455	68.0
Final work	0	246 752	246 752	10.0
Overall cost of the project	1 369 934	1 104 207	2 474 141	100.0
Ratio 56 : 44	1 385 519	1 088 622		
Construction	78 64	21 56		
Other	9 45	90 55		

The sharing of the project costs by both sides depended on:

- ratio of watershed area of the stream,
- ratio of diminished damage cost resulting from the project,
- ratio in validation of previous constructions preventing floods on both sides and of benefits for both sides.

According to the analysis of interests from both sides, the ratios of sharing the estimated cost is as follows: Slovenia covers 56% of the cost and Hungary 44% of the cost (Table 5.5.2). The cost of the project is 2.5 million ECU, which was partly supported by the Interreg III fund with 900 000 ECU. The project has recently finished and the costs will be returned in 30 years with diminishing damage.

5.5.4
Conclusions

The Kobilje stream or Kebele stream (in Hungarian) flood protection project presents an example of good practice in bilateral cooperation between water management authorities from Hungary and Slovenia. The project required a long period of preparation, with work done by technical staff from both countries. Well-prepared documentation allowed a successful agreement on cost sharing projects to be reached.

References

1 Brilly, M. (2001) The integrated approach to flash flood management, in *Coping with Flash Floods*, (eds E. Gruntfest and J.W. Handmer), NATO Science Series 2, Environmental Security, Kluwer Academic Publishers, Boston, pp. 103–113.
2 Novak, J. (2008) Kebele Árvíztározó, Zadrževalnik na Kobiljskem potoku, A

Szlovénia-Magyarország-Horvátország Szomszédsági Program 2004–2006 Irányító Hatósága Szlovéniában: Sluzba Vlade Republike Slovenije za lokalno samoupravo in regionalno politiko, Government of the Republic of Slovenia, Government Office for local self-government and regional development.

5.6
Climate Change Impacts on Dams Projects in Transboundary River Basins. The Case of the Mesta/Nestos River Basin, Greece

Charalampos Skoulikaris, Jean-Marie Monget and Jacques Ganoulis

5.6.1
Introduction

Based on the concept that greenhouse gas augmentation due to human activities could be responsible for the recent rise in temperature and thus drive changes in climate [1], the IPCC has developed several assessment reports to describe the effect of anthropogenic activities on the evolution of greenhouse gas emissions. The results of the climate simulation reveal a non-uniform augmentation of global temperature from 1.4 to 5.8 °C during this century. Furthermore, the simulation results from the GCMs also indicate that the global predicted average water vapour concentration and precipitation should increase during the twenty-first century in the northern hemisphere and particularly in mid-high latitudes [1]. At lower latitudes results are less contrasted. Year to year variations in precipitation could also increase. As with the change in climate variables, the changes in hydrological systems are not expected to be uniform around the globe. Thus, large impacts may occur on renewable energy sources, such as hydropower generation, as this is sensitive to the amount, timing and geographical pattern of precipitation as well as to temperature [2].

This chapter examines the climate change impacts under the A1B and B1 climate scenarios on the existing dams on the Mesta/Nestos River basin. The Mesta/Nestos River basin is a transboundary river basin in South Eastern Europe (SEE) (Figure 5.6.1) almost equally shared between Bulgaria, the upstream country, and Greece, the downstream country. The basin orientates from the north (where the headwaters are located) to the south east (river outlet). The river, before discharging

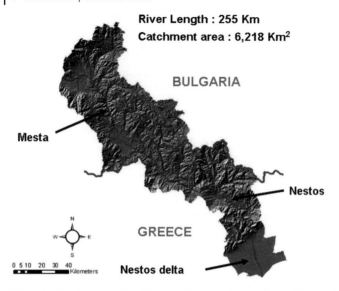

Figure 5.6.1 Characteristics of the transboundary Mesta/Nestos River basin.

into the Aegean Sea, forms the Nestos delta. In the Greek part of the basin there are two hydroelectric power plants, which are located in the mountainous part of the basin: the upstream dam of Thissavros and the downstream dam of Platanovryssi. The reservoir of the Thissavros dam has a surface area of $18\,\mathrm{km^2}$ and stores $565 \times 10^6\,\mathrm{m^3}$ of water and the reservoir of Platanovryssi has a surface area of $3.25\,\mathrm{km^2}$ and stores $11 \times 10^6\,\mathrm{m^3}$ of water. The installed capacity of the Thissavros dam is 384 MW and the total electricity generation in 2002 was $568 \times 10^6\,\mathrm{kWh}$ and in 2005 was $440 \times 10^6\,\mathrm{kWh}$, while the installed capacity of the Platanovryssi dam is 116 MW and the total electricity generation in 2005 was $240 \times 10^6\,\mathrm{kWh}$ [3]. The Thissavros dam was selected to illustrate the output results of this research study.

5.6.2
Numerical Models and Tools

5.6.2.1 **MODSUR-NEIGE Runoff Model**
The spatially distributed hydrologic model MODSUR ('modélization des transferts de surface'), which was used for the simulation of Mesta/Nestos River flow, consists of the surface modelling component of the coupled surface and groundwater MODCOU simulation model, which was developed at the Ecole Nationale Supérieure des Mines de Paris [4]. MODSUR is based on a densely spaced grid and uses a progressive quadtree structure with varying cell sizes ranging from 250 m for the smallest to 2000 m for the largest (Figure 5.6.2). Each cell follows a four-connectivity rule, that is, it may be connected to equal sized cells or to a cell that is four times larger or four times smaller. The ensemble of connected cells builds a runoff network, which gathers the flow down to the catchment outlet.

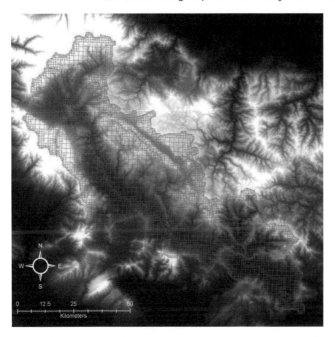

Figure 5.6.2 Topology of the MODSUR grid in the Mesta/Nestos River basin over an Topography Mission Digital Elevation Model (SRTM DEM) at 100 m resolution.

The NEIGE component of the MODSUR-NEIGE model is used for the simulation of the snow cover regime and is based on the principle of 'degree days', which was originally developed by the United States Army Corps of Engineers [5]. It was later adapted to the distributed model principle [6] using an approach that separates the process of snow melting between forested and non-forested areas.

The MODSUR-NEIGE was successfully calibrated from the period July 1991 until December 1995 [7], which was the only period during which a uniform set of precipitation measurements were publicly available for both Bulgaria and Greece. One advantage of the specific model is that it enables the computation of the flow for any 'river' cell rather than only at the river basin outlet. Thus, the produced daily results of the runoff were computed for the 'river' cells corresponding to the location of the dams.

5.6.2.2 Dams Simulation Model

The dams' simulation was conducted using the HEC-ResSim (U.S. Army Corp of Engineers Hydrologic Engineering Centre-Reservoir Simulation) tool [8]. HEC-ResSim was developed for flood control operation and hydropower analysis. Its hydropower simulation capabilities include an analysis of run-of-river generation, peak power generation, pumped storage and system power operation. To simulate hydropower operation, the reservoir releases are determined to meet power production goals, which may vary on a monthly, daily or hourly basis. Additionally, the hydropower component takes into account the penstock capacity and losses, as well as leakage parameters.

In the case of the Mesta/Nestos basin, the model was fed with the outputs of the hydrological model to determine the produced energy under climate change conditions. The definition of the operational and technical parameters, such as the geometric properties of the pool and the capability of the hydropower plant, as well as the definition of the various constraints regarding the electric power production, the regime of released flow and the operation in conditions of flooding, were based on data published by the Power Public Corporation (PPC) of Greece.

5.6.2.3 CLM Climate Model

The results obtained from GCMs are very useful in order to evaluate potential climate changes at a global scale. However, due to their coarse spatial resolution, they do not adequately take into account the local land surface characteristics, for example, the orography. Thus, specific methods have been developed to adequately transfer the GCMs results to regional scales. These 'downscaling' methods are divided into two major categories: the statistical and the dynamical. Statistical downscaling involves the development of analytical relationships between large-scale atmospheric variables, also known as predictors, and local surface variables, also known as predictands. Dynamical downscaling occurs when the results produced by a Global Circulation Model (GCM) are used as boundary conditions (idem, 'forcing') to dedicated Regional Climate Model (RCM) [9]. This study was based on a dynamical downscaling technique derived from the CLM model of the Max Plank Institute.

CLM, which stands for Climate version of the 'Local Model', simulates the climate of the European region and can be used for simulations on time scales up to centuries and spatial resolutions between 1 and 50 km. The boundary conditions of the CLM are provided by the simulation results of the coupled atmosphere-ocean global climate model ECHAM5/MPIOM [10] at six hourly intervals. The overall output data of the CLM regional model is provided in netCDF format and is generated at two spatial resolutions: a 0.165° spatial resolution on a rotated Gaussian grid and a 0.2° (~25 km) spatial resolution on a regular geographical grid. The simulation work in the Mesta/Nestos basin has been conducted with the use of the latter for a time period of 50 years (2016–2065) and for the results of the A1B and B1 IPCC climate scenarios. A reference climate flow (RF) regime, which projects to the future a measured flow regime of the past and represents a stable climate, was used to compare the simulations between climate change and non-climate change conditions.

5.6.3
Simulation Results

5.6.3.1 Reference Climate (RF)

According to the simulation results for this scenario, the average reservoir elevation is 369.0 m and the derived produced power for the period of simulation (2015–2065) is equal to 44.3 MW, which corresponds to 1062.8 MWh of generated energy (Figure 5.6.3).

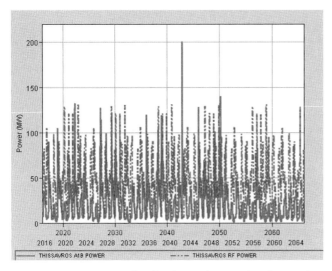

Figure 5.6.3 Comparison of produced power between the reference climate (RF, dotted grey) and climate change scenario A1B (solid line) for the Thissavros dam.

5.6.3.2 Climate Scenario A1B

Variations are present both in the generated power and the reservoir elevation: the average reservoir elevation is 341.0 m, which corresponds to a storage capacity of $560 \times 10^6 \, m^3$ and the minimum is 303.0 m, which corresponds to a storage capacity of $379 \times 10^6 \, m^3$. As far as power generation is concerned, the estimated generated power is illustrated in Figure 5.6.3. More specifically, the average produced is equal to 19.1 MW, which corresponds to 459.0 MWh of generated energy. Because of water releases from the Thissavros dam to meet the agriculture demand for water, the generated power during summer time is close to zero.

5.6.3.3 Climate Scenario B1

The obtained results for this scenario are slightly more optimistic. The average reservoir elevation is 346.0 m, which corresponds to a storage capacity of $563 \times 10^6 \, m^3$, and the minimum is 288.0 m, which corresponds to a storage capacity of $288 \times 10^6 \, m^3$. As regards power generation, the average generated power was estimated at 22.6 MW, which corresponds to 542.0 MWh of produced energy (Figure 5.6.4).

5.6.4
Conclusions

The coupling of the different modelling tools indicated the impacts of climate change on the Nestos River water balance in both climate scenarios. Variations in water availability and power generation are presented for the whole set of dams. The simulation results demonstrate that according to the PPC data the average foreseen produced energy for the period from 2015 to 2065 is smaller than that produced in

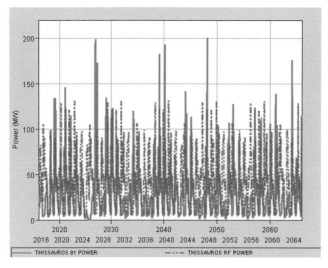

Figure 5.6.4 Comparison of produced power between the reference climate (RF, dotted grey) and climate change scenario B1 (solid line) for the Thissavros dam.

2002, but greater than that produced in 2005. In any case, both the climate change scenarios used in this study present a future decrease in the water regime in Greece with consequences for the existing hydroelectric dams. These results coincide with similar studies that have been implemented in different river basins in Greece by different research teams [11, 12], where a diminution of the produced power could occur in the long run during the twenty-first century.

References

1 IPCC (2001) *Climate Change 2001: The Scientific Basis. Contribution of Working Group I to the Third Assessment Report of the Intergovernmental Panel on Climate Change*, Cambridge University Press, Cambridge, United Kingdom, 881 pp.

2 Harrison, G.P. and Whittington, H.W. (2002) Investment in renewable energy: accounting for climate change. *Power Engineering Society Summer Meeting, 2002 IEEE*, IEEE, vol. 1, pp. 140–144. ISBN 0-7803-7518-1.

3 Greek Ministry of Development (2005) 3rd National Report for the Use of Renewable Sources in Athens.

4 Ledoux, E., Girard, G., de Marsily, G. and Deschenes, J. (1989) *Spatially Distributed Modeling: Conceptual Approach, Coupling Surface Water and Ground Water in*

Unsaturated Flow Hydrologic Modeling - Theory and Practice (ed. H.J. Morel-Seytoux), NATO ASI Series S 275, Kluwer Academic, Boston, pp. 435–454.

5 U.S. Army Corps of Engineers (1956) Snow Hydrology: Summary report of snow investigations, U.S. Department of Commerce Office of Technical Services, PB 151660, 437 pp.

6 Charbonneau, R., Fortin, J.P. and Girard, G. (1971) Précision et sensibilité des modèles paramétriques. Paper presented at the Proceedings of Canadian Hydrology Symposium, 26–27 May, 1971, Québec.

7 Skoulikaris, Ch. (2008) Mathematical modelling applied to the sustainable management of water resources projects at a river basin scale: the case of the Mesta/

Nestos. PhD thesis, Aristotle University of Thessaloniki, Greece.

8 USACE (2003) US Army Corps of Engineers (USACE) Hydrologic Engineering Center –reservoir system simulation. User Manual, Version 2.0, September 2003.

9 Mearns, L.O., Giorgi, F., Whetton, P., Pabon, D., Hulme, M. and Lal M., (2003) Guidelines for use of climate scenarios developed from regional climate model experiments, DDC of IPCC TGCIA.

10 Roeckner, E., Brokopf, R., Esch, M., Giorgetta, M., Hagemann, S., Kornblueh, L., Manzini, E., Schlese, U. and Schulzweida, U. (2006) Sensitivity

of simulated climate to horizontal and vertical resolution in the ECHAM5 atmosphere model. *Journal of Climate*, **19**, 3771–3791.

11 Mimikou, M. and Baltas, E. (1997) Climate Change impacts on the reliability of hydroelectric energy production. *Hydrological Science Journal*, **42** (5), 661–678.

12 Baltas, E.A. and Mimikou, M. (2005) Climate change impacts on the water supply of Thessaloniki. *International Journal of Water Resources Development*, **21** (2), Selected Global Water Issues, 2005, 341–353.

Further Reading

Ganoulis, J. and Skoulikaris, C. (2011) Impact of climate change on hydropower generation and irrigation: a case study from Greece, in *Climate Change and its Effects on Water Resources* (eds A. Baba et al.), NATO Science for Peace and Security Series C: Environmental Security, vol. **3**, Springer, Ch. 5

Skoulikaris, Ch., Ganoulis, J. and Monget J.M. (2009) Impact of climate change on river water flow: the case of the transboundary Mesta/Nestos river between Bulgaria and Greece. *Proceedings of the 33rd International Association of Hydraulic Engineering & Research (IAHR) Biennial Congress, Vancouver, British Columbia, August 9–14, 2009*, International Association of Hydraulic Engineering and Research (IAHR), pp. 1108–1115. ISBN 978-94-90365-01-1.

Skoulikaris, Ch., Ganoulis, J. and Monget, J.M. (2009) Integrated modelling of a new dam: a case study from the "HELP" Mesta/Nestos River, in *18th World IMACS Congress and MODSIM09 International Congress on Modelling and Simulation* (eds R.S. Anderssen, R.D. Braddock and L.T.H. Newham), Modelling and Simulation Society of Australia and New Zealand and International Association for Mathematics and Computers in Simulation, pp. 4050–4056. ISBN 978-0-9758400-7-8.

Ganoulis, J., Skoulikaris, Ch. and Monget, J.M. (2008) Involving stakeholders in transboundary water resources management: the Mesta/Nestos "HELP" Basin. *Water SA Journal*, **34** (4), 461–467.

5.7

Assessment of Climate Change Impacts on Water Resources in the Vjosa Basin

Miriam Ndini and Eglantina Demiraj

5.7.1

Introduction

Recent studies carried out in the field of climate change in Albania show that climate change will affect Albania with less precipitation, an increase in temperature and a rise in sea level [1–3]. More specifically, the impact of climate change is found to be significant on water resources and Albania's power sector, which is more than 90% dependent on hydropower and consequently upon climate conditions. Albania is currently receiving less rain than ever before, resulting in dry reservoirs.

Despite this, the existing policy and strategic documents do not address climate risk.

Projects carried out in Albania concerning the impact of climate change on the hydrological cycle are:

- *Implications of climate change for the Albanian coast* (MAP Technical Reports Series No. 98, UNEP, 1996);
- *Vulnerability assessment of climate change and adaptation options in the "First National Communication of Albania to the United Framework Convention to Climate Change (UNFCCC). Tirana, July 2002.*
- Assessment of Climate Change Impact on the Hydrological Cycle Elements in the South Eastern European Countries, UNESCO 2006.

The projects were undertaken to assess the impact of climate change not only in the hydrological field but also in other fields such as lithosphere, hydrosphere, natural ecosystems, managed ecosystems, energy and industry, tourism, transport and related services, sanitation and health aspects, population and settlement.

In this chapter, we will concentrate mostly on the vulnerability assessment of climate change on surface water. The current trends of temperature, precipitation and runoff are identified by calculating the seasonal and yearly anomalies from the long-term averages (1961–2000).

5.7.2
Surface Water Assessment

The present work brings an overview of a study of climate impact on basic hydrological balance elements for seven profiles distributed all over the Vjosa River catchment and for the period 1961–2000. The runoff and precipitation values for this period were measured. The climate impacts have a consequence on water flow, but man's influence cannot be ignored, although the profiles used in this study were selected in such a way so as to be free of human influence.

Results of the analyses showed that the long-term mean runoff, precipitation and evapotranspiration are distributed throughout the seasons as follows:

runoff: 38.6% occurs in winter, 35.6% in spring, 16.1% in autumn and 9.7% in summer;
precipitation: 40.3% occurs in winter, 28.3% in autumn, 20.2% in spring and 11.2% in summer;
evapotranspiration: 5.7% occurs in winter, 20.7% in spring, 27.1% in autumn and 46.5% in summer.

The Vjosa River is a transboundary river, shared between Albania and Greece. The total area of the basin is about $8100 \, \text{km}^2$. The Albanian part of the catchment of the Vjosa River has an area of $6706 \, \text{km}^2$ and a length of 272 m. Only $^2/_3$ of the entire basin is situated in the territory of Albania, the rest of the basin belongs to Greece.

A characteristic feature of the catchment of the Vjosa River is the presence of deep karst aquifers, which assure an abundant underground supply during the dry season. In this hydrographic basin there are also a large number of springs, some of which have a very strong impact on the water resources of the basin.

The basic characteristics are as follows:

- annual discharge volume: 5550 million m^3;
- specific discharge: $26 \, l \, s^{-1} \, km^{-2}$;
- ratio wettest month (February) to driest month (August–September) = 7.3, a low value for a river without regulating structure;
- one in ten years high flow: about 21 times the river module;
- storage capacity on the rivers: none; the hydropower plants do not include any significant storage.

The economic development of many important districts of Albania is closely connected to the water resources of this basin.

The available water in the watershed shows significant dependence on precipitation and evapotranspiration. For two seasons (winter and summer) the seasonal variability of runoff conforms to the seasonal variability of precipitation and evapotranspiration; in that sense, the highest and the lowest values of runoff follow the highest and lowest values of precipitation and the lowest and highest values of evapotranspiration, respectively. In the two other seasons (spring and autumn) we see that the value of runoff in spring is higher that that in autumn, whereas precipitation in spring is lower than that in autumn. In this case the antecedent condition of the soil should be taken into account. During the spring season the soil is saturated and, therefore, low infiltration rates increase the surface runoff. During the autumn season the soil is unsaturated and very dry after the summer season, resulting in higher infiltration and lower surface runoff. Furthermore, evapotranspiration is higher in autumn than in spring.

To achieve the goal of this study, the first step taken was to identify the present situation trends. Then, for every profile, the seasonal and yearly runoff anomalies from the long-term (1961–2000) mean were calculated. The runoff anomalies were expressed as a ratio of the long-term mean. The Spearman test was applied to determine whether the trend is significant.

According to the measured data, for the time period 1961–2000, it can be seen that all the profiles showed a decreasing trend. Figure 5.7.1 shows two out of all the stations considered in the study.

5.7.3
Vulnerability Assessment of Surface Water

5.7.3.1 Evaluation of Impact of Climate Change on the Mean Annual River Runoff
An analysis of the climate change scenarios for Albania shows that an increase in temperature and decrease in precipitation are expected. Consequently, we may

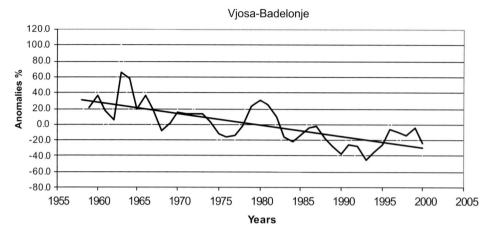

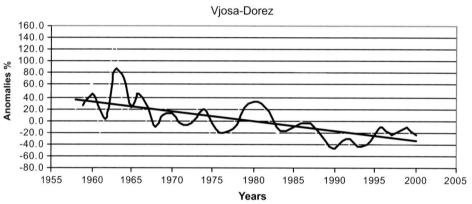

Figure 5.7.1 Yearly anomalies of two stations in the Vjosa River basin.

expect milder winters, warmer springs, hotter and drier summers and drier autumns. It is important to emphasize that the outputs of the scenarios should be considered as indicators of changes that might occur and not as absolute values of changes [5, 6].

To determine the impact of a changing climate on the annual runoff, the model runoff–forming factors (annual precipitation and annual evapotranspiration) in relation to the long-term annual runoff were tested. This is a predictive model, which is based on the concept of the Turc model, which relates precipitation and temperature to runoff according to the following formulas:

$$R = P \times \left[1 - \frac{E_{\mathrm{PI}}}{(c \times E_{\mathrm{PI}}^{n} + P^{n})^{\frac{1}{n}}} \right] \tag{5.7.1}$$

The long-term values of potential evapotranspiration (E_{PI}) are computed with the Thornwait formula and the resulting relationship with the temperatures is given by the expression:

$$E_{PI} = a_1 + b_1 T + c_1 T^2 + d_1 T^3 \qquad (5.7.2)$$

where

R (mm) is the annual runoff;
P (mm) is the annual precipitation;
E_{PI} (mm) is the index value of the potential annual evapotranspiration;
T (°C) is the mean annual air temperature;
c, n, a_1, b_1, c_1, d_1 are coefficients that can be calibrated.

Table 5.7.1 Runoff change for annual and seasonal periods derived from CCSA.

Table 5.7.1 Runoff change for annual and seasonal periods.

	2025	2050	2100
Annual			
Temperature (°C)	0.8 to 1.1	1.7 to 2.3	2.9 to 5.3
Precipitation (%)	−3.4 to −2.6	−6.9 to −5.3	−16.2 to −8.8
Runoff (%)	−12.9 to −9.9	−26.1 to −18.3	−48.2 to −28.9
Winter			
Temperature (°C)	0.7 to 0.9	1.5 to 1.9	2.4 to 4.5
Precipitation (%)	−1.8 to −1.3	−3.6 to −2.8	−8.4 to −4.6
Runoff (%)	−1.2 to −0.3	−2.3 to −0.8	−6.9 to −0.75
Spring			
Temperature (°C)	0.7 to 0.9	1.4 to 1.8	2.3 to 4.2
Precipitation (%)	−1.2 to −0.9	−2.5 to −1.9	−5.8 to −3.2
Runoff (%)	−16.8 to −13.5	−30.5 to −21.0	−66.7 to −43.9
Summer			
Temperature (°C)	1.2 to 1.5	2.4 to 3.1	4.0 to 7.3
Precipitation (%)	−11.5 to −8.7	−23.2 to −17.8	−54.1 to −29.5
Runoff (%)	−29.1 to −22.5	−64.8 to −47.6	−89.0 to −67.4
Autumn			
Temperature (°C)	0.8 to 1.1	1.7 to 2.2	2.9 to 5.2
Precipitation (%)	−3.0 to −2.3	−6.1 to −4.7	−14.2 to −7.7
Runoff (%)	−20.2 to −14.6	−34.9 to −22.5	−47.0 to −30.9

The trends in the changes remain the same. For 2025 the differences are negligible, for the second time horizon, 2050, more significant and for the third time horizon, 2100, they are considerable.

The main result is that the scenario causes a decrease of long-term annual runoff. This means that a decrease in precipitation and an increase in temperature lead to a decrease in runoff. The degree of decrease varies according to the scenarios employed but both exhibit acceleration in the rate of decrease towards 2100. The extremity of the decrease in runoff will increase with the rise in the mean annual temperature, and thus with the broadening of the time horizon.

5.7.3.2 Evaluation of Impact of Climate Change on the Seasonal River Runoff

This section evaluates the potential impact of a changed climate on the seasonal distribution of river runoff in Albania. A simple hydrological rainfall–runoff model was chosen for modelling river runoff in a seasonal time step. The precipitation and temperature input into the model was spatially averaged over the basin using the Thiessen method for precipitation and arithmetic mean for temperature [7, 8]. The model was calibrated with data for the standard period 1961–2000.

The procedure for evaluating the hydrological seasonal scenarios was as follows:

- Calibration of the coefficients of Equation (5.7.3):

$$E_{PI} = a_1 + b_1 T + c_1 T^2 + d_1 T^3 \tag{5.7.3}$$

where T (°C) is the mean seasonal air temperature and a_1, b_1, c_1 and d_1 are coefficients, which are calibrated.

- Calibration of the coefficients of Equation (5.7.4):

$$R = a + bP + cE_{PI} \tag{5.7.4}$$

where

$-R$ (mm) is the seasonal runoff,

$-P$ (mm) is the seasonal precipitation,

$-E_{PI}$ (mm) is the index value of the seasonal evapotranspiration, which is calculated in the same way as in the annual model; a, b and c are coefficients that are calibrated.

The CCSA anticipates an increase in air temperature and decrease in precipitation for the time horizons chosen (2025, 2050 and 2100). Table 5.7.1 shows the average changes in seasonal discharges in (%) according to the CCSA.

The winter period represents a period of decrease in runoff but for very low values. It is expected to reduce by a maximum of 0.75% by 2100. It may be concluded that there are no significant changes for the winter period for all time horizons.

After summer, the spring period shows the second highest levels of expected decrease, which may reach values as high as −66.7% when the temperature increases by 4.2 °C and precipitation decreases by −5.8% in 2100.

During the summer we may expect the highest decrease compared with the other seasons. In comparison with the standard (baseline) period, river water resources can markedly change during the summer period, when the anticipated decrease for the horizon 2025 may vary from −22.5 to −29.1%. In the Vjosa River basin the minimum discharge occurs during the summer and the low-flow frequency will increase across most of the catchment area. The season of lowest flow will still be the summer. During summer we may expect the highest decrease compared with the other seasons, because of the climate changes (temperature increase and precipitation decrease), but the Vjosa River basin has a very low ratio between maximum and minimum discharge. This is because its underground watershed is karstic and calcareous, which creates the perfect condition for a consistent nourishing of the river flow during the dry period.

During the autumn period the runoff will be probably gradually decrease from −30.9 to −47.0% for the farthest horizon. The highest decrease −47.0% is observed in 2100 when the temperature will increase by 5.2 °C and the precipitation will decrease by −14.2%.

5.7.4
Conclusions

Comparing the results of the changes in river runoff obtained for the CCSA, which expects an increase of the long-term mean annual and seasonal air temperature and a decrease of mean annual and seasonal precipitation, it can be stated that a decrease in the long-term mean annual and seasonal runoff has to be expected for the whole territory and for all the three time horizons with acceleration in the decrease towards 2100 (Figure 5.7.2).

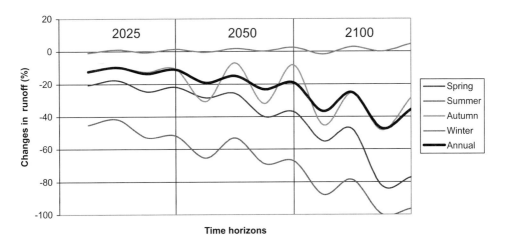

Figure 5.7.2 Average change in mean runoff according to the CCSA for three time horizons (2025, 2050 and 2100).

As contributors to hydrological systems, snow and ice and potential changes to them in a warmer climate will have profound impacts on streams and rivers [9]. Higher temperatures will shift the snowline upwards and for this reason the winter floods will still occur in winter but spring floods will shift towards the winter. The seasonal patterns of snowfall are likely to change with the snow season beginning later and ending earlier. Consequently, the spring runoff is expected to reduce noticeably. The maximum reduction amounts to 30% and 66%, respectively, by 2050 and 2100. Albania, which is dependent on hydropower and industry, should therefore take into consideration the expected reduction in runoff, which will affect the production of electricity.

The highest decrease in runoff is expected in summer and the lowest in winter. Therefore, the runoff in summer will be more vulnerable than the runoff in the winter period. The summer droughts may be serious. Spring shows the second-highest decrease in runoff and autumn the third highest.

Although the results of the predictive equations seem satisfactory for annual, spring, autumn and winter runoff, for the summer runoff the predictive equation gives unrealistic results. The value of the runoff decrease during summer is expected to be the highest compared with the other seasons but not for those values predicted by the equation. None of the equations are intended to be definitive and there is considerable scope for refining them. However, they can provide information for preliminary studies of water resources planning [10, 11].

An implication of simulated changes in stream flow is that flood risk generally would increase; the time of greatest risk would move from spring to winter. Effects on groundwater recharge (a major resource of this catchment's area) could be increased by climate change.

References

1 Demiraj, E., Bicja, M., Gjika, E. and Gjiknuri, L. (1996) Implications of climatic changes on the Albanian coast, Mediterranean Action Plan, Technical Reports Series No. 98, UNEP Athens.

2 UNDP (2003) Disaster risk assessment in Albania.

3 UNDP (2002) The first national communication of Albania to the united Nations Framework Convention on Climate Change.

4 Institute of Hydrometeorology, IHM Archives.

5 EEA (2005) Vulnerability and adaptation to climate change in Europe. Technical report No. 7/2005.

6 Hennessy, K. (2003) Climate change and its projected effects on water resources, in Proceedings of the Symposium of the Australian Academy of Technological Sciences and Engineering, Australian Academy of Technological Sciences and Engineering, pp. 174–191.

7 EEA (2007) Climate change and water adaptation issues. Technical report No. 2/2007.

8 FRIEND (2003) Processes and Estimation Methods for Stream flow and Groundwater. Developments in Water Science, Nr. 48.

9 Intergovernmental Panel on Climate Change (IPCC) (1994) Technical guidelines for assessing climate change impacts and adaptations.

10 WMO (2004) Expert meeting on hydrological sensitivity to climate conditions, Centre for Ecology and Hydrology (CEH), United

Kingdom. Publication WCASP-67
TD 1242.

11 IPCC (2001) *Climate Change 2001:
Impacts, Adaptation, and Vulnerability*

(eds J.J. McCarthy, O.F. Canziani, N.A.
Leary, D.J. Dokken and K.S. White),
Cambridge University Press.

5.8
Identification and Typology of River Water Bodies in the Hellenic Part of the Strymonas River Basin, as a Transboundary Case Study

Ioannis Chronis, Maria Lazaridou, Efthalia Lazaridou, Thomas K. Alexandridis, George Zalidis and Nikolaos Tsotsolis

5.8.1
Introduction

Monitoring according to the Water Frame Directive (2000/60/E.C.) guidelines requires the identification of river water bodies, typology and an investigation of reference conditions within each river basin. In the framework of the Community initiative INTERREG III A/PHARE CBC GREECE-BULGARIA 2000–2006, the Water Directorate of the region of Central Macedonia in Greece has been the first to fund an integrated study on Greek territory to assess the quality of and the pressures on the surface waters of the transboundary Struma/Strymonas River basin, in order to facilitate the implementation of the Water Frame Directive (WFD).

The identification of 'water bodies' based on geographical and hydromorphological determinants enables their status to be accurately described and compared to the environmental objectives of the Directive. A surface water body should be a discrete element of surface water, so that no water body overlaps with another or is composed of elements of surface water that are not contiguous. Heavily modified water bodies may be identified and designated where good ecological status is not being achieved because of impacts on the hydromorphological characteristics of surface water resulting from physical alterations. Related to this requirement, there are considerations regarding pressures and impacts [1].

Typology is the next step for the categorization of water bodies, and this is based on two systems, System A or System B [2] The two most common approaches for the structure of the typology are the 'top-down' approach, where the environmental variables (abiotic characteristics) are selected according to the geomorphology of the area [3], and the 'bottom-up' approach, where the data of the biocommunities are used to stream the types of the rivers and the combination of the two approaches. The WFD uses only abiotic characteristics for the classification of rivers bodies into types (Annex II, 1). Two systems are proposed, of which the first (System A) is a combination of the 'top-down' and 'bottom-up' approaches, and which defines the typology according to 25 predefined ecoregions (based on the distribution of the biocommunities in the European surface water). Apart from 'System A', the WFD proposes an alternative classification system, the so-called 'System B' as a 'top-down' approach.

The aim of the present study is to apply the best typology system for the determination of river types in the Hellenic part of the Strymonas river basin.

5.8.1.1 Study Area

The Greek part of the Strymonas River basin is the downstream part of a 17.730 km^2 river basin. This is a transboundary river between Greece (34.6%) and Bulgaria (51.3%) and a small part (14.1%) belongs to the Former Yugoslav Republic of Macedonia (FYROM) and this flows directly to the Bulgarian part. According to the data of the Ministry of Agriculture and Food Affairs in Greece, the mean annual discharge of the river is 3.440 × 106 m^3. The length of the river in Greek territory is 118 km. The three major side rivers of Aggistro/Bistritsa, Mpelitsa and Aggitis, which have a constant flow in the river basin, were also taken into consideration. The Aggistro comes from Bulgaria and its main flow originates from the Greek springs at the village of Aggistro. The Mpelitsa consists of two units, one is the Krousovitis, which crosses through the town of Sidirokastro, and the other is the Mpelitsa, which drains the central part of the Serres plain. The Aggitis consists of three major units. The smallest is called Aggitis and originates north of the city of Drama and joins the stream of Agia Varvara and the Tafros Fillipon before entering the Strymonas River (Figure 5.8.1).

Figure 5.8.1 Hellenic part of the Strymonas River basin – satellite image.

5.8.2
Methodology

River water bodies in the basin were identified and recognized as coherent units, according to the EC's Guidance document No.2 'Identification of water bodies' [1], so that monitoring and the implementation of WFD can be fulfilled. Morphological elements and river confluences were examined thoroughly so as to achieve the best division grade amongst the river water bodies. System B was chosen for the typology and the type specific conditions, to address the huge variety and differences in nature in Greece. For its implementation, four factors from System B were selected as being the most representative, three of which were obligatory factors (catchment size, geology, altitude) and one optional (slope). Other researchers [4] have applied System B with extra optional factors (temperature, rainfall) but no other type was discovered because the same hydro-geomorphologic conditions exist in the study area.

All factors were categorized in three classes, either after a statistical analysis [5] or after a slight modification of the existing bibliography [6, 7] (Table 5.8.1), achieving, however, the same degree of differentiation as in System A. Although data for the upstream Bulgarian part of the river were not available, in the calculations the whole transboundary catchment size was taken into consideration for the different water bodies/types of the main river. A four digit numerical system was adapted to present the types, with each digit representing one of the three classes of each factor (Table 5.8.1).

Table 5.8.1 A four digit numerical system was adapted to present the types.[a]

Digit	Category
1st	Altitude category [6] $1 = 0–150$ m $2 = 150–600$ m $3 \geq 600$ m
2nd	Catchment area category [5] $1 = 0–500$ m^2 $2 = 500–5000$ m^2 $3 \geq 5000$ m^2 (including transboundary river basin)
3rd	Geology category [8] $1 = Ca$ $2 = Si$ $3 = C$
4th	Slope [7] $1 = 0–5°$ $2 = 5°–15°$ $3 \geq 15°$

a) Example: type $1211 =$ altitude 0–150 m, catchment 500–5000 m^2, geology Ca, slope 0–5°.

Table 5.8.2 Recognized water bodies in the Strymonas River basin.

Number and type of water body
Three artificial water bodies (AWBs) artificial water body (AWB) (human construction)
Five heavily modified water bodies (HMWBs) heavily modified water body (HMWB) (severe channel transformations)
Nine natural water bodies (NWBs) natural water body (NWB) (mostly upstream parts)

5.8.3
Results

In the Greek part of the river basin, 17 water bodies were recognized, seven of which flow in the Strymonas river. The upper parts of the river after the Greek–Bulgarian border and the downstream part before the estuary were recognized as natural water bodies The central parts of the Strymonas River, together with the Mpelitsa River, were recognized as heavily modified because of the creation of Lake Kerkini in 1932, the diversion of the main river course and the alteration of the river banks. The Aggistro and the major flow of the Aggitis were also recognized as natural water bodies (Table 5.8.2).

The five types that were recognized in these 17 river water bodies differed in their catchment size, geology or slope. The seven water bodies of the main flow of the

Table 5.8.3 Types of river systems (R-) according to System B.

Type	Name	Water body	Length (m)
1121	Aggistro	R-NWB	7 033
	Aggitis Upstream	R-NWB	20 394
	Krousovitis	R-NWB	4 383
	Tafros Ag. Varbaras	R-AWB	10 850
	Tafros Fillipon 1	R-AWB	19 986
	Tafros Fillipon 2	R-AWB	5 268
1221	Aggitis 2	R-NWB	23 500
	Mpelitsa	R-HMWB	41 500
	Tafros Fillipon 3	R-HMWB	3 000
1223	Aggitis 1	R-NWB	4 905
1311	Strymonas Upstream 2	R-NWB	3 979
1321	Strymonas Upstream 1	R-NWB	1 241
	Strymonas Upstream 3	R-NWB	6 465
	Strymonas Upstream 4	R-HMWB	17 590
	Strymonas Downstream 1	R-HMWB	45 902
	Strymonas Downstream 2	R-HMWB	17 889
	Strymonas Downstream 3	R-NWB	14 613

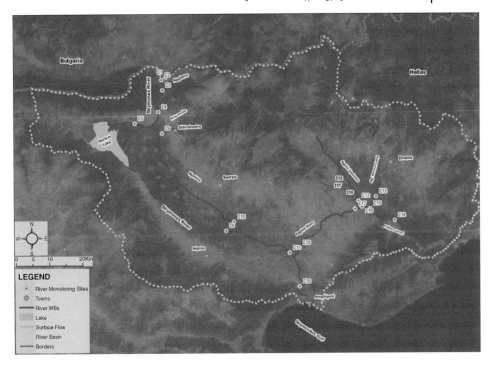

Figure 5.8.2 Monitoring river sites in the Strymonas River basin.

Strymonas River belong to the same type except for one, and this is due to its geology. No type-specific reference conditions were found in the river basin. Checks were performed to see whether in central and northern Greece there are types with specific reference conditions similar to the existing types in the Strymonas River but none were found [4]. The above procedure resulted in the 17 monitoring sites for all water bodies and types in the basin (Table 5.8.3, Figure 5.8.2).

5.8.4
Conclusions and Discussion

According to the international bibliography, most researchers use 'System B' because it allows a level of flexibility in the choice of parameters. For example, Bossche and Usseglio-Polatera [9] applied 'System B' for the determination of types in Belgium (Wallonia) with optional factors (slope, natural area).

Type-specific reference conditions were not found because of a high level of human intervention in the Strymonas River basin. The following methodology helped with the selection of monitoring sites in the basin. The application of System B resulted in 17 proposed monitoring sites for monitoring the 17 identified river water bodies (one station per water body). System B responded to the particularities (slope, clough of Aggitis) of the transboundary basin, including types of unique river systems

(Aggitis 1) in the basin, which would have been eliminated if system A was followed. The implementation complied fully with the requirements of the WFD [2].

Because of the absence of reference conditions, the multimetric indices were used for the types RM1, RM2 [10], which provide reference conditions [11] and/or the Hellenic Evaluation System (HES) [12] for RM3 type. The HES and its Interpretation Index were produced based on samples from (i) all seasons reflecting any kind of seasonal effect, (ii) different substrate types and (iii) with different habitat characteristics representing some of the System B parameters of the WFD characterizing river typology.

Acknowledgements

This work formed part of the project 'Monitoring of surface waters of the Strymonas River basin' funded by the Community Initiative INTERREG III A/PHARE CBC GREECE-BULGARIA 2000–2006. The authors are grateful to the Water Directorate of the Region of Central Macedonia for their support.

References

1 EC (2003) Common Implementation Strategy for the Water Framework Directive (2000/60/EC). Guidance document No. 2, Identification of water bodies.

2 EC (2003) Common Implementation Strategy for the Water Framework Directive (2000/60/EC). Guidance document No. 10, River and lakes – Typology, reference conditions and classification systems.

3 Ehlert, T., Hering, D., Koenzen, U., Pottgiesser, T., Schuhmacher, H. and Friedrich, G. (2002) Typology and type specific reference conditions for medium-sized and large rivers in north Rhine-Westphalia: methodical and biological aspects. *International Review of Hydrobiology*, **87** (2–3), 151–163.

4 Kanli, L. (2008) Comparison of typological river systems in Greece. Master thesis, Library of Department of Biology, Aristotle University of Thessaloniki, Greece.

5 Kemitzoglou, D. (2006) Characterization of river types in Central and Northern Greece and determination of benthic macroinvertebrates communities according to WFD. Master thesis, Library

of Department of Biology, Aristotle University of Thessaloniki, Greece.

6 Dikau, R. (1989) *The Application of a Digital Relief Model to Landform Analysis*, Taylor and Francis, London.

7 Demek, J. (1972) *Manual of Detailed Geomorphological Mapping*, Academia, Prague.

8 Aristotle University (AUTH) (1952) Geological Map of Greece, Library of the Geology Department, Aristotle University of Thessaloniki, Greece.

9 Bossche, J.P. and Usseglio-Polatera, P. (2005) Characterization, ecological status and type-specific reference conditions of surface water bodies in Wallonia (Belgium) using biocenotic metrics based on benthic invertebrate communities. *Hydrobiologia*, **551**, 253–271.

10 Van de Bund, W., Cardoso, A.C., Heiskanen, A.S. and Noges, P. (2004) Common Implementation Strategy for the WFD (2000/60/EC).

11 Buffagni, A., Erba, S., Birk, S., Cazzola, M., Feld, C., Ofenbock, T., Murray-Bligh, J., Furse, M.T., Clarke, R., Herring, D., Soszka, H. and Van de Bund, W. (2005) *Towards European Inter-calibration for the Water Framework Directive: Procedures and Examples for Different*

River Types from the E.C. Project STAR, Instituto di Ricerca Sulle Acque, Rome.

12 Artemiadou, V. and Lazaridou, M. (2005) Evaluation score and interpretation index of the ecological quality of running waters in Central and Northern Hellas. Environmental Monitoring and Assessment, 110, 1–40.

Further Reading

EC (2003) Common Implementation Strategy for the Water Framework Directive (2000/60/EC). Guidance document No.4, Identification of Heavily Modified Water Bodies.

EC (2000) Directive 2000/60/EC of the European Parliament and of the Council of 23rd October 2000 establishing a framework for Community action in the field of water policy. Official Journal of the European Communities, L327, Luxembourg, 22 December 2000.

Water Directorate, Region of Central Macedonia,(2008) Final deliverable of the project "Monitoring of surface waters of Strymonas River Basin", Thessaloniki, Greece.

5.9
Calculation of Sediment Reduction at the Outlet of the Mesta/Nestos River Basin caused by the Dams

Manolia Andredaki, Vlassios Hrissanthou and Nikolaos Kotsovinos

5.9.1
Introduction

The Nestos River flows through two European countries, Bulgaria and Greece, and discharges its water into the Aegean Sea. The basin area of the Nestos River, considered in this study, is about $5100 \, \text{km}^2$. In the Greek part of the river, two dams, the Thisavros dam and the Platanovrysi dam, have already been constructed and started operating in 1997 and 1999, respectively. The construction of the dams implies a reduction of sediment yield at the outlet of the Nestos River basin and the alteration of the sediment balance of the basin in general, which results in coast erosion. Thus, the main aim of this study is to calculate the sediment yield at the outlet of the Nestos River basin before and after the construction of the dams.

To reduce sedimentation in the Thisavros reservoir, measures against soil erosion should be taken in both the Greek and Bulgarian parts of the Thisavros reservoir basin. This means that, apart from the issues regarding water quantity and quality of the Nestos River in both countries, sediment transport issues should also be confronted.

5.9.2
Description of the Simulation Model

The simulation model RUNERSET (RUNoff–ERosion–SEdiment Transport) [1] was applied to the Nestos River basin for the calculation at certain locations of the mean annual value of sediment yield, due to rainfall and runoff. The above simulation model consists of three submodels: a rainfall-runoff submodel [2], a soil erosion submodel [3] and a sediment transport submodel for streams [4].

The rainfall-runoff submodel is a water balance model for soil moisture in the root zone. The available soil moisture increases via rainfall and decreases via potential evapotranspiration, deep percolation and runoff. Both runoff and deep percolation can be estimated from the water balance equation. The potential evapotranspiration included in the rainfall-runoff submodel is calculated using the Thornthwaite method [5].

Runoff estimated from the rainfall by means of the first submodel serves as input to the soil erosion submodel. By means of the second submodel, soil erosion due to rainfall and runoff is computed. According to Schmidt [3], the erosive impact of droplets and overland flow is proportional to the momentum flux contained in the droplets and the overland flow respectively. The sediment supply to the main stream of a basin is estimated by means of a comparison between the available sediment in the basin area and the sediment transport capacity by overland flow. Gully erosion is not taken into account in the above submodel.

Sediment yield at the outlet of the main stream of a basin can be computed by comparing the available sediment in the main stream and the sediment transport capacity by streamflow. The latter quantity is computed using the relationships of Yang [4], which are based on the concept of unit stream power.

5.9.3
Application of the Simulation Model

For more precise calculations, the Nestos River basin was divided into 60 sub-basins (Figure 5.9.1). Specifically, the basin of the Thisavros dam (Bulgarian and Greek parts) was divided into 31 sub-basins, the basin of the Platanovrysi dam (Greece) into nine sub-basins and the basin downstream of Platanovrysi dam into 20 sub-basins. The outlet of the last basin is named Toxotes.

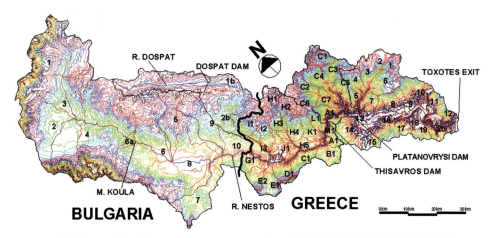

Figure 5.9.1 Altitude contours map of the Nestos River basin divided into sub-basins.

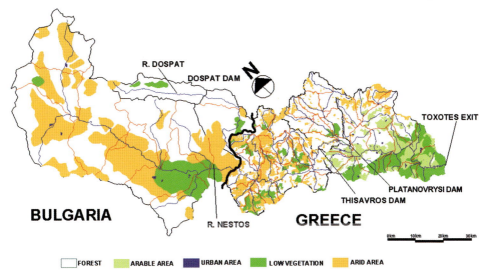

Figure 5.9.2   Main streams and soil cover map of the Nestos River basin.

Meteorological data (monthly rainfall data and mean monthly temperature data) from 22 meteorological stations in Greece and Bulgaria were used as input data for the simulation model. The following thematic maps were constructed for the accurate computation of sub-basin parameters: altitude contours map with sub-basins (Figure 5.9.1), main streams map (Figure 5.9.2), Thiessen polygons map, soil cover map (Figure 5.9.2) and geological map (Figure 5.9.3). Soil maps were not available.

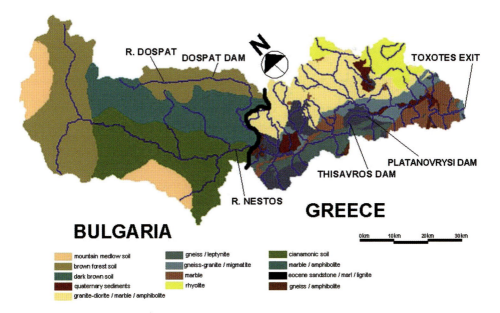

Figure 5.9.3   Geological map of the Nestos River basin.

The values of the sub-basin parameters (e.g. sub-basin area, mean slope gradient of soil surface, mean slope gradient of the main stream of a sub-basin, percentage of Thiessen polygons, percentage of soil cover etc.) were also used as input data for the model. The calculations were performed for each sub-basin on a monthly time basis.

This means that cumulative monthly values for runoff, soil erosion and sediment transport were calculated for each sub-basin; thus the impact of extreme rainfall events can be recognized indirectly by the relatively high arithmetic values of the corresponding monthly runoff and soil erosion.

Table 5.9.1 gives the physiographic characteristics of the sub-basins.

5.9.4
Model Testing

Sediment measurements (suspended load) for 53 years (1937–1989) were available for the location 'Momina Koula' [6] in the Bulgarian part of Nestos River (Figure 5.9.1).

According to the measurements, the mean annual suspended sediment yield for the time period given above is $202\,t\,km^{-2}$ [6]. RUNERSET was applied to the basin corresponding to this location for the same time period [7]. The basin area is $1511\,km^2$, which is about 30% of the entire basin area of the Nestos River. Bed load measurements were not available; therefore, the following assumption was made: the ratio of bed load to suspended load amounts to 0.25. According to this assumption, the measured mean annual sediment yield at Momina Koula is $252.5\,t\,km^{-2}$. As a final result, the simulation model used underestimates the measured mean annual sediment yield by about 18%. The arithmetic results of the model for the different years, as well as the results of the field measurements, are given in Table 5.9.2.

5.9.5
Main Computations

The relatively low deviation between computation and measurement results for the mean annual sediment yield at the location 'Momina Koula' was an encouraging indication for the further application of the simulation model to other parts of the Nestos River basin. The following calculations were therefore performed for a time period of eleven years (1980–1990):

1) calculation of mean annual sediment amount inflowing into the Thisavros reservoir from the Bulgarian part ($3052\,km^2$) and from the Greek part ($804\,km^2$) of the Nestos River basin [7, 8];
2) calculation of mean annual sediment amount inflowing into the Platanovrysi reservoir from the corresponding basin ($405\,km^2$, Greece) [9];
3) in a previous study [1], the mean annual value of sediment yield at the outlet of the Nestos River basin (Toxotes) was calculated; this sediment yield originates mainly from the part of the Nestos River basin which lies downstream of the Platanovrysi dam ($840\,km^2$, Greece).

Table 5.9.1 Physiographic characteristics of the sub-basins.

Hydrologic basin	Thisavros reservoir				Basin downstream of Dospat dam		Platanovrysi reservoir		Downstream of Platanovrysi dam (basin after the construction of the dams)	
	Bulgarian part of the basin		Greek part of the basin							
Physiographic characteristics	Max value	Min value	Max value	Min value	Max value	Min value	Max value	Min value	Max value	Min value
Mean slope gradient of soil surface (%)	23.59	16.78	50.90	18.00	22.75	21.57	57.46	45.45	58.03	22.69
Mean slope gradient of the main stream of the sub-basins (%)	14.79	0.23	23.51	1.31	7.70	3.12	13.30	3.38	20.00	2.50
Area of the sub-basins (km²)	477.63	89.65	53.76	5.96	208.12	61.16	55.76	14.49	67.60	13.30
Vegetation										
forest (%)	94.53	14.41	90.24	22.40	88.62	71.38	90.24	22.40	97.95	82.51
arable area and low vegetation (%)	47.31	0.00	41.40	0.84	6.44	4.86	15.30	0.00	99.13	0.00
urban area (%)	—[a]	—[a]	—[a]	—[a]	—[a]	—[a]	—[a]	—[a]	—[a]	—[a]
arid area (%)	67.26	5.47	49.69	0.00	22.18	6.52	14.96	2.05	23.81	0.00
Rocks										
permeable (%)	27.52	0.00	85.02	0.00	0.00	0.00	21.83	0.00	100.00	0.00
semi-permeable (%)	88.81	0.00	0.00	0.00	100.00	100.00	0.00	0.00	100.00	0.00
impermeable (%)	79.49	0.00	100.00	0.00	0.00	0.00	100.00	78.17	86.91	0.00

a) Relatively negligible percentages.

Table 5.9.2 Computation and measurement results of sediment yield for the location 'Momina Koula'.

Year	Annual sediment yield (t)	Arithmetic simulation results				Year	Annual sediment yield (t)
		Year	Annual sediment yield (t)				
1937	366 000	1955	381 000			1973	318 000
1938	300 500	1956	458 000			1974	225 000
1939	271 500	1958	413 000			1975	55 500
1940	447 500	1959	274 000			1976	319 500
1941	144 500	1960	555 500			1977	85 500
1942	313 500	1961	160 500			1978	280 500
1943	36 000	1962	798 500			1979	315 000
1944	327 000	1963	705 500			1980	314 000
1945	485 000	1964	91 000			1981	200 500
1946	338 000	1965	276 500			1982	209 500
1947	517 500	1966	475 500			1983	68 500
1948	72 000	1967	107 500			1984	185 000
1949	230 500	1968	252 500			1985	239 000
1950	278 500	1969	634 500			1986	511 000
1951	428 500	1970	152 500			1987	288 500
1952	374 000	1971	511 500			1988	253 000
1953	359 000	1972	179 000			1989	7500
1954	786 000						

Arithmetic values

Sum (t)	Mean annual value (t)	Mean annual value (t km^{-2})	Max value (t km^{-2})	Min value (t km^{-2})	Stand. deviat. (t km^{-2})
16 377 500	315 000	207.87	527.01	4.95	120.36

Field measurements

Mean annual suspended sediment yield (t km^{-2})	Bed load (t km^{-2})	Mean annual sediment yield (t km^{-2})	Deviation from computational results (%)
202	50.5	252.5	17.7

Table 5.9.3 Computational results of sediment yield at various locations of the Nestos River basin.

Year	Hydrologic basin: Thisavros reservoir			Hydrologic basin: Platanovrysi reservoir (t)	Hydrologic basin: downstream of Platanovrysi Dam (basin after the construction of the dams) (t)	Entire hydrologic basin of the Nestos River (t)
	Bulgarian part of the basin (t)	Greek part of the basin (t)	Basin downstr. of Dospat dam (t)			
1980	1084 000	184 000	154 500	366 000	278 000	2066 500
1981	968 000	128 000	128 500	344 000	588 000	2156 500
1982	850 000	312 500	112 500	409 000	426 000	2110 000
1983	309 500	114 000	32 500	99 500	73 000	628 500
1984	678 500	360 500	107 500	277 000	494 000	1917 500
1985	991 000	94 500	127 500	54 500	131 000	1398 500
1986	1495 500	613 000	162 000	303 500	198 000	2772 500
1987	1021 000	875 500	131 500	761 500	673 000	3462 500
1988	884 000	357 500	130 000	241 000	383 000	1995 500
1989	73 500	121 000	XX[a]	192 500	207 000	594 000
1990	545 500	552 500	46 500	289 500	64 000	1498 000
Arithmetic values						
Mean (t)	809 000	337 500	113 000	314 500	331 000	1873 000
Max (t)	1495 500	875 500	162 000	409 000	673 000	3462 500
Min (t)	73 500	94 500	32 500	54 500	64 000	594 000
St. dev. (t)	391 725	252 278	42 344	186 329	208 229	765 845
St. dev./mean	0.48	0.75	0.37	0.59	0.63	0.45

a) XX: No results due to very low values.

Table 5.9.3 summarizes the arithmetic results for the annual sediment yield for different years at certain locations of the Nestos River basin (Thisavros reservoir, Platanovrysi reservoir and Toxotes).

5.9.6
Conclusion

According to Table 5.9.3, the mean annual value of sediment yield at the outlet of the Nestos River basin before the construction of the dams is about 1.9×10^6 t, while after the construction of the dams this amounts to 0.33×10^6 t. This shows that the construction of the dams implies a dramatic decrease (about 83%) in the sediments supplied directly to the basin outlet and indirectly to the neighbouring coast. This reduction influences the seashore sediment balance, resulting in an analogous increase in the erosion rates of the Nestos River mouth as well as the adjacent shorelines. In fact, the reduction percentage that is calculated in the present chapter is directly comparable to the increase of the coastal erosion rates that are stated in the work of Xeidakis *et al.* [10].

Acknowledgements

Thanks are due to the European Union for their financial support of the research project INTERREG IIIC-BEACHMED-e.

References

1 Hrissanthou, V. (2002) Comparative application of two erosion models to a basin. *Hydrological Sciences Journal*, **47** (2), 279–292.

2 Giakoumakis, S., Tsakiris, G. and Efremides, D. (1991) On the rainfall-runoff modelling in a Mediterranean island environment, in *Advances in Water Resources Technology* (ed. G. Tsakiris), Balkema, Rotterdam, pp. 137–148.

3 Schmidt, J. (1992) Predicting the sediment yield from agricultural land using a new soil erosion model. Presented at the Proceedings 5th International Symposium on River Sedimentation, Karlsruhe, Germany, April 6–10, 1992, University of Karlsruhe, Germany.

4 Yang, C.T. (1973) Incipient motion and sediment transport. *Journal of*

the Hydraulics Division, ASCE, **99** (10), 1679–1704.

5 Thornthwaite, C.W. (1948) An approach towards a rational classification of climate. *Geographical Reviews*, 38, 55–94.

6 Gergov, G. (1996) Suspended sediment load of Bulgarian rivers. *GeoJournal*, **40** (4), 387–396.

7 Andredaki, M. (2008) Calculation of mean annual sediment yield of Nestos river before and after the construction of the dams using RUNERSET model. Technical Report Number T-2-08, Hydraulics and Hydraulic Structures Laboratory, Department of Civil Engineering, Democritus University of Thrace, Greece.

8 Kapona, E. and Tona, E. (2003) Computation of the inflowing sediments into Thisavros reservoir of Nestos River. Diploma thesis, Department of Civil

Engineering, Democritus University of
Thrace, Xanthi, Greece.

9 Klisiari, A. (2002) Computation of the
inflowing sediments into Platanovrysi
reservoir of Nestos River. Diploma thesis,
Department of Civil Engineering,
Democritus University of Thrace, Xanthi,
Greece.

10 Xeidakis, G., Georgoulas, A., Kotsovinos,
N., Delimani, P. and Varagouli, E. (2010)

Environmental degradation of the
coastal zone of the west part of Nestos
River Delta, N. Greece, in *Proceedings
12th International Congress of the Geological
Society of Greece, Patras, Greece,
19–22 May, 2010*, Geological Society of
Greece, vol. **2**, pp. 1074–1083.
Available at http://www.geosociety.gr/
GSG_XLIII_%202.pdf (accessed 31
March 2011).

5.10
Methodologies of Estimation of Periodicities of River Flow and its Long-Range Forecast: The Case of the Transboundary Danube River

Alexey V. Babkin

5.10.1
Introduction

With a length of 2857 km, the Danube River is the second largest river in Europe (after the Volga). The Danube originates in the Black Forest mountains of western Germany. Along its course, it passes through nine countries: Germany, Austria, Slovakia, Hungary, Croatia, Serbia, Bulgaria, Romania and Ukraine.

Discontinuous time series (TS) of the Danube River runoff are available from 1840, Orsova, Romania. The runoff from the river is significantly variable, a minimum level of 105 km^3 was recorded in 1863 and maximum levels of close to 250 km^3 yr^{-1} were recorded in 1915 and 1941. This means a spread of 145 km^3 yr^{-1}.

The variation in runoff from the Danube River has an impact on different branches of a modern economy, such as fisheries, water transport and tourism, and industrial and communal water consumption [1]. The hydrological problem of developing a methodology of analysis of TS of the Danube River runoff for long-range forecasting is actually closely related to the problems of developing the local and regional economy and integrated transboundary water resources management.

Traditionally, the methods of Fourier transform, correlation and spectral analysis are used for analysis of periodicities in time series of hydrological characteristics [2]. The present study develops a methodology for presenting the periodicities in TS of the Danube's runoff and applies it to long-range forecast estimations. Periodicities are estimated using the method of "least squares".

Annual time series were analysed and modelled for the time interval from the beginning of instrumental observations up to 1978. The training forecasts for the period 1979–1988, and for the intervals 1979–1983 and 1984–1988, were computed and tested by the new data.

5.10.2
Methodology for Presenting Periodicities in Time Series of River Runoff

The time series of runoff for the Danube River were approximated by sine functions using the method of the least squares [3, 4]:

$$S_Q = \sum_{1}^{n} (Q_i - Q_0 - b \sin \omega t_i - c \cos \omega t_i)^2 = \sum_{1}^{n} \left[Q_i - Q_0 - \frac{\delta Q}{2} \sin(\omega t_i + \varphi_Q) \right]^2$$

(5.10.1)

where

S_Q is the sum of the square differences, respectively, between the time series of the river runoff and the sinusoids of their approximation;
Q_i is a value of the river runoff variable from time series in the year t_i;
i is the number of the year in the observation range of n years;
Q_0 is an additional item of approximation;
b and c are constant parameters that reflect the amplitude δQ and phase φ_Q of approximation sine:

$$\delta Q = 2\sqrt{b^2 + c^2}$$

(5.10.2)

$$tg\varphi_Q = \frac{c}{b}$$

(5.10.3)

The approximation sine is best when the sum (5.10.1) is smallest [5]. To determine the best sine for any fixed period it is necessary to find the particular derivatives of Equation (5.10.1) by parameters Q_0, b and c. These derivatives should be equated with 0 and combined into the system:

$$\frac{\partial S_Q}{\partial Q_0} = -2 \sum_{1}^{n} (Q_i - Q_0 - b \sin \omega t_i - c \cos \omega t_i) = 0$$

(5.10.4)

$$\frac{\partial S_Q}{\partial b} = -2 \sum_{1}^{n} [(Q_i - Q_0 - b \sin \omega t_i - c \cos \omega t_i) \sin \omega t_i] = 0$$

(5.10.5)

$$\frac{\partial S_Q}{\partial c} = -2 \sum_{1}^{n} [(Q_i - Q_0 - b \sin \omega t_i - c \cos \omega t_i) \cos \omega t_i] = 0$$

(5.10.6)

Solution of the system of Equations (5.10.4)–(5.10.6) permits evaluation of the amplitude, phase and additional item of the approximation sine for any period (T) and for the succession of the periods. The sum of square differences between the approximation sine and values from the time series may be calculated using Equation (5.10.1).

The time series of the Danube River were approximated by sine functions successively with the unitary period step. Table 5.10.1 shows the parameters of the approximation sinusoids.

Table 5.10.1 Sine approximation of time series of runoff from the Danube River (1859–1978).

T, year	Q_0 (km^3 yr^{-1})	$\delta Q/2$ (km^3 yr^{-1})	ϕ_Q (radian)	S_Q (km^3 yr^{-1})^2
3.0	1 717 272	35 372	45 427	125 472.1
4.0	1 717 502	31 091	22 580	125 664.2
5.0	1 717 163	99 217	0.9766	119 453.6
6.0	1 717 223	38 616	26 116	125 304.4
7.0	1 717 422	12 293	41 748	126 231.9
8.0	1 718 295	53 874	39 732	124 306.5
9.0	1 715 987	85 354	33 328	121 272.4
10.0	1 717 359	31 968	37 783	125 621.6
11.0	1 717 688	53 886	25 909	124 301.5
12.0	1 715 867	71 465	35 100	122 796.0
13.0	1 718 779	69 047	−0.0128	123 044.6
14.0	1 717 533	78 421	43 506	122 037.9
15.0	1 718 974	79 309	-0.7625	121 949.6
16.0	1 716 274	43 588	−10 143	125 025.5
17.0	1 717 598	11 786	26 613	126 238.9
18.0	1 716 149	41 941	39 369	125 133.2
19.0	1 715 722	44 786	17 261	124 918.4
20.0	1 717 748	64 393	24 656	123 463.7
21.0	1 720 400	92 245	−0.2248	120 443.4
22.0	1 720 561	103 683	0.6158	118 870.5
23.0	1 717 898	93 232	−0.9298	120 283.9
24.0	1 715 374	66 560	17 840	123 311.5
25.0	1 715 883	27 326	27 385	125 825.7
26.0	1 718 046	15 608	−0.4982	126 166.7
27.0	1 718 769	53 116	37 391	124 338.8
28.0	1 716 753	85 687	0.6195	121 269.2
29.0	1 712 958	109 611	27 244	118 280.4
30.0	1 709 848	119 497	38 593	116 746.2
31.0	1 709 174	115 970	40 961	117 050.1
32.0	1 710 739	104 950	35 048	118 531.6
33.0	1 713 371	91 304	21 573	120 406.0

It can be seen that there are some local minima of sums of square differences between the time series of Danube River runoff and their approximation in dependence from the period (highlighted in italics in the table). These minima are revealed near the periods of 5, 9, 12, 15, 22 and 30 years.

5.10.3
Long-Range Forecast of Runoff from the Danube River

The minima of sums of square differences of time series and their approximation are smallest near the periods of 30, 22 and 5 years. The correlation of each of these harmonics with the time series is larger then 23%. When these sinusoids were added together the correlation of their sum and the TS of the Danube River runoff

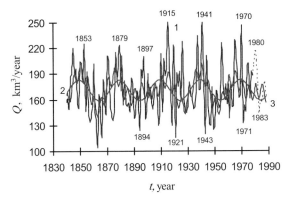

Figure 5.10.1 Variation of runoff from the Danube River (Orsova): 1 – time series (dotted line – training forecast interval, 1979–1988); 2 – approximating sine for the period of 30 years; 3 – the sum of periodicities of 30, 22 and 5 years.

were increased. The correlation of the TS of runoff from the Danube River with the sum of these three harmonics is equal to 0.419. The TS of runoff from the Danube River, the 30 year sine and the sum of harmonics are illustrated in Figure 5.10.1.

The annual long-range forecast is considered as true if the difference between the real and forecasted runoff does not exceed 67.4% from the standard deviation of its TS [6]. The long-range forecast of water resources may be estimated as successful if its results are no worse than the mean value prediction.

The analysis of TS of the Danube River runoff has shown that for 1840–1978 its mean value (Q_m) and standard deviation are equal to 171.7 and 30.1 $km^3 yr^{-1}$, respectively. Consequently, the annual long-range forecast of runoff from the Danube River is successful if it is no more than 20.3 $km^3 yr^{-1}$ out.

The results of the forecasting of runoff from the Danube River for 1979–1988 are analysed in Table 5.10.2. The first column of the table presents the years of the training forecast interval, and the second and third columns the real water runoff from the Danube River (Q) and its values computed by the sum of periodicities of 30, 22 and 5 years (Q_s). The last rows illustrate the mean values of the real and predicted runoff for the five and ten year intervals.

The difference $(Q_s - Q)$, presented in the fourth column facilitates the estimation of the number of true forecasts for the first and second five year periods as well as for the whole forecasting interval. The fifth column illustrates the squared annual forecast margins of error, which were added together for the first and second five year periods and for the whole ten year period. The last two columns reflect the forecast results computed by the mean value.

It can be seen (Figure 5.10.1, Table 5.10.2) that seven out of ten of the values of the Danube River runoff were correctly predicted (two true values from 1979 to 1983 and five true values from 1984 to 1988). The forecast results of the Danube River runoff computed by the method of periodicities are somewhat better than the forecast using the mean value. The forecasting of runoff from the Danube River for

Table 5.10.2 Estimation of results of the training forecast of the runoff from the Danube River.

t (years)	Q (km³ yr⁻¹)	Q_s (km³ yr⁻¹)	$Q_s - Q$ (km³ yr⁻¹)	$(Q_s - Q)^2$ (km³ yr⁻¹)²	$Q_m - Q$ (km³ yr⁻¹)	$(Q_m - Q)^2$ (km³ yr⁻¹)²
1979	201.5	166.4	−35.1	1233	−29.8	887
1980	216.9	177.9	−39.0	1523	−45.2	2046
1981	205.4	178.0	−27.4	754	−33.7	1136
1982	177.2	167.0	−10.2	104	−5.4	29.7
1983	144.0	160.3	16.3	267	27.7	770
1984	157.7	167.1	9.4	88	14.1	197
1985	157.1	177.3	20.2	407	14.6	213
1986	163.3	175.9	12.6	160	8.5	71.4
1987	180.6	163.6	−17.0	289	−8.9	79.4
1988	164.0	156.0	−8.0	64	7.7	59.3

Arithmetic results for five- and ten-year intervals

	Mean (Q)	Mean (Q_s)	No. of true forecasts	Sums of squares of margins of error	No. of true forecasts	Sums of squares of margins of error
1979–1983	189.0	169.9	2	3882	1	4869
1984–1988	164.6	168.0	5	1010	5	621
Total	176.8	169.0	7	4892	6	5489

the same years using its mean value for 1840–1978 produced six out of ten true values (one true value from 1979 to 1983 and five true values from 1984 to 1988). The sum of the squared margin of error in the forecast when using the method of periodicities is somewhat smaller than that the mean value for the first five years and the whole ten year interval.

The shortcoming of the forecast is that for 1979–1983 the mean value of predicted runoff is significantly lower than its real mean value. The real mean runoff for 1979–1983 is much larger, while the computed runoff is a bit smaller than its mean value for 1840–1978. However, the years of maxima of the Danube River runoff (1980 and 1981), causing its increased mean value, and the year of its minimum (1983) were successively predicted by the respective maxima and minimum of the sum of periodicities. Consequently, the result of the training forecast using the method of periodicities may be considered satisfactory.

5.10.4
Conclusion

The methodology of long-range forecasting of the Danube River runoff has been developed. The time series (excluding the last verification interval of 10 years) were approximated by the sine functions successively with the unitary period step. The periods with the minimum sums of square differences respectively between the time series of the Danube River runoff and their approximation were determined and the sinusoids of 30, 22 and 5 years, whose correlation with time series is larger than 23%, were taken into consideration. When the sinusoids with these periods were successively summed the correlation of their sums with the time series were increased. These sinusoids show some indication of periodicities.

The sum of periodicities was integrated into the forecast equation and the forecast estimations of the Danube River runoff with the lead time up to 10 years were produced. The verification of this prediction by independent data shows a good correlation between the forecasted and observed values of the Danube River runoff.

References

1 Pencev, P.G. (2008) Danube River, Encyclopaedia Britannica (DVD).

2 Shlaymin, B.A. (1962) Super long range forecast of the Caspian Sea level. *Proceedings of the VGO*, **94** (1), 26–33 (in Russian).

3 Babkin, A.V. (2005) An improved model of the assessment of periodic changes in the level and elements of the water balance of the Caspian Sea. *Meteorology and Hydrology*, **N11**, 63–73 (in Russian).

4 Babkin, A.V. (2008) A long-term forecast technique for lake Ladoga water level and the Neva River runoff. *Proceedings of the Russian State Hydrometeorological University*, **N8**, 31–37 (in Russian).

5 Linnik Iu, V. (1962) *Method of the Least Squares*, Nauka, Moscow (in Russian).

6 Apollov, B.A., Kalinin, G.P. and Komarov, V.D. (1974) *The Manual for the Hydrological Forecasts*, Gidrometeoizdat, Leningrad (in Russian).

Part Three
Legal, Socio-Economic and Institutional Approaches

Transboundary Water Resources Management: A Multidisciplinary Approach, First Edition.
Edited by Jacques Ganoulis, Alice Aureli and Jean Fried.
© 2011 Wiley-VCH Verlag GmbH & Co. KGaA. Published 2011 by Wiley-VCH Verlag GmbH & Co. KGaA.

6
Legal Approaches

6.1
The Law of Transboundary Aquifers: Scope and Rippling Effects

Lilian Del Castillo-Laborde

6.1.1
Introduction

The management of groundwater entails special requirements. These should be taken into consideration when planning water use, assessing water quantity, controlling water quality and in issues concerning environmental protection, interrelation with recharge areas, effects on and resulting from surface waters and lakes, and the management of non-rechargeable aquifers. These features should be incorporated in the formulation of legal rules applicable to the proper management of transboundary aquifers, as far as they are shared resources whose utilization in one State impinges upon the volume and quality of the aquifer in another State or States.

6.1.2
Legal Principles for Transboundary Aquifers

The legal status of principles is particularly important, because the binding or nonbinding nature is the essence of any provision with the aspiration to become a 'legal' provision. Actually, it is inappropriate to consider a non-binding formulation 'legal'. When stating that a principle of behaviour is a 'legal' principle, its binding nature is implicit. In environmental and natural resources documents, the terminology used is sometimes not quite consistent and refers to 'principles' when it is uncertain whether or not they are legal rules. However, the content and legal status of 'principles' applicable to natural resources in general and transboundary aquifers in particular are of the utmost relevance for their utilization and management.

Principles, in a broad sense, are the common feature of State practice dispersed in space and time; they are an abstract formulation of various conducts from different origins, that is, national statutes, international agreements, judgments

Transboundary Water Resources Management: A Multidisciplinary Approach, First Edition.
Edited by Jacques Ganoulis, Alice Aureli and Jean Fried.
© 2011 Wiley-VCH Verlag GmbH & Co. KGaA. Published 2011 by Wiley-VCH Verlag GmbH & Co. KGaA.

and awards, resolutions of international organizations, contributions of academic bodies and so forth.

When accepted as mandatory, principles become legal rules, and they prevail over other rules because they cannot be derogated, as is the case with treaties. Some general customary rules are also regarded as legal principles, but their content relates more often to specific international legal matters. Principles and custom are the fabric that supports an agreement. Agreements are usually the application of general principles and customary rules to specific situations and for that reason are a significant contribution to their acceptance and consolidation.

Many efforts have been made to build international rules for the access and utilization of natural resources, a subject that traditionally has not been regulated by international law. Water is an exception because local rules and inter-jurisdictional compacts have been adopted from former times, when such rules had very specific purposes in terms of, for example, water allocation, navigation, irrigation, hydropower generation and timber floating. They were not meant to create general rules, although they could generate trends in State practice, which in turn could become customary law.

More recently, state practice at the national level, the works of relevant international organizations and the not so abundant number of agreements addressing transboundary aquifers were systematized through the codification and progressive development process developed by the International Law Commission (ILC), the United Nations General Assembly specialized body in charge of carrying out this task, which culminated in August 2008.

6.1.3
The Scope of the Draft Adopted by the UN General Assembly

In 2008, at its 60th session, the ILC adopted the Draft Articles on 'Shared Natural Resources: Transboundary Aquifers' [1]. The ILC developed the subject in several sessions, chaired by the Special Rapporteur Ambassador Chusei Yamada. The Draft Articles established rules for transboundary aquifers or aquifer systems (Article 1.a). Once approved by the ILC, the Draft Articles with their commentaries were submitted to the General Assembly. The legal provisions consist of a preamble and nineteen articles organized into four parts, the first being the 'Introduction', which deals with the scope and the use of terms and incorporates the technical definition of aquifers and aquifer systems, followed by the substantive sections addressing the 'General principles' in the second part and the 'Protection, preservation and management' of transboundary aquifers in the third part, and bringing together different issues in the fourth part under 'Miscellaneous provisions'.

The ILC recommended that the General Assembly should: (i) take note of the Draft Articles on the Law of Transboundary Aquifers in a resolution and annex the articles to the resolution; (ii) recommend that States concerned make appropriate bilateral or regional arrangements for the proper management of their transboundary aquifers on the basis of the principles enunciated in the articles; and (iii) also consider, at a later stage, and in view of the importance of the topic, the elaboration of a convention on the basis of the Draft articles.

On 11 December, 2008, the General Assembly took note of the Draft Articles on the Law of transboundary aquifers, commended them to the attention of governments, encouraged the States concerned to make appropriate bilateral or regional arrangements, and decided to include in the provisional agenda of its 66th session the item of 'The law of transboundary aquifers' [2].

The set of rules formulated by the ILC merges codification and progressive development of international law. The Draft articles are a joint effort of scholarly elaboration and governmental commentaries on the subject, roughly including accepted legal principles in their essential features.

6.1.4
Provisions Concerning Access

Above all, water-related agreements dealt with specific water uses and not with water access. Water access refers to the entitlement to use the resource and it was tackled at a later stage of legal evolution. Likewise, taking for granted that permanent sovereignty over natural resources is a well-established legal principle, the right of access was not addressed by the Convention on the Law of the Non-navigational Uses of International Watercourses opened by the United Nations General Assembly in 1997 for the signature of governments, which refers exclusively to the non-navigational *uses* of international watercourses.

It is only more recently, in the Draft articles on the Law of Transboundary Aquifers, that access to the resource is addressed. The sovereignty of aquifer States over the resources in their territories is reaffirmed. National sovereignty, however, does not denote unregulated access, but a type of access conditioned by the general rules of international law applicable to shared natural resources.

Entitlement to access must be shaped by the international legal framework, and the formulation in the Draft articles mirrors the previous resolutions of the United Nations on the subject, that is, General Assembly Resolution 1803 (XVII) of 1962 on 'Permanente Sovereignty over Natural Resources'. This rule is not opposed to joint projects that could be arranged between States sharing a transboundary aquifer or between an aquifer State and other non-aquifer States. Thus, Article 3 of the Draft articles states that:

Sovereignty of aquifer States: Each aquifer State has sovereignty over the portion of a transboundary aquifer or aquifer system located within its territory. It shall exercise its sovereignty in accordance with international law and the present articles.

6.1.5
Provisions Concerning Utilization

The international legal framework for the utilization of shared natural resources abides by the general rule of State sovereignty over its natural goods. Its sources are agreements, recommendations of international organizations, international and municipal case law and the contribution of doctrine. The enumeration of those rules, from the most general to the most specific, can be summarized as follows:

1) **Co-Operation between States**
    The general obligation to cooperate is the most widely stated principle, and one of the purposes of the United Nations which commits itself 'to achieve international cooperation in solving international problems of an economic, social, cultural or humanitarian character' (Article 1.3.UN Charter). The principle of cooperation implies a set of autonomous rules which, although not excluding other rules not mentioned in this paragraph, certainly include the set of rules described *ut infra*. Cooperation should be promoted in order 'to attain equitable and reasonable utilization and appropriate protection of their transboundary aquifers or aquifer systems' (Article 7).

2) **Equitable and reasonable utilization**
    The Draft describes the contents of this rule and enumerates what conditions aquifer States should fulfil to achieve equitable and reasonable utilization (Article 4). These conditions are rounded off by the description of the relevant factors that should be taken into account in order to determine in each specific case the equitableness and the reasonableness of the use (Article 5).

3) **Obligation not to cause significant harm**
    The duty not to cause significant harm to other States, especially but not exclusively aquifer States, is a duty of prevention to avoid any damage being caused, and it is a duty of reparation for any damages already caused (Article 6).

4) **Exchange of data and information**
    The regular exchange of data and information is a continuous process and aquifer States should establish a proper methodology for its implementation, either by means of a permanent transnational body or without it (Article 8).

5) **Institutional cooperation through bilateral or regional agreements**
    Agreements could set up permanent mechanisms supporting the exchange of data and information and such an organization would be a commendable achievement for aquifer States (Article 9).

6.1.6
Provisions Concerning Protection, Preservation and Management

The set of rules on the subject incorporates the general duty of aquifer States to protect their aquifers and preserve their ecosystems. Additionally, there are explicit duties relating to recharge and discharge zones as well as to pollution prevention and control. Furthermore, the obligations to monitor, to draw up management plans and to enter into consultation processes are also foreseen.

6.1.6.1 Provisions on Duties

1) **Protection and preservation of ecosystems**
    States should be aware of the quantity and quality of the water retained and released in the portion of the aquifer in their territory in order to protect the ecosystems dependent on them (Article 10).

2) **Recharge and discharge zones**
Aquifer States should monitor the recharge and discharge zones in view of the detrimental effects on other aquifer States and on the ecosystems (Article 11).

3) **Prevention, reduction and control of pollution**
The vulnerability and difficult remediation of aquifers impose on States the duty to prevent pollution sources and to adopt a precautionary approach on this issue (Article 12).

6.1.6.2 Provisions on Implementation Mechanisms

1) **Monitoring**
Aquifer States should monitor their aquifers in order to keep a record of their situation regarding quality and quantity, and they should also exchange that information with other aquifer States. This monitoring should be carried out in accordance with generally accepted parameters (Article 13).

2) **Management**
It is foreseen that management plans for transnational aquifers or aquifer systems should be made accessible to other aquifer States, and that consultations could be entered into regarding those plans. States could also consider the joint management of the resources (Article 14).

3) **Planned activities**
When aquifer States intend to carry out activities that could adversely affect the use of the resource by other aquifer States, they should enter into consultations, providing the other States with proper information about the future use. The notified States should be able to formulate observations and, as a consequence, a process of consultation and negotiation would develop to provide technical assessment (Article 15).

6.1.7
Provisions Concerning Technical Cooperation, Emergency Situations and Armed Conflict

The fourth and last chapter of the Draft articles entrusts States with the duty to promote scientific, technical and legal cooperation, amongst other matters, with developing aquifer States (Article 16) and concomitantly assist them in emergency situations (Article 17.4).

Aquifers and the installations and facilities for their utilization are protected by the rules of international law and international humanitarian law applicable to armed conflicts of international and non-international character (Article 18). Aquifer States, however, should not provide any data or information that could be 'vital to its national defence or security' (Article 19).

6.1.8
Other Rules Applicable to Transboundary Shared Resources

The ILC Draft articles do not include all the rules developed by international law for the utilization of transnational resources. Still there are, for instance, specific rules applicable to biotic or living resources, and others applicable to abiotic or non-living resources, although it is not always possible to isolate the living resource from the element in which it is immersed. These rules are dealt with in several conventions, such as the 1972 UNESCO Convention on the Conservation of Nature and Living Resources, the 1979 UNECE Geneva Convention on Long-range Transboundary Air Pollution and subsequent specific eight protocols, the 1991 Protocol to the Antarctic Treaty on Environmental Protection, the 1992 UNECE Helsinki Convention on the Transboundary Effects of Industrial Accidents, the 1992 UNECE Helsinki Convention on the Protection and Use of Transboundary Watercourses and Lakes and 1999 Protocol on Water and Health, the 1992 UN Convention on Biological Diversity, the 1994 Convention to Combat Desertification, the 1998 UNECE Aarhus Convention on Access to Information, Public Participation in Decision-making and Access to Justice in Environmental Matters, the 2002 Dakar Senegal River Water Charter, the 2003 Maputo African Convention for the Conservation of Nature and of Natural Resources, amongst others, which for brevity will not be listed in this outline. Nonetheless, some general rules will be mentioned here as they may be considered implicit in the provisions already described.

1) **The principle of public participation and governance of decision-making mechanisms**
   The fact that international organs are necessary for the utilization of shared international resources has already been incorporated in different agreements for the establishment of working groups and permanent commissions.

2) **The principle of compliance with international commitments and peaceful settlement of disputes**
   Joint commissions or other agencies have been set up for monitoring the implementation of agreed quantity and quality standards. Access to national courts and to international adjudication bodies and tribunals is also part of different agreements for the utilization of transboundary resources. Different mechanisms for the peaceful settlement of disputes, including negotiation and other means of dispute avoidance are included in instruments for shared resources regulation.

6.1.9
Case study: The Guarani Aquifer System

The Guarani Aquifer System (GAS) spans the territory of Argentina, Brazil, Paraguay and Uruguay over an area of approximately 1 200 000 km^2 and is made up of one or more different aquifers. It was subject to several studies under an internationally supported programme (Global Environment Facility, Organization of American States) that concluded in February 2009. In 2004, the Draft Declaration of Basic Principles and Action Guidelines was approved (Resolution N° 9/2004, Steering Committee – Guarani Aquifer System Project). It was subsequently decided to draw up an agreement, aiming at the protection of the resource, which was concluded on August 2, 2010. The Agreement invokes, as relevant backgrounds, Resolutions of the United Nations General Assembly – A/RES/1803(XVII) of December 14, 1962, on 'Permanent Sovereignty over Natural Resources' and A/RES/63/124 of December 11, 2008, on 'The Law of Transboundary Aquifers – as well as the 1972 Stockholm

Declaration on Human Environment, the 1992 Rio Declaration on Environment and Development, and the Environmental Framework Agreement adopted in 2001 by the Parties of the Common South American Market (MERCOSUR).

The 22 provisions of the Guarani Agreement highlight the right of access to the resource by the aquifer territorial States (Articles 1 and 2) and the management capacity of the member Parties of the Guarani resources at the national level, based on rational and sustainable utilization criteria, within the framework of the 'no harm to other States' principle (Article 3).

Above all, the Parties agree to promote the conservation and environmental protection of the Guarani Aquifer System and multiple, rational, sustainable and equitable use of its water resources (Article 4). The Parties commit themselves to abide by the rules and principles of international law with regard to the studies, activities and works they could carry out in the GAS (Article 5) and specifically with regard to the rule of 'not to cause significant harm' to other aquifer States (Article 6). In case harm occurs, the State where the harm originates shall eliminate or reduce the damage (Article 7). The duty to exchange information on studies, activities or works having potentially transboundary effects is also established (Articles 8 and 9), as well as the mechanism for consultation and negotiation, which are not to last more than six months, if one or more State Parties consider that significant harm could arise (Articles 10 and 11).

A Commission of the Parties for the implementation of the GAS agreement is to be set up within the institutional framework of the La Plata Basin Treaty (1969) (Article 15). Scientific cooperation amongst aquifer States is also foreseen (Articles 12 and 14) notwithstanding the local projects each Party could develop at the national level (Article 13). A procedure for dispute settlement is established, including an arbitral mechanism that will be set out in an additional protocol (Articles 16, 17, 18 and 19). The agreement admits no reservations (Article 20), is unlimited in time (Article 21.2) and may be denounced at any time (Article 22). The agreement requires ratification by the Parties (Article 21.1), Brazil being the depositary of the original instrument (Article 21.3). The agreement does not aim to establish a joint management of the GAS, but to create through the future Commission a permanent coordination body with a legal framework.

The Guarani Aquifer agreement provisions contain the legal principles addressed by the Draft Articles concerning access and utilization of the resource. Regarding the right of access, Articles 1 and 2 of the Guarani Agreement state the access principle described by Article 3 of the Draft Articles. Furthermore, the driving principle of co-operation between States established by Article 7 of the Draft Articles is adapted in Article 12 of the Guarani Aquifer Agreement. With regard to the rule of equitable and reasonable utilization drawn by Articles 4 and 5 of the Draft Articles, this is included in Article 4 of the Guarani Agreement. In addition, the duty not to cause significant harm to other States or to the environment underlined by Article 6 of the Draft Articles is established as well by Article 6 of the Guarani Agreement. Likewise, the duty to exchange data and information expressed in Article 8 of the Draft Articles is also established by Articles 8, 13 and 14 of the Guarani Agreement. Some principles of the Draft Articles may be traced in the Guarani Agreement provisions, while some others are not included. However, the parallelism of the core legal principles adopted by both instruments is remarkable.

6.1.10
Conclusions

The description of the above provisions is merely one of several possible system-atizations of rules for transboundary resources. The crystallization of principles into a set of rules is the challenge that any codification effort faces. Statements of international organizations and academic bodies as well as specific agreements incorporate legal rules, but these scattered illustrations can only be transformed into general rules of worldwide acceptance if the practice is coherent, compatible and unchallenged. Thus, the relevant conclusion is that legal rules exist for accessing, utilizing and protecting transboundary aquifers, rules that are not dependent on but would benefit from the culmination of the codification process.

However, the weak edge of any international rule is that it has to be implemented at a national level. Furthermore, the coordination of different national administrations requires constant and progressive efforts to be made. Reaching agreement on shared natural resources is a difficult target to achieve and much has to be done before results are accomplished. In the specific case of transboundary aquifers, priority should be given to raising public and private awareness with the cooperation of national and international agencies and academia.

Concluding the Draft Articles on the Law of Transboundary Aquifers has been a significant achievement of the ILC in particular, and of the UNESCO commitment as a United Nations Specialized Agency and catalysing centre on transboundary aquifers in general. States, international organizations and courts can now rely on a set of rules adopted by the highest codification organ of the United Nations and annexed to General Assembly resolution A/RES/63/124 for further consideration. The Draft Articles will have immediate effects on the specialized doctrine and will help to develop improved national policies in aquifer States. The recent Guarani Aquifer System Agreement supports this view, incorporating in a binding instrument concomitant legal principles for transboundary aquifers. Moreover, the Guarani Aquifer System Agreement implements institutional co-operation through regional agreements, as promoted by Article 9 of the Draft Articles. In fact, not only this provision but a considerable number of the Draft's rules have now become the international law rules for the Guarani Aquifer System, one of the world's biggest regional aquifers, thus reaffirming both the rippling effects of the Draft Articles and the legal consistency of its rules.

References

1 International Law Commission (ILC) (2008) Sixtieth Session, Draft articles on 'shared natural resources: transboundary aquifers, http://untreaty.un.org/ilc/sessions/60/60docs.htm (accessed 09 November 2010).

2 United Nations (2009) Resolution A/RES/63/124, http://daccess-dds-ny.un.org/doc/UNDOC/GEN/N08/478/23/PDF/N0847823.pdf?OpenElement (accessed 09 November 2010).

Further Reading

McCaffrey, S.C. (2009) The international law commission adopts draft articles on transboundary aquifers. *American Journal of International Law*, **103** (2), 272–293.

Mechlem, K. (2009) Moving ahead in protecting freshwater resources.

Leiden Journal of International Law, **22** (4), 801–822.

Vick, Margaret J. (2008) International water law and sovereignty. A discussion of the ILC Draft Articles on the Law of Transboundary Aquifers. *Pacific McGeorge Global Business & Development Law Journal*, **21** (2), 191–221.

6.2
Water Management Policies to Reduce over Allocation of Water Rights in the Rio Grande/Bravo Basin

Samuel Sandoval-Solis, Daene C. McKinney and Rebecca L. Teasley

6.2.1
Introduction

The Rio Grande/Bravo basin has a severe problem of over allocation of water rights, especially in the Rio Conchos tributary of Mexico, due to a misunderstanding of the basin's hydrology and incorrect water rights allocation policies in the 1970s and 1980s. Historically, the hydrology of the basin has shown periods of 25 years with plenty of water followed by extended drought periods of about ten years. During the 1970s and 1980s, which was a wet period, there was an increase in the allocation of water rights in the basin, mostly in the irrigation districts. Later, from 1992 to 2002, an extended drought period occurred, threatening not only the water users, but also the international obligations of water delivery in the Treaty of 1944 between Mexico and the USA. In fact, during the drought period of 1992 to 2002, there was a deficit of water delivery for the treaty obligations from Mexico to the USA. In 2007, this deficit of water was completely paid back, but now there is a consciousness about the over allocation of water [1] and the necessity to solve this problem before another drought period. Nowadays, there is not enough water in the basin to meet the institutional and international obligations that the Rio Grande/Bravo basin is subject to.

6.2.2
Buying Back of Water Rights

In 2003, the Mexican Ministry of Agriculture, Livestock, Rural Development, Fisheries and Food (SAGARPA from its acronym in Spanish) announced the PADUA programme [2]. The objective of this programme is to preserve the productivity and competitiveness of irrigation districts through the permanent buy-back of water rights conferred to irrigation districts that under drought conditions would be impossible or hard to supply, for either economic or hydrological reasons. The PADUA programme reduces the water rights in the basin, trying to match the demand with the availability of water under different hydrologic conditions. In 2008, according to NAFTA (North America Free Trade Agreement) regulations, free commercial tariffs exist between Mexico and the USA for agricultural products. Because of this, productivity and competitiveness are key objectives in the PADUA programme. Table 6.2.1 shows the volume of water bought back [3, 4]. Surface and groundwater rights were bought at $148 and $185 US dollars per 1000 m^3, respectively. Figure 6.2.1 shows the location of these irrigation districts.

6.2.3
Scenarios

Two scenarios are analysed in this chapter, a baseline and a PADUA programme scenario. The baseline scenario represents the actual water management in the basin considering the water demand before the PADUA programme; this scenario can also be considered a no-action scenario. In the PADUA scenario, the water demands after the PADUA programme are used to evaluate the efficiency of the PADUA programme. A basin simulation model has been developed to perform the evaluation of this policy.

6.2.4
Simulation Model

A simulation model of the Rio Grande/Bravo basin has been developed using the Water Evaluation and Planning System software (WEAP) [5]. The model was constructed jointly with the collaboration of the Natural Heritage Institute (NHI), the Stockholm Environment Institute (SEI), the Centre for Research in Water Resources (CRWR), the Texas Commission on Environmental Quality (TCEQ), the International Boundary and Water Commission (IBWC), the National Water

Table **6.2.1** Buying back of water in the PADUA programme.

Irrigation district	Water source	Water demand per year		Water bought back (million m^3)	Investment ($million USD)[a]
		Before PADUA (million m^3)	After PADUA (million m^3)		
005 Delicias	Surface	941.6	850.3	91.3	13.53
	Groundwater	189.0	170.7	18.3	3.39
090 Bajo Rio Conchos	Surface	85.0	63.7	21.3	3.16
	Groundwater	—	—	—	—
			Total	130.9	20.07

a) Monetary exchange 13.5 Mexican pesos per US dollar.

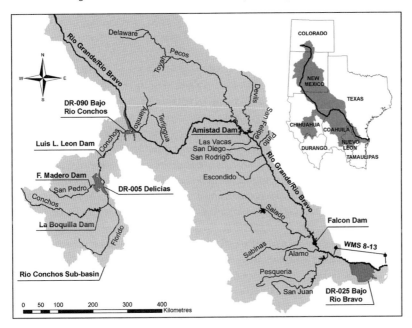

Figure 6.2.1   Rio Grande/Bravo Basin.

Commission of Mexico (CONAGUA) and the Mexican Institute of Water Technology (IMTA). The model contains the logic to separate the water between the USA and Mexico as required in the Convention of 1906 [6] and the treaty of 1944 [7]. An international reservoir accounting system was created in the model for tracking the inflows, outflows and storage of both countries in the international dams, Amistad and Falcon, in accordance to the 1944 treaty. For the water users in the USA, water is allocated according to the Texas Watermaster logic [8] implemented for the USA storage in the international reservoirs. For water users in Mexico, water is allocated according to the National Water Regulation [9] implemented for storage in the international reservoirs and reservoirs located on Mexican territory. A 60 year period of hydrologic analysis is considered, from October 1941 to September 2000. Further information on construction, calibration and validation of the model can be found in Danner *et al.* [5].

6.2.5
Performance Criteria

Four performance criteria are used to evaluate water management in the Rio Grande/ Bravo over the simulation period: reliability, resilience vulnerability and maximum deficit [10]. These performance criteria are selected because they express the period of time the system is working properly (reliability), the severity of the system's failures (vulnerability), the worst failure (maximum deficit) and how fast the system recovers from a failure (resilience).

Reliability is the frequency that water demand was fully met during the simulation period, or in other words, the probability of no-deficit in the water demand during the hydrologic period of analysis [11, 12].

Resilience is the probability that the system recovers from a period of failure, which in this case is a deficit in water supply. According to Hashimoto *et al.* [11], resilience is the probability that a year of no-deficit follows a year of deficit in the water supply for a particular water user.

Vulnerability is the expected value of the deficits, in other words, it is the average of the water supply deficits experienced over the simulation period [11]. Vulnerability expresses the severity of the failures. We use the dimensionless vulnerability, dividing the average annual deficit by the annual water demand for any particular water user [12].

The maximum deficit is of particular importance as a performance criterion in this basin. This parameter allows a comparison of changes during the worst water supply period over the simulation period.

There are 6989 water users in the Rio Grande/Bravo basin. For simplicity we focus our attention on the 1944 treaty obligations and the five largest water users in the basin, which are irrigation districts 004 Delicias (DR-005 Mexico), 025 Bajo Rio Bravo (DR-025 Mexico), 090 Bajo Rio Conchos (DR-090 Mexico) and Watermaster Section 8-13 (WMS 8-13 USA);

The treaty of 1944 between the USA and Mexico specifies the obligations for water delivery from Mexico to the USA. In summary, Mexico has to deliver $^1/_3$ of the flow reaching the Rio Grande/Bravo from six Mexican tributaries (Conchos, Arroyo Las Vacas, San Diego, San Rodrigo, Escondido and Salado), provided that this third shall not be less than 431.721 million m^3 per year, as an average amount in cycles of five consecutive years. Treaty cycles can expire earlier than five years if the conservation capacity assigned to the United States in both international dams, Amistad and Falcon, is filled with water belonging to the United States, in which case all the previous deficits are considered fully paid. Deficits can occur only when a cycle lasts the full five years and the $^1/_3$ of the flow reaching the Rio Grande/Bravo from the six Mexican tributaries is less than 2158.605 million m^3 per cycle (5×431.721 million m^3 per year). Any deficit must be made up in the following treaty cycle. The performance criteria are evaluated according to the previous description of treaty obligations.

6.2.6
Results

Table 6.2.2 shows the results of the performance criteria for the selected water users for both the baseline and the PADUA scenarios. For irrigation district DR-005 Delicias, there is an increase in reliability from 62% to 70%; the recovery from deficit periods (resilience) is also faster from 30% to 33%; the expected deficit (vulnerability) decreased from 50% to 47% and the maximum deficit does not change.

For irrigation district DR-025 Bajo Rio Bravo, there is a slight decrease in reliability from 88% to 87%; recovery from deficit (resilience) is faster from 14% to 25%; the expected deficit (vulnerability) decreased from 37% to 33% and the maximum deficit also decreased, from 56% to 55%.

Table 6.2.2 Summary of results.

Water user	Scenario	Demand (million m³ per year)	Reliability (%)	Resilience (%)	Vulnerability (% demand)	Max. deficit (% demand)
DR-005	Baseline	1130.5	62	30	50	97
	PADUA	1021.0	70	33	47	97
DR-025	Baseline	860.5	88	14	37	56
	PADUA	860.5	87	25	33	55
DR-090	Baseline	85.0	98	100	0.2	0.2
	PADUA	63.7	100	100	0	0
WMS 8-13	Baseline	1116.9	95	40	18	29
	PADUA	1116.9	95	40	20	29
Treaty obligations	Baseline	2158.6[a]	67	75	28	42
	PADUA	2158.6[a]	83	75	31	55

a) Demand per cycle.

For irrigation district DR-090 Bajo Rio Conchos, there is an increase in reliability from 98% to 100%; recovery from deficit periods (resilience) is the same (100%); the expected deficit (vulnerability) decreased from 0.2% to 0% and the maximum deficit also decreased from 0.2 to 0%. These are all positive results for DR-090.

For irrigation district Watermaster Section 8-13, reliability, recovery from deficit periods (resilience) and the maximum deficit are all the same; the expected deficit (vulnerability) increased from 18% to 20%.

For the treaty obligations, there is an increase in reliability from 67% to 83%; recovery from deficit periods (resilience) is the same; the expected deficit (vulnerability) increased from 28% to 31% and the maximum deficit increased from 42% to 55%.

6.2.7
Conclusions

The PADUA programme improved the water supply not only for the water users located where the programme was applied, but also for water users downstream. Benefits for the water users (DR-005 and DR-090) where the PADUA programme was applied include more water over time, fewer expected and smaller maximum deficits and faster recovery from deficit periods. As a result of the PADUA programme, water users downstream were subject to changes in their water supply. Benefits for water users downstream (DR-025 and WMS 8-13) include smaller deficits and faster recovery from deficit periods. For treaty obligations there will be fewer periods of deficit, but the expected and maximum deficits will be more frequent and bigger.

Furthermore, this analysis showed that policies should not be analysed locally where the policies are applied. In this case, local actions such as the buying back of water rights resulted in both benefits and detriments for other water users. In complex systems, such as the Rio Grande/Bravo basin, analysis must be done on the basin level due to the close interaction and dependence between all water users.

Acknowledgements

Special acknowledgements are given to the National Council of Science and Technology of Mexico, CONACYT, for sponsoring the first author of this chapter. Partial funding for this research has been provided by the US Department of Agriculture, the Mexican Institute of Water Technology (IMTA) and the National Heritage Institute (NHI).

References

1 CONAGUA – Comisión Nacional del Agua (2008) Acuerdo por el que se da a conocer el resultado de los estudios de disponibilidad media anual de las aguas superficiales en la cuenca del Rio Bravo, Diario Oficial de la Federación, 29 de Septiembre de 2008.

2 SAGARPA – Secretaria de Agricultura Ganadería, Desarrollo Rural, Pesca y Alimentación (2003) Reglas de operación

del programa de adquisición de derechos de uso del agua, Diario Oficial de la Federación, 12 de Agosto de 2003.

3 SAGARPA – Secretaria de Agricultura Ganadería, Desarrollo Rural, Pesca y Alimentación (2005) Informe de Beneficiarios del programa PADUA 2004.

4 SAGARPA – Secretaria de Agricultura Ganadería, Desarrollo Rural, Pesca y Alimentación (2007) Informe de Beneficiarios del programa PADUA 2005-2006.

5 Danner, C.L., McKinney, D.C., Teasley, R.L. and Sandoval-Solis, S. (2006) Documentation and Testing of the WEAP Model for the Rio Grande/Bravo Basin, CRWR Online Report 06-08, revised February 2009. http://repositories.lib. utexas.edu/bitstream/handle/2152/7011/ crwr_onlinereport06-08.pdf?sequence=2

6 IBWC – International Boundary and Water Commission (1906) Convention between the United States and Mexico. Equitable distribution of the waters of the Rio Grande, May 21st, 1906, Washington DC.

7 IBWC – International Boundary and Water Commission (1944) Treaty between the United States and Mexico. Utilisation of waters of the Colorado and Tijuana Rivers and of the Rio Grande, February 3rd, 1944, Washington DC.

8 TCEQ – Texas Commission on Environmental Quality (2006) Operation of the Rio Grande. Texas Administrative Code, Title 30, Part 1, Chapter 303, Subchapter C.

9 CONAGUA – Comisión Nacional del Agua (2004) Ley de Aguas Nacionales y su Reglamento, México DF.

10 Loucks, D.P. and van Beek, E. (2005) Water resources systems planning and management, in *Studies and Reports in Hydrology*, UNESCO Publishing, 321 pp.

11 Hashimoto, T., Stedinger, J.R. and Loucks, D.P. (1982) Reliability, resiliency and vulnerability criteria for water resource system performance evaluation. *Water Resources Research*, **18**(1), 14–20.

12 McMahon, T.A., Adeloye, A.J. and Sen-Lin, Z. (2006) Understanding performance measures of reservoirs. *Journal of Hydrology*, **324**, 359–382.

Further Reading

FAO-SAGARPA Food and Agriculture Organisation – Secretaria de Agricultura Ganadería, Desarrollo Rural, Pesca y Alimentación (2005) Informe de Evaluación Nacional. PADUA 2004, Octubre 2005.

6.3
Interstate Collaboration in the Aral Sea Basin – Successes and Problems

Viktor A. Dukhovny and Galina Stulina

6.3.1
Introduction

The region of Central Asia is a typical example of an arid and semi-arid region with a serious water deficit, where for thousands of years the well-being and survival of the population has been based on irrigated agriculture and complicated water systems.

In the twentieth century the trend was to utilize natural resources including water as fully as possible and huge engineering and social infrastructures for the growth of irrigated lands, hydropower production and common water use were developed. All these system existed in a single state, namely the USSR, which was administrated in a 'top-down' manner. Nowadays, the new conditions of independent states, the transfer to market economies, the destruction of previous connections and the need to create new ones have resulted in the need to establish mutual collaboration for the use of transboundary water in the Aral Sea Basin being largely ignored, which has had tragic environmental consequences for the Aral Sea coast and the sea itself.

6.3.2
Achievement of Collaboration

Collaboration between just two states is no simple matter, but when five players are involved and the principal goal is to manage water in such a way that the needs of millions of people are respected, then the situation is even more complicated. Having understood the huge weight of the common responsibility in this matter, the five national water organizations, all members of ICWC, gave priority to creating tools and mechanisms that would support successful collaboration. In this particular case, mutual understanding and cooperation are facilitated by the fact that these countries have the same water management roots, as in pre-independent times they shared a common administrative framework and educational system and also worked together as one in the past. The leaders of the five states reconfirmed their decision to work together on water management issues in their agreements of 1993, 1999 and 2002. The situation today, however, requires formally specified common action plans and tools of support:

1) In its initial agreement and later in its declaration of principal provisions of mutual work, the ICWC agreed to maintain the same water allocation as had been the case between the republics of the former Soviet Union, to make decisions only with a strong consensus, to uphold obligations to agree on measures concerning transboundary rivers, and to collaborate and cooperate on relevant issues to prevent damage to neighbouring countries. The executive bodies of the ICWC (Basin Water Organizations (BWOs) Amudarya and Syrdarya and the Scientific Information Centre (SIC) ICWC) were appointed to be responsible for implementing this declaration.

2) The ICWC elaborated and implemented a single system of planning, monitoring and operation of water flow in transboundary rivers that was accepted by all states. Water allocation between states, based on hydrological forecasts and agreed shares, is approved on an annual basis by the ICWC following the approved proposal from the executive bodies (BWOs and SIC ICWC).

3) Together with the Swiss Development Cooperation (SDC), the ICWC created a regional information system called the Central Asia Regional Water Information Base Project (CAREWIB) that accumulated a huge scope of knowledge and information, including publications and experience exchanges, on a specially designated web portal (www.cawater-info.net). Part of this portal is an analytical window that shows on a ten-day basis the exact situation concerning a particular river, and how executive bodies and other principal stakeholders (owners of reservoirs) are following the appointed plan and schedule. This helps to build trust and transparency and allows partners to check up on each other. This

Information System (IS) was so successful, with more than 1000 hits each day, that States were encouraged to use the same approach to create a National Information System (NIS) to which staff of SIC ICWC transferred their own software, interface, methods and accumulated data, and subsequently organized training sessions for national providers.

4) The establishment of a broad training system is the next tool needed for closer interaction between representatives of the different states. Courses held at the headquarters of ICWCs Training Centre (TC) in 1999 in Tashkent with the support of the Canadian International Development Agency (CIDA) and McGill University focused on increasing the level of knowledge of high- and middle-level specialists. Over a five year period more than 2400 people attended specific courses on IWRM, international water law, conflict regulation and advanced irrigation methods amongst others. On this basis a training network was developed, located in seven different places suitable for spreading this education to local staff, to the Water User Association (WUA) and to farmers. The number of people receiving training in 2007 reached 9000. The training focuses on interactive education that allows representatives from different states to develop a single approach and vision.

5) Tools facilitating IWRM implementation are needed. As SIC ICWC's experience in the Ferghana Valley shows, IWRM is based on a number of principles, which can produce an impressive result only if applied altogether. The well-known hydrographic principle cannot by itself be effective unless:

   – it addresses all kinds of water (surface, ground and return);
   – representatives of water and nature users at all water hierarchical levels and in all sectors take an active part in management, including water delivery planning, its correction, financing, repair and maintenance and, finally, improvement.

Basin Water use Councils, Water use Councils for Canal and their Sections, and WUA Councils and their groups will work together with water management organisations (WMOs) to determine appropriate regimes and order of water use, establish control, set up a systems for enforcement and arbitration amongst water users, and facilitate the contribution of their creative capacity and knowledge to WMO activities. To achieve this end, the process would benefit from expert assistance. Social mobilization is an integrated part of water management and should be accompanied by the training and education of water users. However, it is necessary to apply other mechanisms as well, such as blocking payments for water, bonuses for water saving both for water users and WMOs, control of public water supply costs, subsidies to farmers for application of new irrigation techniques, and so on.

Ways to unite farmers and WUAs may be developed, by recommending methods of water use and distribution, accounting and watering in strict accordance with climatic parameters.

It is particularly important to aim at coordinated management and development at the interstate level, since any effort at the local level may fail if water supply from the interstate sources is unstable and depends on various subjective factors. In this context, special attention should be paid to intensified construction of waterworks facilities for energy and runoff regulation purposes. Undoubtedly, they play a positive role in increasing water supply, controlling floods, guaranteeing water supply during dry years, and producing energy that ensures sustainable electricity supply and creates the possibility of energy export. Along with the positive results from establishing efficient management and regulation rules, a considerable contribution may be expected from the implementation of a Supervisory Control and Data Acquisition (SCADA) system in all hydraulic structures and gauging stations on transboundary rivers, as well as from the updating of short- and long-term forecasts that still produce a wide variation of flow, thus making it very difficult to draw up tentative water distribution plans.

The results of such implementation in the Fergana Valley are impressive. From 2003 to 2008 water delivery to the South Fergana canal at the 86th km decreased from 1060 million m^3 to 680 million m^3. All measures undertaken within the framework of water cooperation were tangible and yielded results. Despite the mass media's monthly predictions of water wars in the region, in reality water management and governmental organizations successfully and jointly handle and ensure the normal supply and use of water in the basin. ICWC's system of unified annual planning and control, information exchange, training and management improvement, including the application of new automatic tools, efficient work by executive bodies and close interaction with the national Ministries of Agriculture and Water Resources and the Committees for Water Resources, have made it possible over the last 15 years to successfully handle three extremely dry years and five humid years without any conflict arising. A continuous reduction of unit water use in irrigation is character-istic. Gross withdrawal for irrigation was 14 000 m^3 per hectare in 1990, which decreased to 12 300 m^3 in 2006 and more recently to 11 500 m^3. Undoubtedly, activities related to transboundary water resources face difficulties, and sometimes meet with failure. There are many issues to be decided upon at the interstate, national and local levels. It is necessary to pay constant attention to enhancing the transition from the management of water alone to the joint management of both water and its demand. However, the most important issue is to focus on the future stable functioning of the water sector in Central Asia, taking into account both regional and nationwide tendencies and national and intersectoral changes in water use.

6.3.3
Future of the Region

Future equilibrium in the region is jeopardized by substantial destabilizing factors:

1) population growth, which means that in 2030 the region will cross the UNs line indicating water shortage with 1700 m^3 of water per capita a year compared to almost 2500 m^3 at present;
2) expected climate change, which, according to recent predictions, will abruptly change the current optimistic forecasts of water availability, especially for the Amudarya river, with a probable 30% decrease in surface water resources by 2030;
3) development of Afghanistan and growth of withdrawals in this country.

The transformation of water into a good through hydropower production poses a particular risk. Huge energy potential in flow formation zones may create favourable conditions for overcoming problems of decline in dry years by shifting to long-term regulation. On the other hand, this may create problems in cases where the regime is purely for energy generation, as shown by ICWC in the operation of Roghun. To avoid such problems, the joint use of reservoirs within the region and the development of potential hydrocapacities for the increased production of electric energy are recommended. Moreover, one-sided use of reservoirs for energy production through the lowering of water levels down to dead-storage capacity (Nurek practices) leads to inefficient use of water in reservoirs. This again raises an issue about the establishment of a water-energy consortium to provide a financial mechanism for sustainable supply of fuel and energy resources in order to ensure the required water regimes. Lessons may be learned from South African countries, which organized the development of their hydro-energy resources in this way.

The future survival of the region's countries lies, first of all, in strict pursuance of the international water law. An excellent example was shown by Uzbekistan, which was the first country in the region to ratify two international conventions of 1992 and 1997. Apart from Uzbekistan, only Kazakhstan ratified the 1992 convention. This indicates that some countries in the region are trying to reserve the right to act autonomously regarding the future management of water in the region. This is corroborated by the artificial delays in signing several already approved interstate documents on the Syrdarya River. On the one hand, this demonstrates a misunderstanding of international law itself, which is effective irrespective of whether one country has ratified a global legal instrument or not, and, on the other hand, it indicates short-sightedness since, given the region's closely interlinked and interdependent water system, no one country can consider itself absolutely untouchable. Moreover, the principles 'do not harm' and 'those who do harm should pay' are in force under common law and not only under international water law.

Guidance on the water sector in general and on water protection should create suitable conditions for creating a solid basis of equitable water allocation through:

- elaboration by all countries of the details of the agreements and rules for regulation and management of water resources in Amudarya and Zarafshan, including small transboundary rivers, and completion of the agreements on the Syrdarya River; moreover, similar to the Water use Councils for Canals, Basin Water use Councils are established at BWOs and are made up of major stakeholders (Hydroelectric Power Plants, Delta management, etc.);
- development and approval of national water strategies setting a national approach to water development and conservation focusing on national priorities, IWRM implementation, regional limitations and general objectives;
- forming of a financial mechanism for efficient water use, including charges for services and pollution, direct and cross-cutting subsidies, bonuses and incentives

for rational use, water market elements, and priorities of investments and their feasible sources;

- institutional measures for efficient water use, including the establishment of a large-scale network of extension services.

6.4
Kidron Valley/Wadi Nar International Master Plan
Richard Laster

6.4.1
Introduction

The Kidron Valley/Wadi Nar runs from the city of Jerusalem through the Judean Desert to the Dead Sea. For thousands of years, the valley has supplied local peoples with water, prompting the growth of several civilizations in this area. Many of the Middle East's most famous cultural and historic sites can be found here, including religious sites, ancient tombs, underground watercourses, monasteries and beautiful desert landscapes.

But pollution and neglect, together with an increase in population, has had major health, environmental and economic consequences. Today one of the great centres of civilization serves as a conduit for raw sewage. Water and drainage infrastructure is primitive at best. Historic sites are neglected; their cultural value depreciated by poor environmental practices and strife in the region. Endemic species are disappearing. Much of the fertile land in the valley is not being farmed and landowners have been forced to find other ways to make a living. The resulting pollution of the groundwater in the area endangers existing water sources. The polluted surface water constitutes a health hazard to Dead Sea bathers and local residents.

For the past 30 years, Israelis and Palestinians have deliberated solutions to the rehabilitation and beautification of the Kidron Valley/ Wadi Nar area, including building and operating a joint sewage purification plant. There have been a number of 'historic meetings' such as the one between Teddy Kollek, the Mayor of Jerusalem and Elias Freig, the Mayor of Bethlehem in 1991. Contracts have been drafted and signatures appended but no solution has been reached. As decided in the Oslo Accords of 1995, Israel and the Palestinian Authority set up a Joint Water Committee (JWC) to resolve water issues in a formal setting. Suggestions have been made to the JWC for a solution to the sewage problem in the valley, but a decision has yet to be made. In addition, unlike Europe where borders are coming down, here borders are going up, making basin planning even more difficult.

6.4.2
Development of the Master Plan

Things are changing, however. During the last few years, scholars who have studied the entire basin have met with professionals and politicians from both sides of the border to discuss the problems of the valley and to present the idea of a Master Plan. A joint Master Plan steering committee has been set up, made up of Palestinian representatives from the Palestine Hydrology Group (PHG), Bethlehem University and Al-Quds University, and Israeli representatives from the Peres Centre for Peace, the Jerusalem Institute for Israel Studies, the Milken Institute, the City of Jerusalem and the Dead Sea Drainage Authority.

The proposed collaborative work between Israeli and Palestinian experts in the Kidron Valley/Wadi Nar represents an unprecedented opportunity for setting up a framework for collaborative integrated basin management between the two parties for a shared water resource in a place of enormous historical, cultural and ecological importance and beauty. There is no alternative to collective action when managing a shared international basin. This was recently recognized by the EU Water Framework Directive, which requires water management on the basin scale regardless of political borders. Nature, ecosystems and the intricate interaction between water, climate, soil, flora and fauna can only be protected by collective action.

This is not theoretical research, but an actual planning process, based on the successful model of the Yarqon River Master Plan in the Tel Aviv region, which converted the Yarqon River and its environs from an environmental nuisance into a beautiful recreational area. The steering committees will create and guide a team of Israeli and Palestinian professionals, who will act in concert to develop the Master Plan. The team will consist of a plan coordinator, planners, hydrologists, engineers, archaeologists, ecologists, economists, jurists, sociologists, agronomists and public awareness specialists. Additional experts will be brought in as needs change. The selection of the team members will be done jointly by the steering committee in order to create a balanced, realistic and useable Plan.

The major beneficiaries of the Master Plan are the valley residents themselves. Improved planning, land usage and infrastructure will make the Kidron Valley/Wadi Nar a better place to live and visit. When implemented, local farmers will benefit from effluent which, when treated sufficiently, can provide irrigation for crops and animals in an area with little freshwater resources and limited jobs. Tourists, hikers and other visitors will benefit from the restoration of the natural beauty of the region, its cultural and historic sites, and from increased access. Increased tourism will in turn create additional employment opportunities.

6.4.3
Descriptions of the Master Plan and its Effectiveness

The Master Plan will present a number of possible scenarios based on current and expected land and water use and will detail how best to utilize the resources of the valley while preserving its historic value. In addition, these discussions will bring together local and regional stakeholders, as well as planners and local authorities, to discuss their visions of the valley. This joint effort will exemplify the ability of both parties to coexist and work together to solve mutual problems.

Like most things in the Middle East, solving problems always has political undertones. Creating a master plan for the valley is interpreted by some people on both sides as recognition of Israeli or Palestinian sovereignty. For 40 years, a temporary political situation has been in place in the region, and deterioration of the Kidron Valley/ Wadi Nar is just one result. The researchers, both Israeli and Palestinian, recognize that a master plan based on ecological, historical, physical, economic and geographical terms agreed upon by both sides will serve the best interests of the valley, regardless of present or future political sovereignty. The creation of the first master plan for a joint Israeli–Palestinian water source will be a breakthrough in Palestinian–Israeli cooperation. Once completed, it can serve as a blueprint for similar plans for the other 15 cross-boundary waterways that are desperately in need of rehabilitation.

As a starting point for reclamation, a survey of the Israeli and Palestinian stakeholders involved is of vital importance. The survey will identify the issues, interests, priorities and characteristics of the various stakeholders and stakeholder representatives, and the relationships between the different stakeholders and stakeholder groups. The information gathered by this research will inform the planners as to the needs of the stakeholders.

A comprehensive study of the valley's geography, history and ecology will be carried out. Every portion of the valley will be studied, from drainage and runoff to sewage and solid waste disposal. A study will be made of the endemic species of fauna and flora in the valley, agricultural methods and crops, industry and arts and crafts. A study is also needed of the potential economic benefits of a properly managed basin, showcasing the area's cultural, scientific and ecological heritage to the world. All this information will be compiled into an outline plan that includes the region's historical, archaeological, ecological, tourist and agricultural aspects and identifies the issues, concerns and problems of the Kidron Valley/Wadi Nar that need to be addressed by the Master Plan.

One of the major problems to be tackled is sewage treatment. Without purifying the effluent flowing in the valley there can be little environmental enhancement. For years, Israel and the Palestinian Authority have debated the creation of sewage treatment plants in the valley. In the past, several plans for sewage treatment were developed under the auspices of different government authorities. These plans will be collected and brought to the attention of the stakeholders for their evaluation. The team will work towards achieving agreement on a workable plan.

A subsequent plan for proper distribution of the treated water to the various riparians in the valley will then be developed. This includes the issue of a 'dry' or 'wet' wadi; should water continue to flow year-round, or should the Kidron Valley/ Wadi Nar return to its previous state as an intermittent stream?

The planners will then work to coordinate a Master Plan that takes into consideration future land use planning. The Master Plan will consider: reduction of environmental degradation; water quality and effluent treatment; use of the water that flows in the stream; rehabilitation of the river's ecosystem including flora and fauna; development of parks and pathways for leisure and recreation; reconstruction and rehabilitation of historic sites; designating land usage including protected areas, farmland, open spaces, parks, business and residential areas; and the establishment of a visitors' centre.

It will detail investment opportunities as well as the roles and responsibilities of the various agencies that will be involved in the implementation of the plan and the management of the resources of the valley. One major focus will be the recognition of the Kidron Valley/Wadi Nar as a UN World Heritage Site.

The Master Plan will introduce Integrated Water Resource Management (IWRM) to regional planning, whereby land use, water use and the environment will be managed holistically, social and economic needs will be balanced with the needs of the environment, and decision-making will be coordinated between the local, regional and national levels.

Our goal is not only to develop a Master Plan, but to do so while incorporating the principles of basin management, including transparency, dialogue and public involvement at and between all levels. This is not something to be taken for granted in an area of conflict without much experience in using these principles. Suspicion and resistance may arise at different levels, and frustration and disappointment when expectations and reality collide. The Master Plan is a blueprint for the common future of the Kidron Valley/Wadi Nar developed together by Israelis and Palestinians in coordination and in consultation with stakeholders from both entities and from all levels. It reflects a common vision that will improve the quality of life for the residents of the valley and visitors, both environmentally and economically.

This will be the first Master Plan for a transboundary stream in the region. When completed, the Kidron Valley/Wadi Nar Master Plan will serve as a model for other Palestinian–Israeli transboundary streams and rivers, as well as being a starting point for continued cooperation within the basin itself.

6.4.4
Conclusions

When process supersedes action, glaciers move faster than people. In the Middle East, the amount of coffee drunk during the negotiation processes could replenish the depleted rivers and streams that cross between Israel and the Palestine Authority. Sometimes it seems that the parties enjoy meeting each other over coffee so much, preferably in some lovely spot in Europe, that it would be a pity to conclude the negotiations. At the same time, serious environmental degradation continues apace, as if by slowing down the process, one can reduce man's impact on nature. The Kidron Valley serves as an example of how prolonged negotiations reduce the value of a historical site, impact on the lives of thousands of people, reduce the amount of reusable water available for a parched area and demonstrates how the peace process delays environmental rehabilitation. To break this impasse, without resorting to a resolution of all disagreements between the parties, several scholars joined together to develop a Master Plan for the Kidron Valley.

In December of 2009 a steering committee was created that consists of six institutions: the Dead Sea Drainage Authority, the City of Jerusalem, the Peres Centre for Peace, the Jerusalem Institute for Israel Studies, the Israel Environment Ministry and the Milken Institute. The Chair of the Steering Committee has met with the Director General of a Palestinian environmental NGO, who has expressed an interest in serving on the steering committee, once funds are available, for a study of all transboundary streams in Israel and the Palestinian Authority. The steering committee has appointed an interdisciplinary team, headed by an architect and composed of experts in the fields of planning, water and drainage, environment, history, archaeology, economics, transportation and law. In January 2011 the Master Plan was presented to interested stakeholders and government officials from both Israel and the Palestinian Authority.

Acknowledgements

I would like to acknowledge the help of the Water Commissioners of Israel and the Palestine Authority as well as the stakeholders who make up the steering committee for the Kidron Valley/Wadi Nar Master Plan Project.

Further Reading

Almog, R. (2007) *Geographical and Environmental Characteristics of the Kidron/ Nar Basin, From Conflict to Collective Action: Institutional Change and Management Options to Govern Transboundary Watercourses*, BMBF Germany and Israel Ministry of Science.

Becker, N. (2007) *Cost Benefit Analysis of Alternative Wastewater Options in the Kidron/ Nar Basin, From Conflict to Collective Action: Institutional Change and Management Options to Govern Transboundary Watercourses*, BMBF Germany and Israel Ministry of Science.

Feitelson, E. and Haddad, M. (eds.) (1994) *Management of Shared Groundwater Resources: The Israeli- Palestinian Case with an International Perspective*, International Development Research Centre and Kluwer, Academic Publishers, Ottawa.

Laster, R. and Livney, D. (2007) *The Applicability of the Governance Scheme for the Elbe Catchment to the Kidron Valley/Wadi Nar, From Conflict to Collective Action: Institutional Change and Management Options to Govern Transboundary Watercourses*, BMBF Germany and Israel Ministry of Science.

Laster, R. and Livney, D. (2007) *The Nahal Kidron/Wadi Nar Governance Institutions and Legal Structure: Israel and the Israeli – Controlled West Bank, From Conflict to Collective Action: Institutional Change and Management Options to Govern Transboundary Watercourses*, BMBF Germany and Israel Ministry of Science.

Laster, R., Livney, D. and Holender, D. (2005) The Sound of One Hand Clapping: Limitations to Integrated Water Resource Management in the Dead Sea Basin. *Pace Environmental Law Review*, **22** (1), 123–149.

Mostert, E. (1999) River Basin Management. Presented at the Proceedings of the International Workshop on *River Basin Management*, The Hague, 27–29 October 1999.

Rahamimoff, A. (ed.) (1996) *Yarqon River Master Plan (Hebrew)*, Yarqon River Authority, Tel Aviv.

6.5
The Development of Transboundary Cooperation in the Prespa Lakes Basin

Daphne Mantziou and Miltos Gletsos

6.5.1
Introduction

Prespa is a mountainous basin situated in Southeast Europe and shared by Albania, Greece and the Former Yugoslav Republic (FYR) of Macedonia (Figure 6.5.1). The basin stretches over 1519 km² and combines two Lakes – Greater and Lesser

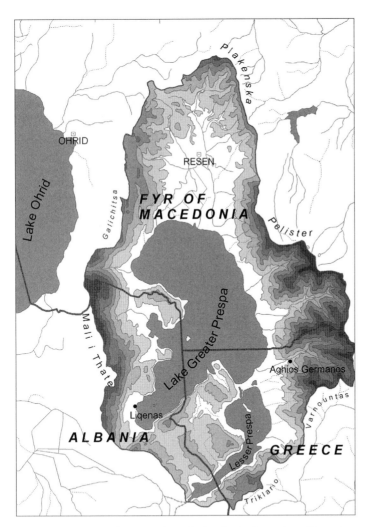

Figure 6.5.1 Prespa Lakes Basin (Map: Society for the Protection of Prespa).

Prespa – encircled by high mountains (over 2000 m above sea level) [1]. The area is widely known for its natural beauty and its globally significant biodiversity. The mosaic of habitats found here shelters a high number of plant and animal species, including numerous endemics. Notable is the presence of waterbird species that are rare or endangered at European or global level: Prespa hosts the world's largest breeding colony of Dalmatian Pelican (*Pelecanus crispus*), a species classified as Vulnerable in the IUCN red list.

Acknowledging the ecological importance of Prespa, all three littoral countries have granted protection status to parts of the basin. The whole Prespa basin in Albania and Greece is declared a National Park, while in the FYR of Macedonia three separate national protected areas are designated. Moreover, parts of the Prespa basin are assigned international (Ramsar) and European (Natura 2000) protected status [2].

6.5.2
Prespa Park: The Early Years of Transboundary Cooperation

The international ecological significance of the area, and especially the need for sustainable water management for the benefit of both nature and the inhabitants, led to the declaration of Prespa as the first transboundary protected area in Southeast Europe. In February 2000, on World Wetlands Day, the Prime Ministers of the three neighbouring states met in Prespa (Ag. Germanos, Greece) and jointly declared the establishment of the Prespa Park [3]. Their initiative, which followed a proposal of the environmental NGOs Society for the Protection of Prespa (SPP) and WWF-Greece, was held under the auspices of the Ramsar Convention on Wetlands and enunciated the states' intention to join forces across borders for the protection and sustainable development of the region.

Following the Prespa Park Declaration an interim joint body, the Prespa Park Coordination Committee (PPCC), was established by the three governments for the coordination of planning and implementation of joint activities in the area. The PPCC was set up as a non-legal entity of representatives of the Ministries of Environment, the local municipalities and the NGO community of each state party, as well as a permanent observer from the MedWet/Ramsar Initiative. Within the first decade of transboundary cooperation, the PPCC has held 13 meetings and has promoted the implementation of joint projects and activities in Prespa. The work of the PPCC has been technically supported by a Secretariat consisting of three officers from the collaborating NGOs.

Stakeholder participation permeates many aspects of the trilateral cooperation. The development of a joint vision for the Prespa Park, with the contribution of experts and key stakeholders from the three sides of the basin, was one of the main early aims of the PPCC. Within the first year of transboundary cooperation, working groups with specialists from the three countries were formed in order to produce the Strategic Action Plan (SAP) for the Sustainable Development of the Prespa Park. The SAP records the ecological and socio-economic situation in the transboundary protected area, identifies the main management issues and lays down objectives and measures for the region [2]. Following an extensive period of consultations with local, regional and national stakeholders on each side of the basin, in May 2004 the SAP was adopted by the PPCC as the guiding document for the planning of activities in the Prespa basin. Regarding water management, the SAP proposes a range of measures addressing water quality and quantity, aiming at both ecosystem conservation and satisfaction of human water demands.

In the following years, a series of sectoral collaboration initiatives were carried out in the Prespa region, including meetings of the national protected area authorities, the competent veterinary services and the fire-fighting authorities of the three sides of the basin. Trilateral collaboration of the local municipalities has particularly advanced as the Mayors signed a Memorandum of Understanding (2006) and a Cooperation Protocol (2007) and hold regular cross-border meetings.

Soon after its establishment, the PPCC initiated action for the submission of a funding proposal to the Global Environment Facility (GEF) aimed at the promotion of ecosystem management in the region. The efforts were successful and in 2004 GEF approved a PDF B grant for the development of a full-size proposal. This led to the approval of a five-year regional project on integrated ecosystem management in the Prespa Lakes basin which commenced in 2006 [4]. The project, which is co-funded by GEF, the governments of the three respective states and other international donors, is implemented by the United Nations Development Programme (UNDP) and has provided financial and technical support at both transboundary and national levels (Albania and FYR of Macedonia being the recipient countries). The main areas of focus include conserving biodiversity, mitigating pollution, encouraging sustainable resource use, strengthening stakeholder participation, maturating institutional cooperation and promoting integrated water management.

6.5.3
Advances on Integrated Water Management

Transboundary water management is a priority issue in the Prespa Park process, as affirmed in various meetings of the PPCC. Differing perceptions on the causes of water problems in the basin have gradually converged over time and through consultations, and relevant joint studies addressing water management issues have been carried out. During the early years of cooperation, consensus was reached amongst the PPCC stakeholders on the necessity of cross-border cooperation on water management on the basis of the European Union Water Framework Directive (WFD, 2000/60/EC). The WFD provides for coordination of efforts within shared basins across administrative and national boundaries. Where basins extend beyond the European Union's territory, the WFD requires Member States to establish appropriate cooperation mechanisms with the third parties. Albania and the FYR of Macedonia are indirectly bound by the WFD, as they are candidate countries at present and therefore in a process of gradual approximation with the requirements of the *acquis communautaire* and the harmonization of their national legislations with EU Directives and standards [5].

A series of steps and actions for the promotion of integrated water management have been undertaken in the ten years of trilateral cooperation for the Prespa Park at both institutional and scientific level. In autumn 2006 the water management authorities of the three littoral states held their first joint meeting in Korcha, Albania. The meeting concluded with the representatives stating the need for the establishment of an *ad hoc*, trilateral working group on water management and the development of a transboundary monitoring system in the Prespa basin. The GEF/UNDP project contributed to the development of a draft proposal on the specifications and composition of the working group [5], which were discussed in detail at the second meeting of the water management authorities (2008, Pyli, Greece). In addition, the identification of common water monitoring indicators as well as of respective measurement protocols is pursued within the frames of a project for the development of a transboundary environmental monitoring system for the Prespa Park. The project (2007–2011) is implemented by the Society for the Protection of Prespa in collaboration with experts from the littoral countries and the Tour du Valat research centre (France) and in coordination with the GEF/UNDP Prespa project [6].

Nevertheless, despite the significant progress achieved over the years in the Prespa Park the lack of formalization of the transboundary collaboration has undoubtedly hindered further evolution. For instance, the adoption of joint management measures in sectoral areas, such as water management, faced difficulties due to the lack of a binding commitment amongst the states.

6.5.4
The Prespa Park International Agreement

In November 2009, a major step in the advancement of transboundary cooperation was taken, when on the tenth anniversary of the Prespa Park the Prime Ministers of the three countries met in Prespa and announced their intention to adopt an intergovernmental Agreement on the protection and sustainable development of the area. In their Joint Statement the Prime Ministers acknowledged the progress in trilateral cooperation and the experiences gained in the framework of the Prespa Park, and made a commitment to establish mechanisms for the development of joint strategies, plans and measures for the effective conservation and management of the region, with special reference to integrated water management [7].

On 2nd February 2010, at a ceremony held in Pyli, Greek Prespa, the Ministers of Environment of the three littoral states and the European Commissioner for the Environment signed a legally binding Agreement for the Prespa Park, reiterating their commitment for the conservation of this unique ecosystem [8]. The 2010 Agreement is a significant step forward as it sets up a solid legal frame for cooperation and integrated water management. The adoption of this high-level political Agreement vindicated the efforts of the Prespa Park stakeholders, who for the past several years had been working for the institutionalization of the trilateral cooperation.

The participation of the European Union as a party in the Prespa Park Agreement underlines the importance of the Prespa basin for Europe, and the European Commission's will to support cooperation in the region. Furthermore, the Agreement falls within the Commission's priorities according to the Council Decision of 27 June 2006 'on the participation of the EC in negotiations aiming at the conclusion of international river basin agreements to improve cooperation in European river basins shared between certain Member States and Third Countries'.

The Prespa Park Agreement of 2010 provides for mechanisms of cooperation at several levels. A Prespa Park Management Committee (PPMC) is established in order to substitute and enlarge the previous Prespa Park Coordination Committee. The PPMC will consist of representatives of the Ministries of Environment, the local communities, the protected area management authorities, the environmental NGOs and the European Union, as well as of two permanent observers, one from the MedWet/Ramsar Initiative and the other from the Ohrid Management Committee. The Agreement foresees that the PPMC convenes twice a year on a rotating basis. To fulfil its duties the PPMC is assisted by a subsidiary technical organ, the Secretariat, which will consist of three members appointed by each state and will be headed by an expert on transboundary cooperation and river basin management. A high-level segment consisting of the Ministers of Environment and the EU representative will hold regular meetings for setting up the agenda, providing political guidance and reviewing respective progress.

The Agreement specifies transboundary cooperation on water management. Namely, Article 5 refers to the parties' commitment to promote integrated water management on the basis of the EU legislation and the WFD in specific. To this effect it stipulates that the states work for the provision of sufficient water quantity and appropriate quality for meeting both ecosystem needs and human demands, the rehabilitation of past negative impacts, protection against detrimental effects, and the effective control of the water management regime to be established. The Agreement also provides for the establishment of a trilateral working group on water management (Article 14) for the accomplishment of the above. The working group will consist of representatives of the competent authorities and stakeholders of each state and will serve as a technical expert body of the PPMC to facilitate coordination of efforts for integrated water management, as reflected in the EU WFD.

6.5.5
Supporting Trilateral Cooperation: Other Actors

The role of NGOs and international initiatives (Ramsar Convention/MedWet Initiative for the Mediterranean Wetlands) was catalytic to the establishment of the Prespa Park and the development of the process [9]. The Declaration of the Prespa Park in 2000 derived from a proposal by the Society for the Protection of Prespa and WWF Greece, with the contribution of MedWet/Ramsar. Over the years, the NGOs and MedWet/Ramsar have essentially backed up transboundary cooperation in the region, by supporting the work of the PPCC, lobbying at national and international levels, securing sources of financing and carrying out joint projects. Since 2001 the seat of the PPCC Secretariat was hosted by the SPP, which has provided continuous technical support to the Committee over the years. Most importantly, the NGOs have contributed to the engagement of local and regional stakeholders and to keeping the momentum alive.

Additionally, the contribution of the international donor community has been fundamental for the development of the Prespa Park initiative, the support of the process and the realization of joint projects and activities [9]. German bilateral development assistance funds have been supporting conservation work in Prespa since the mid-1990s, such as the institutional and technical support provided to the National Parks in Albania and the FYR of Macedonia by the KfW (German Bank for Reconstruction). In addition, the multi-donor GEF/UNDP Prespa Lakes project is expected to contribute to the further promotion of integrated water management, sustainable development practices, stakeholder involvement and awareness raising.

6.5.6
Conclusion

The establishment of the transboundary Prespa Park by a high level political declaration amongst Albania, Greece and the FYR of Macedonia in 2000 was a landmark in transboundary cooperation in Southeast Europe. Within the first decade of trilateral cooperation, the Prespa Park provided a forum for communication and exchange of ideas amongst the littoral countries. The establishment of an institutional structure, the regular contacts and the involvement of key stakeholders in the process have proven to be fundamental elements for building trust and confronting the difficulties and constraints with which transboundary cooperation is unavoidably faced. This is particularly demonstrated with respect to water management, where consensus has been reached amongst Prespa Park stakeholders on important water issues and on the need to apply the EU Water Framework Directive provisions in the shared basin. Ten years after the Prespa Park Declaration, the signing of a binding international Agreement amongst the three neighbouring states and the EU lays down the foundations for a new era of cooperation in this transboundary basin of Southeast Europe. The 2010 Prespa Park Agreement verifies the will of the governments to jointly work for the conservation of the region and establishes the legal frame for future cooperation on ecosystem conservation and sustainable use of its resources. The institutional coordination mechanisms envisaged are expected to enhance integrated water management in Prespa. At the time of writing, the Prespa Park actors and stakeholders anticipate the completion of the ratification procedure by the signatory parties.

References

1 Perennou, C., Gletsos, M., Chauvelon, P., Crivelli, A., DeCoursey, M., Dokulil, M., Grillas, P., Grovel, R. and Sandoz, A. (2009) *Development of a Transboundary Monitoring System for the Prespa Park Area*, Society for the Protection of Prespa, Aghios Germanos, Greece. Available at http://www.spp.gr/ fullstudy_vol1.pdf (accessed 6 May 2011).

2 Society for the Protection of Prespa (SPP), WWF-Greece, Protection and Preservation

of Natural Environment in Albania (PPNEA), Macedonian Alliance for Prespa (MAP) (2005) *Strategic Action Plan for the Sustainable Development of the Prespa Park, Executive Summary,* Society for the Protection of Prespa, Ag. Germanos, Greece, http://www.spp.gr/sap_executive_summary_edition_en.pdf (accessed 6 May 2011).

3 Prime Ministerial Declaration on the Prespa Park (2000), available at: http://www.ypeka.gr/LinkClick.aspx?fileticket=fEBt4RjBNy4%3d&tabid=550 (accessed 6 May 2011).

4 United Nations Development Programme (UNDP) (2005) Integrated ecosystem management in the Prespa Lakes Basin of Albania, FYR-Macedonia and Greece, UNDP Full size Project Document, UNDP.

5 McIntyre, O. (2009) Enhancing transboundary cooperation in water management in the Prespa Lakes basin. Consultant report, GEF/UNDP Prespa Regional Project.

6 Society for the Protection of Prespa (2010) Transboundary Environmental Monitoring System, http://www.spp.gr/monitoring_en (accessed 6 May 2011).

7 Prime Ministers of Albania, Greece and the Former Yugoslav Republic of Macedonia (2009) Joint Statement, 27 November 2009, Pyli, Greece, http://www.spp.gr/prime%20ministers%20joint%20communique_pyli%202009_en.pdf (accessed 15 November 2010).

8 Ministerial Agreement on the Protection and Sustainable Development of the Prespa Park Area, 2 February 2010, Pyli, Greece. Available at http://www.ramsar.org/pdf/wwd/10/wwd2010_rpts_prespa_agreement.pdf (accessed 6 May 2011).

9 Christopoulou, I. and Roumeliotou, V. (2006) Uniting people through nature in Southeast Europe: the role (and limits) of nongovernmental organizations in the transboundary Prespa Park. *Southeast Europe Black Sea Studies,* **6**(3), 335–354.

6.6
International Relations and Environmental Security: Conflict or Cooperation? Contrasting the Cases of the Maritza-Evros-Meriç and Mekong Transboundary Rivers

Sotiris Petropoulos and Anastasios Valvis

6.6.1
Introduction – Conflict or Cooperation in Transboundary River Basins?

Environmental security is becoming a very well-known term nowadays. According to Jon Barnett there are six principal approaches to environmental security that can be discerned from the literature [1]. These approaches can be grouped into those focusing mostly on environmental protection and those focusing on securing the environment and environmental resources in order to maintain the security of the state of its citizens and its institutions. The first, supported by the United Nations (UN), corresponds to 'ecologic' security requiring collective action on the grounds that environmental problems are universal, while the latter supports the belief that the environment should be taken into consideration along with the security of the state.

As far as transboundary river basin management is concerned, the tensions can be described as a dilemma between cooperation and conflict scenarios. In fact, due to the multiple uses a river may have, the management of transboundary rivers becomes quite challenging, and in many cases is a matter of priority in a country's foreign policy, as it may lead either to interstate cooperation or conflict.

Thankfully, cooperation seems to be a more frequent scenario than extended conflict. Hayton, for example, tried to examine the status of cooperative agreements concerning the development of water resources shared by two or more countries. In his research he concluded that while there is a growing concern about the management of shared water resources, this concern is not followed by any equivalent anxiety over the protection of these resources, thus underlying the urgency for a more institutionalized engagement [2].

Third parties can play a crucial role in institutionalized engagements. For instance, Fano underlines the role of third parties in the particular cases of developing countries and water scarcity [3]. An international institution that could play such a role is undoubtedly the World Bank.

Nevertheless, even the involvement of international institutions cannot act as a substitute for states' willingness to give up some of their sovereignty privileges, and even if this happens there are other problems that could be dissuasive. For instance, Hofius mentioned the case of the Rhine basin countries, stressing the administrative problems associated with the implementation of cooperation of several states bordering a large river basin [4]. Furthermore, cooperation also depends on the institutional capacities of the states concerned [5]. This may also explain the large number of treaties that exist mostly in Europe (a region boasting in general high levels of institutionalization), and to a lesser extent in the Americas, versus the small number of treaties that exist in Africa.

To summarize, therefore, transboundary water management is unambiguously a really complicated matter. It is affected to a great extent by states' relations and states' comparative advantage in terms of power. The importance of water itself is a great factor for tensions. Nevertheless, more often than not, riparian states proceed to multilateral negotiations, based on the general principles provided by international water law, to avoid possible conflict. These negotiations are assisted by the involvement of international institutions, such as the UN.

6.6.2
The Maritza-Evros-Meriç Case

6.6.2.1 The Evros River and its Importance

The Maritza/Evros/Meriç sub-basin, including the Arda, Tundja and Ergene tributaries, is one of the major river systems located in the eastern Balkans. About 66% of the basin belongs to Bulgaria, 28% to Turkey and 6% to Greece. The delta area (about 188 km^2) is a very important ecological site, protected under the Ramsar Convention on Wetlands that was signed in Iran in 1971.

Apart from the great ecological importance of the Evros, it is also very important for the economic development of local communities in all three countries. To be more specific, in Bulgaria and Greece the Evros serves as a water source for agricultural use. In Turkey, half of the basin area is used for irrigational and dry farming while some industrialization has occurred [6].

The complexity of the Evros case derives from two main reasons:

- the importance of the river for the three riparian countries;
- the political and military history of the three riparian states (International Network of Water-Environment Centres for the Balkans (INWEB), www.inweb. gr (accessed 03 November 2010)).

6.6.2.2 International Management of the Evros River

Despite the importance of the Evros, it seems that there are no common routes of collaboration between the three riparian states. Bilateral cooperation between Bulgaria and Greece can be traced back to 1964, while both countries ratified the Helsinki Convention (1992) (in force in Greece since 1996) and created a joint monitoring system that included the Evros-Maritza river. In 1971, another agreement was established between Greece and Bulgaria, after the formulation of a joint committee for cooperation in the field of energy and the use of cross-border river waters (Sofia, 1971). Many other protocols have been signed by both countries but only the agreement of 1964 provides specific measures to be taken, mostly concerning flood protection (International Network of Water-Environment Centres for the Balkans (INWEB), www.inweb.gr (accessed 03 November 2010)).

In 1991, another initiative led to the Protocol of the Meeting of the Joint Greek (GR)-Bulgarian (BG) Committee of Experts concerning the preparation of a common proposal to the EU for the joint monitoring and control of water quality and quantity of the transboundary rivers Maritsa/Evros, Mesta/Nestos and Struma/Strymonas. This actually led to the 2000–2006 EU-BG-GR agreement under the umbrella of the Interreg programmes, which supported the installation of hydro-meteorological monitoring stations to deal with floods.

6.6.2.3 Main Issues in Managing the Evros River

The complexity of the Maritza-Evros-Meriç River is mainly due to political and historical factors. First of all, almost 208 km of the river forms the border between Greece and Turkey; thus, the Evros is located in a military-controlled area. This means that a special permit has to be requested from the military authorities if any scientific or other activities are to be carried out. Historical relations between Turkey and Greece may also have a negative influence on possible future cooperation.

Furthermore, the upstream country Bulgaria is only a recent EU member and is undergoing a period of transition with many institutional reforms on the way. Greece, one of the two downstream countries, is a longstanding EU member with a high dependence on upstream transboundary waters, while Turkey, the other downstream country, is negotiating to join the EU. This means that Turkey is not actually obliged to follow the Water Framework Directive of the EU [7]. At the same time, implementation of the Directive by Greece and Bulgaria is quite slow, and indeed the sluggish progress on cooperation between these two EU members may even be perceived as unwillingness to cooperate, especially from the side of the upstream country. Paradoxically, according to Bulgarian experts, the national legislation of their country, issued since 1999, has been strictly harmonized with the EU in the field of water resources management. However, at the same time the Greek authorities seem to question the Bulgarian administration's interpretation of the WFD on the issue of mutual cooperation and the exchange of information between riparian states.

The river's importance for Bulgaria is also a major reason for the expansion of political tensions between the three riparian states. Bulgaria uses the river for electric power generation. To safeguard its energy needs, Bulgaria keeps the level of water in the dams at a high point, which, in periods of extended rainfalls, means that overflow is unavoidable and leads to extended floods in the Greek and Turkish parts of the basin.

To sum up, the main hindrances for comprehensive and cooperative sustainable use of the river can be categorized in two pillars – political and institutional:

- lack of scientific, technical and institutional infrastructure, especially in the field of monitoring;
- differing ranking in national interests and security means a very low placement of cooperation on transboundary water management issues in the political agenda of the riparian states;
- absence of an international institution that could take the lead in the negotiation process as a mediator (such as the UN);
- and last, but not least, the absence or rather the exclusion of the local population and authorities from the decision making process and the negotiations.

6.6.3
The Mekong–Lancang River Case

6.6.3.1 The Mekong River and its Importance

The Mekong River is a very important international river crossing China, Laos, Myanmar, Thailand, Cambodia and Vietnam before emptying into the South China Sea. It is of importance for many reasons, but mostly because 70 million people earn their living through activities closely related to the existence of the Mekong [8].

From an international relations perspective, the Mekong was the first international river for which a special regional body, the Mekong Committee, was established (Mekong River Commission) (Mekong River Commission website http://www.mrcmekong.org/ (accessed 03 November 2010)). Hence, it is important for the international community to derive specific lessons from this first attempt to coordinate riparian countries for the peaceful and successful development of a common resource.

Moreover, interest in the Mekong is reinforced by the attention drawn to it by China:

- The Mekong has a great potential for generating electricity through the construction of hydropower plants and this fits perfectly with China's significant rise in energy consumption [9].
- China wants easier navigation conditions on the Mekong, which would enhance growth in its south-western regions and create a new road to Southeast Asia and the oil-generating countries (e.g. Indonesia, Brunei, South China Sea) [10].

Finally, the Mekong River is also important in a regional context. It is an area where China and Indochina interact not only bilaterally but also regionally through ASEAN and the Mekong River Commission. As ASEAN is in the process of engaging China in cooperation over trade and regional institutional building, a good track record of positive collaboration over the Mekong is welcome [11].

6.6.3.2 International Management of the Mekong River

The commencement point for managing the Mekong was the creation of the Mekong Committee. In 1957, under the guidance of the UN, the states of Laos, Thailand, Cambodia and Vietnam (the Lower Mekong basin states) agreed to create the Mekong Committee, a regional organization for promoting and coordinating water resource development projects [10].

In 1978, the Mekong Committee was transformed into the Interim Mekong Committee (IMC) while after the end of the Cold War and the accession of Vietnam, Laos and Cambodia to ASEAN, there was a new tendency towards increased cooperation. Consequently, the Mekong River Commission (MRC) was created in 1995 [12].

Under the MRC several rules were set relating to consultation and/or notification prior to undertaking national activities with effects on the Mekong, as well as obligations in maintaining flows in the mainstream [12].

6.6.3.3 Main Issues on Managing the Mekong River

The management of the Mekong has been a focus of the international community since as early as the 1950s. However, China, which is a major riparian country, has never participated in any regional initiative for the equitable management of the Mekong, except for the recent Quadpartide Commercial Navigation Agreement, which was a Chinese initiative to ameliorate the navigation of the Mekong.

The significance of China's absence from a binding regional body increases as the need for energy in China develops. The construction of numerous dams and hydropower plants across the Chinese part of the Mekong will have uncertain effects on fisheries, rice plantations and so on in its lower part [13].

Cooperation on expanding the navigational capabilities of the Mekong only came about when China took the initiative. The other riparian countries accepted the proposal, as an enhanced passage for merchandize ships to China can facilitate trade.

In addition to the activities of the MRC, ASEAN, which is a regional cooperation scheme covering the whole Southeast Asian region, has taken several initiatives. The importance of ASEAN for the Mekong is made greater through ASEAN's engagement with China. This process coincides with the Chinese formulation of a more cooperative stance, the so-called 'peaceful rise' policy. In addition, from 1996 onwards, ASEAN has initiated a series of ministerial meetings on ASEAN-Mekong Basin Development Cooperation in which China has the status of a 'core group' member [14].

Unquestionably, China's absence from a binding regional agreement on the management of the Mekong is very significant. China is not only the most powerful riparian country but also the most upstream country. This provides a clear advantage in negotiations for equitable and environmentally friendly management. In the future, the increased need in China for energy could lead to an increased usage of the Mekong's flows, creating severe difficulties downstream. As many poor Indochinese countries depend significantly on the Mekong for food production, any possible disturbance will certainly increase tensions. Even today, occasional protests against China's unilateral actions on the Mekong take place.

6.6.4
Comparing the two Regions

Historical analysis of the two case studies has revealed a set of common features.

First of all, the *transboundary nature* of both rivers has rendered them important sources of cooperative initiatives and tensions while they also have *attracted international and to some level regional interest.*

The Mekong is the first transboundary water resource for which an international organization, namely the UN, tried to implement a pioneer management framework. After the mid-1990s, ASEAN interest in the issue of the management of the Mekong flows increased.

On the other hand, the Evros did not attract so much international interest, and neither Greece nor Turkey relies on the river for their economic growth (International Network of Water-Environment Centres for the Balkans (INWEB), www.inweb.gr (accessed 03 November 2010)).

Additionally, the historical tensions between the two countries were not conducive to a joint utilization initiative. In contrast with most Indochinese countries, Turkey, and especially Greece, experienced significant levels of economic growth after the end of World War II and moved to being economies based on services. As the share of agriculture in their national GDP decreased, so did the importance of the Evros. Regional interest in the Evros is quite a new feature due mainly to the accession of Bulgaria to the EU and the global increase of the river's importance.

Going further, in both cases *the potentiality of conflict or at least interstate tension cannot be ruled out.*

As far as the Mekong is concerned, conflict amongst the Mekong River basin states is minimized through membership to ASEAN and the fair management processes adopted by the MRC. Additionally, the absence of concrete interest in projects with a significant effect on the river flows (e.g. dams) as well as the cooperation of all Indochinese states in the river's management has reduced the possibility of interstate conflict. The above situation is in total contrast to the behaviour of the main upstream countries, China and Myanmar, which cannot be forecasted. Power asymmetry between China and the states of Indochina is large enough for the former to be able to ignore the latter. This asymmetry is not only based on military criteria but also on economic, as all East Asian countries want access to the huge Chinese market.

Peaceful coexistence has been retained so far for three basic reasons. Firstly, the downstream states have proceeded in a regional agreement on the management of the river (MRC) and in joining a regional cooperation scheme that involves other Southeast Asian states (ASEAN). The interest of ASEAN, as a united group of states, in the management of the Mekong and the China–MRC relations has improved the negotiating ability of the whole of Indochina. Secondly, since 1997 China has followed a quite consistent, good-neighbourhood policy and kept a low profile on issues that could raise tensions (the South China Sea, the Mekong, and Vietnam etc.). Thirdly, the need of the Indochina states for unblocked access to the huge Chinese market has minimized expressions of criticism towards the Chinese stance regarding the utilization of the Mekong's flows.

On the other hand, China's rising power might provide the prism for a change in the good-neighbourhood policy. Moreover, as China's need for energy is quickly growing it has become imperative for it to tap all available resources. The Mekong can provide significant energy production sites but this will possibly affect the flows of water downstream. As Indochina is largely dependent on the Mekong, this might trigger a more aggressive stance towards China with unknown outcomes. Under such circumstances, ASEAN–China relations may not be able to preserve peaceful ways of conduct.

As far as the Evros is concerned, two of the three riparian countries, Greece and Turkey, have a long history of tension and conflict. The fact that both countries have left the Evros quite underdeveloped, due to the militarization of this zone, clearly reduces the possibility of conflict regarding its management. This is also the reason why both countries have not been extremely aggressive towards Bulgaria, which on many occasions has acted unilaterally. The role of the EU has also increased since Bulgaria became a member and Turkey became a candidate-member. This is a new feature that might facilitate more concrete regional river management based on fair utilization of its flows. Greece is a longstanding EU member that has so far not initiated an intensive attempt to formulate a regional agreement, mainly because the Evros is of low significance to the Greek economy. The absence of strong regional interest in institutionalizing a river management protocol leaves significant space for political tension to increase, especially in times of financial crisis.

Similarly to the Mekong case, interstate relations could deteriorate as energy consumption in all three countries increases. More specifically, Bulgaria's obligation to close down the Kozlodui nuclear plant, as well as its rapid economic growth due to its accession in the EU, has geometrically increased its needs for energy production. Likewise Greece, which for many years has suffered occasional power shortages, will not be able to utilize the Evros for the production of energy if Bulgaria proceeds with a plan for intensified usage of Evros waters. Consequently, without even calculating the problems caused to the agricultural productions of Greece and Turkey, any future increase in energy consumption may become a destabilizing factor in interstate relations between the three states. In this case, the framework of the EU might be able to fill the hole left by the absence of a concrete regional agreement on the management of the Evros and therefore minimize any potential interstate tensions.

6.6.5
Conclusion

This chapter deals with the evolution of the concept of environmental security within the study of international relations. More specifically, it focuses on international and regional initiatives to institutionalize a framework of cooperative management for one of the most important transboundary natural resources, fresh water (e.g. rivers).

This analysis attempts to assess the past and current situation of two rivers that are important in their respective areas, the Mekong in Indochina and the Evros in the Balkans. It reveals that in both cases potential conflict, or at least increased bilateral and regional interstate tension, can evolve due to the increasing needs of the upstream states (China and Bulgaria) for tapping all available resources mainly for energy generation. The absence of concrete interstate water management leaves much space for unilateral actions that can spur tension in two areas where peaceful coexistence is a relatively new condition and historical tensions are still remembered. To this end, the absence of a regional agreement covering all riparian states in each case may be detrimental to future deflection from normality.

On the bright side of the story lies the fact that in both cases a regional cooperation scheme exists and can operate as a potential substitute for a river basin organization, at least for a short time. More specifically, ASEAN seems to be a very helpful component in the management of the Mekong issue, but at the same time seems too small to handle a potential crisis between Indochina (as a whole or just particular states of Indochina) and China. On the contrary, while the EU may become the decisive factor in handling the management of the Evros issue, both the accession of Bulgaria to the EU and the award of candidate-member status to Turkey have not yet been seen as stimulating increased cooperation with Greece.

References

1 Barnett, J. (2007) Environmental security, in *Contemporary Security Studies* (ed. A. Collins), Oxford University Press, Oxford, pp. 182–203.

2 Hayton, R.D. (1993) The matter of public participation. *Natural Resources Journal*, **33** (2) 275–281.

3 Fano, E. (1977) The role of international agencies, in *Water in a Developing World-The Management of a Critical Resource* (eds A.E. Utton and L. Teclaff), Westview Press, Boulder, Colorado, pp. 219–230.

4 Hofius, K. (1991) Co-operation in hydrology of the Rhine basin countries, in *Hydrology for the Water Management of Large River Basins* (eds F.H.M. Van de Ven et al..), International Association of Hydrological Sciences.

5 Young, O. (1982) *Resource Regimes: Natural Resources and Social Institutions*, University of California Press, Berkeley and Los Angeles, 276 pp.

6 International Network of Water-Environment Centres in the Balkans (2010) Inventory of Internationally Shared Water Services, www.inweb.gr/index.php?option=com_content&task=view&id=60&Itemid=151 (accessed 03 November 2010).

7 European Union (2000) Directive 200/60/EC of the European Parliament and of the Council of 23 October 2000 establishing a framework for Community action in the field of water policy, Water Framework Directive of the EU (WFD 2000/60), http://eur-lex.europa.eu/LexUriServ/LexUriServ.do?uri=CELEX:32000L0060:EN:NOT (accessed 31 March 2011).

8 Mekong River Commission (December 2008) MRC Work Program 2009, http://www.mrcmekong.org/download/

programmes/work_program_09.pdf (accessed 15 November 2010).

9 Buntaine, M.T. (2008) Trade, interdependence and bargaining with China over environmental cooperation in the Lancang-Mekong River basin, Paper presented at the annual meeting of the ISA's 49th Annual Convention, Bridging Multiple Divides, Hilton San Francisco, San Francisco, USA.

10 Jacobs, J.W. (2002) The Mekong River Commission: transboundary water resources planning and regional security. *The Geographical Journal*, **168** (4), 354–364.

11 Alice, D. (2006) Who's socializing whom? Complex engagement in Sino-ASEAN relations. *The Pacific Review*, **19** (2), 157–179.

12 Mekong River Commission (1995) Agreement on the Cooperation for the Sustainable Development of the Mekong River Basin, http://www.mrcmekong.org/agreement_95/agreement_95.htm (accessed 31 March 2011).

13 Goh, E. (2004) China in the Mekong River basin: the regional security implications of resource development on the Lancang Jiang, Insitute of Defence and Strategic Studies of Singapore, Working papers, 069/04, 26 pp.

14 Association of Southeast Asian Nations (1996) Highlights of the first ministerial meeting on ASEA-Mekong Basin Development Cooperation, Kuala Lumpur, 17 June 1996, http://www.aseansec.org/6351.htm (accessed on 31 March 2011).

Further Reading

Myers, N. (1987) Population, environment, and conflict. *Environmental Conservation*, **14** (1), 43–56.

6.7
Delineation of Water Resources Regions to Promote Integrated Water Resources Management and Facilitate Transboundary Water Conflicts Resolution

Ana Carolina Coelho, Darrell Fontane, Evan Vlachos and Rodrigo Maia

6.7.1
Introduction

The lack of uniform and integrated water resources regions is a critical issue, especially in transboundary water regions and federative countries. Overlaying levels of planning and management, as a result of uncoordinated water resources regions, hampers Integrated Water Resources Management (IWRM). In addition, the process of delineating those regions has often been executed without sufficient scientific support. While this is usually a result of political and historical circumstances, it is possible to improve results by using knowledge from prior experiences, modern techniques and improved decision support systems (DSS). To harmonize multiple objectives and better represent the interaction between environmental, socio-economic, political and historical aspects, it becomes imperative to define appropriate territorial limits for water resources planning and management.

To deal with such a complex and ill-structured problem, a doctoral research project is being conducted by this chapter's primary author. This research introduces an approach to support the process of delineating water resources planning and management regions based both on recognition of more comprehensive aspects and incorporation of those aspects into a DSS. The proposed Water Resources Planning and Management Regions (WARPLAM) DSS is designed to be used by federal and state governments, international commissions and water councils. It intends to promote a common understanding about the logic behind this process, to reinforce the principles of IWRM and to facilitate the resolution of transboundary water conflicts.

The process of developing the WARPLAM DSS can be summarized in three main phases: Phase 1 – understanding the important aspects related to the delineation of water resources planning and management regions; Phase 2 – building the DSS through the definition of a suitable approach; and Phase 3 – validating the system through a case study. This chapter focuses on describing the first phase of the presented approach, which refers to the conceptualization of the problem and the recognition of more comprehensive aspects related to IWRM regions. For that, a comparative analysis was performed based on the adopted water resources regions in different countries in the European and the American Continents.

6.7.2
IWRM and Water Resources Regions

IWRM is a relatively recent practice being adopted by water managers because it reflects the necessity of planning and management of water systems in a way that all relevant objectives are harmonized [1]. According to Vlachos [2] the term appeared early in the 1930s as a new paradigm that reinforces the importance of considering the world's complexities, including new approaches for planning and organizational structures that represent the interaction between environment, society and technology. In this context, geographic integration is an important area since it reflects a wide range of activities, such as planning, management, controlling, data organization, monitoring and water allocation.

Many authors consider river basins the most suitable geographic unit for IWRM. Dourojeanni *et al.* [3] justify the use of river basins as they correspond to the: (i) principal terrestrial form of the hydrologic cycle; (ii) interrelationship and interdependence between water uses and users; and (iii) region where water and physical and biotic systems interact, including the socio-economic system. Therefore, some countries define their water resources regions using solely river basin classification based on topological relationships, such as proposed by Pfafstetter [4]. Most recently, however, different countries are aggregating other criteria for defining IWRM regions, including historic development, cultural and environmental aspects and strategic water uses, representing the 'problemshed' concept, as defined by Vlachos [5] and Allan [6].

Additionally, an important aspect to be considered is the fact that political boundaries, which are generally not coincident with the hydrological limits, can represent a strong barrier to using river basin areas as territorial units for IWRM.

Those political boundaries can be characterized not only by international limits but also by boundaries between different regions in the same country [7]. Political limits, depending on the degree of permeability, can constitute a unifying influence or an obstacle to IWRM, depending on their scale and jurisdictional power over water [8]. Internal issues within national borders and external issues between riparian countries regarding water sharing [7] can be reduced by defining IWRM regions and respective comprehensive institutional structures [9] with sufficient power to lessen the boundary effects.

6.7.3
Comparative Analysis: Water Resources Regions in Europe and America

A summary of the results of the comparative analysis of water resources regions adopted in different countries in the European and American Continents was developed. The methodology used for this Phase 1 of the research is a qualitative comparative analysis using a simple theoretical framework. The information necessary for the analysis was obtained in three different steps: analysis of documented sources, an online survey and personal interviews with experts. A list of selected countries was formulated based on the information available at each step of the process. It constitutes a broad selection of countries in two different continents, including examples from different types of political organizations such as centralized authorities, decentralized powers and federal states. However, the examples are not exhaustive. The main objectives of Phase 1 are to recognize more comprehensive aspects related to the process of delineation water resources regions, and to identify general lessons learned. Therefore, it is expected that the analysis of the selected countries represents unbiased results, with negligible gaps.

From around September 2008 to June 2009, about 150 documented sources were analysed, approximately 50 experts answered the online survey, and 24 specialists went through a face-to-face question and answer structured process that was intended to track the reasoning process. Figure 6.7.1 presents a general overview of the results obtained in eleven different countries, considered as examples: Portugal, Spain, Greece, England and Wales, Netherlands, Germany, France, United States, Colombia, Mexico and Brazil. The analysis focused on seven aspects listed as row-headers: form and system of the government, established water resources regions in two levels, purposes of those regions, criteria considered when delineating those regions, river basin committees, the existence of real planning and/or management at river basin level, and international river basin commissions.

6.7.4
Recognition of More Comprehensive Aspects

As demonstrated in Figure 6.7.1, the comparative analysis focuses on the existing regions for planning and/or management of water resources in different countries. By the analysis of those regions, some important aspects considered to establish them were recognized. In addition to river basin limits, other socio-economic, political

COUNTRIES	PORTUGAL	SPAIN	GREECE	ENGLAND AND WALES	NETHERLANDS	GERMANY	FRANCE	UNITED STATES	COLOMBIA	MEXICO	BRAZIL
Form / System of Government	Unitary Republic Parliamentary	Unitary Monarchy Parliamentary	Unitary Republic Parliamentary	Unitary Monarchy Parliamentary	Unitary Monarchy Parliamentary	Federalism Republic Parliamentary	Unitary Republic Executive & Parliamentary	Federalism Republic Executive	Unitary Republic Executive	Federalism Republic Executive	Federalism Republic Executive
Water Resources Regions	10 River Basin Districts Under WFD	25 River Basin Districts - RBD Under WFD	14 River Basin Districts Under WFD	11 River Basin Districts Under WFD	4 River Basin Districts Under WFD	10 River Basin Districts Under WFD	12 River Basin Districts Under WFD	21 Regions and 222 Sub-regions	33 CARs - Reg. Environ. Authorities	13 Administrative Basins	12 National Hydrographic Regions
Purposes	Mainly Planning	Planning and Management	Mainly Planning	Mainly Planning	Mainly Planning	Mainly Planning	Mainly Planning	Mainly Data Management	Management and Planning	Management and Planning	Limited Planning
Criteria considered when delineating those regions	Hydrographic Mainly, Political, Transboundary Basins, History, Prior Planning Processes, and Size Limits	Hydrographic, History, Political, Transboundary Basins	Hydrographic + Hydrogeology Basins	Hydrographic	Hydrographic	Hydrographic + Ecoregions + Size Limits	Hydrographic /Administrative Mainly, Political, Geology, History, Socio-Economy, Geography, Finance, Culture, and Size Limits.	Hydrographic + Political + Culture	Biogeography, Hydro-geographic and Geopolitical	Hydrographic + Hydro-Administrative	Hydrographic Mainly + Geographic
Other Established Regions	5 ARHs - Hydrographic Region Administrations	Sub-basins in each RBD. Historic 10 Hydrographic Confederations	13 Regional Water Directorates	129 CAMs and 8 Regions	26 Water Boards - Waterschappen	Sub-regions: Working Groups of State Water authorities	Sub-basins: Local Water Commissions + Other River Basin Territorial Public Establishments	Interstate Compacts Regions + Other Isolated River Basin Commissions	-	102 sub-regions + 37 Hydrological Regions	Federal and State Water Resources Units
Purposes	Management and Planning	Planning and Management	Management	Planning and Management	Management	Planning and management	Planning and management	Planning and management	-	Planning and Monitoring	Planning and Management
Criteria considered when delineating those regions	Hydrographic Mainly, History, Political, Prior Planning Processes, Transboundary Basins, Administrative/Institutional, Financial Efficiency, Similar Problems, Water Quality, Socio-Economic, Size and Distance Limits	Hydrographic, Political, Administrative, Size Limits, Geographic Features, Territorial Organization, User Participation and Historic-Social Processes	Administrative Regions	Surface Water Catchments Mainly, Groundwater, Coastal Areas, Significant Abstractions and Water Transfers, History, Urban Protection, Size Limits, Administrative Structures Efficiency	Hydrographic Mainly, Politico-Administrative, Coastal Areas, Groundwater, Artificial Structures, Climate, Environment Protection, Socio-Economic, Geographic, Census, History, Culture, Size and Distance Limits.	Political-Administrative + Hydrographic	Hydrographic (Small River Basins) or Water Systems (Estuary, Aquifer, etc.)	Hydrographic (River Basin Limits in Critical Areas)	-	Political Jurisdictions + Hydrographic	Hydrographic + Administrative Regions + Socio-Political Aspects
Committees	River Basin Districts Councils at ARHs	Water Council + other boards in most RBDs	Regional Water Committees / Councils	Liaison Panels at the River Basin Level	Water Boards	River Basin Communities in some Rivers	River Basin Authorities	Watershed Groups	Regional Boards of CARs	25 River Basin Councils	River Basin Committees
Real Planning and/or Management at River Basin Level	Yes. ARHs will reinforce this process.	Yes. Competent Water Authorities.	No. Carried out mainly along administrative boundaries.	Yes. It is being promoted by the WFD.	No. It is being reinforced by the WFD.	No. It is mostly performed by the Federal States.	Yes. Competent Water Authorities.	No. It is mostly conducted by State level.	No. There is a dispute among among Central Government CARs and and regions. Municipalites.	Balanced among Central Government and regions.	Balance among State and Federal Governments and regions.
International River Basin Commissions	Bilateral Agreements with Spain at Mino-Lima, Douro, Guadiana and Tejo.	Bilateral Agreements with Portugal at Mino-Lima, Douro, Guadiana and Tejo.	Bilateral Agreements: Aoos, Vardar/Axios, Strimon, Marits/Evros, Prespa.	Cross-border arrangements between England and Wales (Dee; Severn).	International Commissions for the Protection of the Rhine, Meuse, Shelde and Ems.	International Commissions: Rhine, Elbe, Danube, Meuse, Mosel Meuse, Shelde and Saar, Oder and Ems.	International Commissions: Rhine, Garrone, Meuse, Rhine, Lake Geneva and Scheldt.	Mexico US International Boundary and Water Commission (IBWC)	-	Mexico US International Boundary and Water Commission (IBWC) and the Guatemala, Belize and Mexico IBWC	Bilateral Agreements at La Plata and Amazon River Basins

Figure 6.7.1 General overview of the comparative analysis results in eleven countries.

and environmental aspects are identified and valued according to IWRM principles. As observed also, the sets of criteria vary among the examples illustrated in this comparative analysis. It is difficult to generalize one common set because it depends on the way the countries define their priorities in terms of river basin planning and management, and their respective institutional framework.

In the first example, Portugal, there are five hydrographic regions administrations based on hydrographic aspects mainly. Some additional important criteria were also considered, such as: political jurisdictions, including municipal, administrative regions and international boundaries; historical aspects, including prior planning processes and prior institutional and administrative structures; financial efficacy; hydraulic connectivity, including water transfer projects; water quantity and quality aspects; similar kinds of problems and priorities; geological and geomorphologic characteristics; socio-economic aspects, including the territorial units for statistical purposes; and geographical distances [10]. France, including its colonies, is divided into 14 regions, based on administrative and hydrologic aspects, adopting the lines corresponding to the delimitation of the communes' territories – which are the smaller administrative territorial unit – closest to river basins or groups of river basins [11]. The Netherlands' Water Boards are based on groups of water systems, for instance, polders or drainage basins, combined with: administrative regions and political jurisdictions; coastal areas; groundwater limits; artificial structures such as reservoirs, channels, and water transfers projects; climatic characteristics; environmental protection areas; socio-economic areas, including agriculture lands; metropolitan regions; geographic features; census divisions; historical development and cultural factors; and size and distance limits [12]. Furthermore, all the examples from the European Continent have river basin districts (RBDs) because the European Union, through the Water Framework Directive (WFD), requires all Member States to identify those RBDs as the main areas for IWRM [13]. According to guidelines provided by the EC [14], those districts are made up of main river basins or groups of small river basins considering climatic, environmental, socio-economic and administrative aspects, weighted based on particular characteristics of the Member States. In the case of transboundary river basin districts, coordinated planning and management must be ensured.

In the American Continent, the United States selected four levels of hydrologic units in 1987, after a long period of disagreement about subdivisions of the Federal, State and local agencies. These agencies had been using incompatible criteria for names, codes and river basins boundaries, strengthening transboundary water conflicts. The four levels of units were delimited considering drainage area of major rivers or a combination of small drainage areas, hydrograph characteristics, culture and political boundaries [15]. In Mexico, 13 hydrologic-administrative regions were established as Regional Management Units by the National Water Council in 1998. The division is based on hydrologic and administrative aspects, having coincident limits with one or more river basins, according to regional characteristics of water resources. The actuation area of those regions, through the creation of River Basin Organisms, is correspondent to the limits of the municipalities contained in each region [16].

Those and further included examples demonstrate that some aspects, other than solely river basin limits, are being considered in order to define integrated water resources regions, such as political-administrative, socio-economic, cultural, historical, geographical and environmental aspects. It is important to emphasize the value of the European WFD in promoting the consideration of such aspects and motivating the EU member countries to delineate IWRM regions. Significant progress, in terms of implementing more integrated water resources regions as advocated by the WFD, is already noticed in the examples analysed. In fact, it becomes imperative that more comprehensive aspects be incorporated into future decision process regarding water resources regions in order to promote IWRM and facilitate transboundary water conflict resolution.

6.7.5
Conclusion

Based on a comparison among water resources regions adopted in selected countries, this analysis resulted in the identification of criteria to be considered into upcoming delineation processes, in order to increase the quality of future decisions. In the next phase of the research, the proposed WARPLAM DSS will incorporate those criteria into an Expert System that will be available for decision makers to increase their understanding about this process and to improve governance. Furthermore, it is important to emphasize that WARPLAM DSS will constitute a very flexible solution to support the delineation of water resources regions into multiple levels of subsidiarity and to be adaptable to regional circumstances.

The relatively large number of examples used in this analysis was intended to assure that a wide variety of criteria might be analysed and the set of recognized aspects might be applied in different other cases. Despite the fact that some information may be subject to different interpretation, the results are considered valid because of the multiple sources selected in this study.

Finally, as observed in the presented examples, water resources regions should be defined at different scales and related to multi-level governance in order to promote IWRM. Therefore, it is expected that the proposed approach addresses the need for better tools, designed to deal with such a complex and ill-structured problem.

Acknowledgements

The authors would like to thank all the specialists – and respective affiliated institutions – that provided their valuable time and effort to share their knowledge through the online survey and/or personal interviews. Their contribution is extremely important for the results of this study.

The first author would like to thank the Fulbright Program, the Brazilian National Water Agency and the Brazilian Agency for Improvement of Higher Education Personnel – CAPES for their financial support through grant 2276/05-4 to Ana Carolina Coelho Maran Gonçalves.

References

1 Grigg, N.S. (2005) *Water Manager's Handbook: A Guide to the Water Industry*, Aquamedia Publishing, Fort Collins.

2 Vlachos, E. (2008) Lecture notes. Seminar on technology assessment and social forecasting – CIVE639, Colorado State University, Fort Collins, Colorado.

3 Dourojeanni, A., Jouravlel, A. and Chavez, G. (2002) Gestión del agua a nivel de cuencas: teoría y práctica, CEPAL Division de Recursos Naturales e Infraestructura. Serie Recursos Naturales e Infraestructura Nr 47, Santiago do Chile.

4 Pfafstetter, O. (1989) Classification of hydrographic basins: coding methodology, Departamento Nacional de Obras de Saneamento, Rio de Janeiro, available from Verdin, K. L. (1997) A System for Topologically Coding Global Drainage Basins and Stream Networks, Earth Resources Observation Systems Data Center, U.S. Geological Survey. 5pp.

5 Vlachos, E. and Mylopoulos, Y. (2000) The status of transboundary water resources in the balkans: establishing a context for hydrodiplomacy, in *Transboundary Water Resources in the Balkans: Initiating a Sustainable Co-operative Network* (eds J. Ganoulis *et al.*), NATO Science Series Vol. 74, Kluwer, Dordrecht, pp. 213–223.

6 Allan, J.A. (2005) *Water in the Environment/ Socio-Economic Development Discourse: Sustainability, Changing Management Paradigms and Policy Responses in a Global System, Government and Opposition*, Blackwell Publishing, Oxford, UK.

7 Ganoulis, J., Duckstein, L., Literathy, P. and Bogardi, I. (eds) (1996) *Transboundary Water Resources Management: Institutional and Engineering Approaches*, NATO ASI Series, Vol. 7, Springer Verlag, Heidelberg.

8 Matthews, O.P. and Germain, D.S. (2007) Boundaries and transboundary water conflicts. *Journal of Water Resources Planning and Management*, **133** (5), 386–396.

9 Waterstone, M., (1996) A conceptual framework for the intitutional analysis of transboundary water resources management: theoretical perspectives, in *Transboundary Water Resources*

Management: Institutional and Engineering Approaches (eds. J. Ganoulis *et al.*), NATO ASI Series Vol. 7, Springer Verlag, Heidelberg, p. 9.

10 MAOTDR Ministério do Ambiente, do Ordenamento do Território e do Desenvolvimento Regional (2008) Administrações de Região Hidrográfica, MAOTDR, Lisboa.

11 MED Ministere de L'Ecologia et du Developpement (2003) Directive Europeene 2000/60/CE du Parlement et du Conseil etablissant un cadre pour une politique communautaire dans le domaine de l'eau. Mise en úuvre des dispositions de l'article 3-8 et de l'annexe I – Basins hydrographiques et autorites competentes.

12 Havekes, H. (2008) Functional decentralized water governance: guarantees, protection and developments. The institutional changes of the water authority in the past fifty years, Dutch Association of Water Boards, The Hague, The Netherlands, Dissertation, University of Utrecht, The Netherlands.

13 Environment Agency (2004) River basin characterization and the Water Framework Directive – briefing note. Version 1, http://www.environment-agency.gov.uk/ wfd (accessed 29 September 2010).

14 EC European Commission (2002) Project 2.9 Best Practices in River Basin Management Planning. Work Package 1: Identification of River Basin Districts in Member States. Version: 1.1 Date: August 2002, http://dqa.inag.pt/dqa2002/port/ docs_apoio/doc_int/13/ RiverBasinDistrict Guidance.pdf (accessed 29 September 2010).

15 Seaber, P.R., Kapinos, F.P. and Knapp, G.L. (1987) Hydrologic Unit Maps, USGS United State Geological Survey – Water supply Paper 2294. US Government Printing Office.

16 CNA (2007) ACUERDO por el que se determina la circunscripción territorial de los organismos de cuenca de la Comisión Nacional del Agua, Secretaria de Medio Ambiente y Recursos Naturales, http:// www.cna.gob.mx (accessed 29 September 2010).

6.8
Transboundary Water Resources and Determination of Hydrologic Prefectures in Greece

Evangelos A. Baltas

6.8.1
Introduction

Greece is a small country (area of 132 000 km^2) with intense ground relief, limited back land and a large coastline. The result of this geomorphologic structure is the division of the area into a plethora of small basins and a large number of water districts [1] (Figure 6.8.1), a fact which does not reflect its small area [2]. France, which is much larger in area than Greece, consists of four water districts, while Spain consists of nine [3]. Having assessed that there is a diversity of conditions and needs the European Parliament decided that this diversity should be taken into account during the planning phase and the implementation of the measures for the protection and sustainable use of water in a basin framework [4]. Decisions should be taken as closely as possible to the place where the water is used or is affected [5].

As much as 80% of water resources in the Mediterranean region and 90% in the Balkan Peninsula is shared between two or more countries. In Greece 25% of water resources stem from transboundary rivers (Aoos, Axios, Strymonas, Nestos and

Figure 6.8.1 The 14 water districts in Greece. Source: National Data Bank of Hydrological and Meteorological Information.

Evros) that flow through the Greek prefectures of Epirus, Macedonia and Thrace as well as from the Prespes and Doirani lakes. The dependence of Greece on transboundary water amounts to 13 billion m^3 per year. This indicates that high priority should be given to agreements with neighbouring countries so as to ensure integrated management and protection of common resources along with the necessary water supply for the sensitive ecosystems of the area.

This study attempts to determine four hydrologic prefectures, based on the hydrologic identity of the existing water districts, the important management problems regarding internal and transboundary water resources, as well as the existing administrative structure. The smaller number of hydrologic prefectures that is in entire keeping with the Directive would make research and organization easier and more complete, as well as facilitating more systematic control, supervision and management.

6.8.2
Difficulties in Implementing the Directive

Greece has a geomorphologic peculiarity, owing to the intense relief and the great extent of coastlines. It consists of small basins, each of which requires a different management plan. The Directive of the European Parliament and of the Council 2000/60 established new legislation for the sustainable management of water resources on the basis of drainage basin area. The implementation of such policy, especially in Northern Greece, is more difficult due to the transboundary nature of the water resources.

The most important problems relating to the state of water resources are the difficulty and deficiency in the systematic and reliable recording and evaluation of the physical and artificial water systems from a quantitative and qualitative point of view, as well as the deficiency in adequate measurements of hydrologic, meteorological, hydrogeological and qualitative parameters.

The available quantity of water resources decreases for various reasons, such as the unequal spatial distribution of water resources and the distribution of water resources through time, and this complicates their exploitation. In Greece, there are regions with great quantities of water reserves and others with intense deficiencies [6]. The mean hyper-annual surface precipitation constitutes an important characteristic for the evaluation of a region's hydrologic identity and response, so its examination provides reliable results on the hydrologic status of each water district. The mean annual surface precipitation of the 30 year period 1965–1995 is depicted in Figure 6.8.2. One can observe the high variability of precipitation, from 1150 mm in the north-western part of the country to 350 mm in the eastern part. This is attributed to the Pindos mountain range, which interrupts the prevailing eastern movement of weather systems, thus dividing the country into two major parts: the windward, high precipitation western areas and the leeward, low precipitation eastern areas. Moreover, other reasons are the unequal distribution of water demand, the plethora of small torrents, surface flow of small duration, the extension

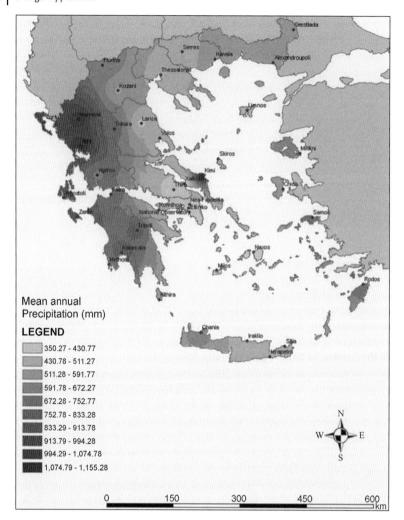

Figure 6.8.2 Mean annual (1965–1995) surface precipitation (mm).

of coasts that results in the intensive exploitation of coastal underground water and sea intrusion, and the large number of islands that are arid or have minimal water resources.

The problems of availability of water resources are increased because of:

- the significant increase in consumption,
- the reduction of water input from neighbouring countries,
- pollution,
- extreme phenomena, such as droughts, which are more frequently observed and are attributed to climatic change.

6.8.3
Determination of the Hydrologic Prefectures

The criteria that were taken into account for the determination of the hydrologic prefectures are:

- The surplus of water resources in the western regions of the country should be utilized for the fulfilment of the needs in the eastern regions.
- The transboundary water resources should be controlled by a single authority.
- All islands, which constitute small independent hydrological units, should be managed by a single authority.

More specifically, the great water potential in western regions that remains unutilized should be integrated into a common management plan for the rational fulfilment of the needs of the eastern part of the country, where the natural supply is insufficient to cover the growing demand. The transboundary water resources of the country, according to the Directive, require a single and special treatment for the implementation of the most rational management policy, so that they are used in the best possible way inside the national boundaries, and that conflict with neighbouring countries is prevented. According to the Directive, countries should collaborate together, complementary agreements should be drawn up and existing agreements be renewed. The feature that characterizes the development identity of the islands is the limited natural supply of water and the increased seasonal demand. To overcome this specific problem, a management policy that aims at the development of infrastructure and at the rational use and protection of the existing water resources is necessary.

The characteristics of each water district, which were examined only for ancillary reasons, proved particularly useful for the evaluation, as they showed the connections between the water districts, as well as their developing peculiarities, so that these can be taken into account in their final integration into a prefecture. The hydrologic prefectures that resulted are the following four (Figure 6.8.3):

- Hydrologic Prefecture of Macedonia-Thrace,
- Hydrologic Prefecture of Sterea-Epirus-Thessaly,
- Hydrologic Prefecture of Peloponnesus,
- Hydrologic Prefecture of Islands.

The management of transboundary water resources, which are necessary for the fulfilment of basic needs, constitutes a crucial issue in the hydrologic prefecture of Macedonia-Thrace. These water bodies include the lake of Big Prespa shared with FYR of Macedonia and Albania and the Small Prespa Lake shared with Albania in the water district of Western Macedonia, the Axios River and the Doirani Lake shared with FYR of Macedonia in the water district of Central Macedonia, the Strymonas River shared with Bulgaria in the water district of Eastern Macedonia, the Nestos River shared with Bulgaria and the Evros River shared with Bulgaria and Turkey in the water district of Thrace. Their management should involve the following actions:

Figure 6.8.3 The four hydrologic prefectures.

promotion of transboundary meetings to ensure the quantity and quality of water, collaboration with the upstream countries on water issues (common networks for monitoring quantitative and qualitative parameters, and common automatic systems warning of floods), minimization of the dependence on other countries and forecasting alternative ways of development.

6.8.4
Conclusions

The above separation constitutes a proposal that is based on scientific data, made available by the use of new technology and mainly geographic information systems, and forming part of the infrastructure of the National Data Bank of Hydrological and Meteorological Information (NDBHM) [7]. This proposal constitutes a hydrologic approach to the complicated issue of the total management of the country's water resources; such management should take into account a plethora of other parameters, such as water consumption and the utilization structures of the country's water potential. Hydrologic data and the resulting conclusions will form the base for policy in that direction.

References

1 Koumouli, V.D. (2001) Geomorphologic and hydrologic investigation for redefinition of water sections in Greece, National Technical University of Athens, Interdisciplinary Interdepartmental programme of

postgraduate studies – Water resources science & technology.

2 Ministry of Development (1987) Law 1739/1987 (FEK A 201) Water Resources Management and other instrumentation.

3 Ministry of Development (1997) Study for the water resources management in Greece.

4 European Parliament (2001) CSI-WFD-Project 2.9 (2001) Guidance on Best Practices in River Basin Management Planning, Work Package 1: Guidance on the Identification of River Basin Districts, Issue Paper, Valencia, 17–18 December 2001.

5 WWF/EC (2001) Elements of Good Practice in Integrated River Basin Management: A Practical Resource for implementing the EU Water Framework Directive. Key issues, lessons learned and 'good practice' examples from the WWF/EC 'Water Seminar Series' 2000/2001, October 2001, Brussels, Belgium, http://assets.panda.org/downloads/WFD-PRD-en.pdf (accessed 30 September 2010).

6 Sofios, S., Arabatzis, G. and Baltas, E. (2007) Policy for management of water resources in Greece. *The Environmentalist*, **28** (3), 185–194.

7 Mimikou, A.M. (2000) National Data Bank of Hydrological and Meteorological Information, http://ndbhmi.chi.civil.ntua.gr/en/index.html

7

Socio-Economic and Institutional Approaches

7.1
Social–Ecological Resilience of Transboundary Watershed Management: Institutional Design and Social Learning

Anne Browning-Aiken and Barbara J. Morehouse

7.1.1
Introduction

International boundaries, indeed even administrative boundaries, can pose serious challenges to managing shared water resources. When issues of sovereignty and jurisdictional authority are at stake, the hurdles can seem insurmountable. Institutional issues in particular may beset the best-intentioned efforts to establish and perpetuate equitable and rational water allocation and utilization. Institutions arise from and are embedded in contexts of particular historical, political and cultural relations; biophysical conditions; and processes of social and environmental variability and change [3]. Water management on the US–Mexico border encompasses all of these contextual factors, but the region is further challenged due to its position between an advanced and a developing country. Here the mix of economic, environmental, political and social factors is notoriously complex [1]. The potential for crises arising from inequitable distribution of water or from ecosystem failure escalates where regulation of the transboundary resource lies within the jurisdiction of two separate and distinct political and legal systems [2] and where society views water both as a right and as an entitlement. Some problems relate to the flow of funds and technology from central areas to communities in borderland areas of one or both countries; other problems may arise due to interagency conflicts within the country. Some institutionalized sources of stress, such as land and related water use practices inappropriate to semiarid environments, tend to be widespread and long-standing. Others, such as climatic change, diminishing water supplies, and increased water demands generated by agriculture, industry, and population growth are of more recent origin. When linkages within ecosystems become strained, when economic growth outstrips existing water availability, and when governance structures and

Transboundary Water Resources Management: A Multidisciplinary Approach, First Edition.
Edited by Jacques Ganoulis, Alice Aureli and Jean Fried.
© 2011 Wiley-VCH Verlag GmbH & Co. KGaA. Published 2011 by Wiley-VCH Verlag GmbH & Co. KGaA.

practices become fragmented or uncoordinated, adaptation to changing conditions becomes increasingly difficult [5].

7.1.2
Issues for Transboundary Institutional Mechanisms

Institutional differences across the Upper San Pedro Basin (USPB) pose a serious threat to the social–ecological resilience of the watershed [6, 7], particularly under escalating stresses posed by climate change. On the biophysical side, border stream flows show marked decline, and drops in the water tables of aquifers on the US side have prompted significant intervention from Fort Huachuca and the City of Sierra Vista. Socio-economically, proxy indicators suggest that resilience is being compromised in the Sonoran Basin: unemployment rates, largely associated with a closure of the mine that has extended for two years now, are projected to increase [8]. Reports of narcotics trafficking-related crime in the Mexican borderlands are also increasing [9].

Tensions between national, state and local levels over water and ecosystem management in the Arizona portion of the basin have increased over the past few decades in tandem with concerns about the impacts of growth on ecosystem functions, particularly in the riparian zone of the river. In 1998 local residents, encouraged by Commission on Environmental Cooperation recommendations, responded to these concerns by developing, the Upper San Pedro Partnership (Partnership), to cope with watershed resource management.

Unlike the USA, Mexico's water resources have traditionally been managed in a top-down fashion from the nation's capital, but more recently the stated intent has been to decentralize water management to state and local levels. The Mexican National Water Commission (CONAGUA) and Sonoran Water Commission (CEA) encourage municipal authorities and water stakeholders to form watershed commissions, and advisory and technical committees. Actual devolution has progressed very slowly and management decisions continue to be made largely at the top then passed down–often with very little funding or technical support–to state and local levels [10, 11]. One bright spot is CONAGUA's effort to form an organization called the San Pedro Binational Commission (SPBC) to collaborate on water issues at a transboundary regional level.

However, boundary issues continue to prevent revisions in institutional design elements that would allow both sides to work together. In the meantime, the Partnership has identified strategies for addressing water quantity issues, established a river monitoring system linked to a groundwater flow model and created a process for conflict resolution–all elements in effective institutional design for natural resource management. Likewise, having realized that management requires a learning process and capacity to integrate new information into its strategic planning, the Partnership moved to adaptive management and the establishment of a water district that could potentially have the power to implement management strategies. Operationalization of the district awaits a public vote of approval.

Also posing problems are assumptions of entitlement to water based on water laws grounded in private property rights [12] and the doctrine of prior appropriation. As for reservation of water for ecosystem support, federal reserved water rights and the Endangered Species Act [13] provide a means for recognizing *in principle* the significance of water for habitat maintenance and by extension for supporting ecosystem services more generally. It was under this proviso that the San Pedro Riparian National Conservation Area could be established under US Public Law 108–136, Section 321 [14].

Both countries face issues with regard to equity between those who benefit and those who bear the costs. In Mexico, local inhabitants are unable to control the adverse impacts of industrial activity on the social–ecological system of which they are a component. Moreover, national and Sonoran state water agencies lack authority to enforce mine infractions of environmental policy. In Arizona, the Partnership is reluctant to control increases in water demand generated by new housing developments. Rather than ecosystem health as essential to provision of water services, land and water are viewed through the lens of individual property rights to be exploited on a purely individual level. Further, Mexico's neoliberal development policy with its privatization of resource management and reliance on markets contradicts institutions aimed at ecological conservation, and poses challenges to Sonorans who believe they have an entitlement to water, based on their historical relations with the mine.

7.1.3
Social Learning

Despite the challenges described above, the transboundary social–ecological system possesses some traits that offer opportunities for attaining system resilience [7]. Social learning processes in river basin management contexts require, according to Tàbara and Postl-Wohl [15], the following attributes: (i) critical mutual reflection and potential modification of assumptions and cultural frameworks; (ii) participatory multi-scale decision-making; (iii) development of polycentric forms of resource assessment and management; (iv) empowerment of a collaborative basin group to improve existing conditions or implement changes; (v) recognition of interdependencies; (vi) capacity to reflect on assumptions about dynamics and cause-and-effect relationships; and (vii) engagement of stakeholders in a collective decision-making process. To address water deficit problems, the Partnership has developed an innovative institutional framework that is based on a social learning process. The group began by first creating a vision of the watershed for the future. During this visioning process members, from the local elected officials to water managers, to NGO representative, to state and federal agency representatives, came to understand each other's values concerning the governance of water and found a common ground upon which to begin the strategic planning process. During this phase members began to realize they needed answers to questions about the geohydrology of the basin. For scientific information to be useful to non-scientists, researchers had to learn to explain the science in a manner appropriate to the watershed audience. Through these phases, members have improved their understanding of the complex hydrology of the San Pedro aquifer and have established levels of mutual trust that have allowed the group to evolve into an effective management institution [16]. The

Partnership's social learning process has resulted in increased understanding of key issues, incorporated participants' issue frames, and achieved integration across scales and policy domains [17]. The Partnership has learned the value of adaptive approaches to addressing unanswered questions about managing the basin's water resources. However, its capacity to overcome lingering sociopolitical barriers, such as limited capacity to create a statutory framework for regulating water use, is currently being tested.

The SPBC in Sonora has a shorter track record and less time to develop trust and build scientific understanding. The Commission has been designated by national and state entities as the institution empowered to address basin management issues, but it lacks the economic and political power to design and implement a new management plan [18, 19]. The SPBC has organized few meetings over the past two-year period and thus has not had an opportunity for critical mutual reflection and modification of assumptions and cultural frameworks, or for developing new management strategies. In addition, since the group has not discussed how the mining company might address serious pollution issues in Cananea, the SPBC has not really recognized its interdependency with the mine. Social learning is a relatively new concept for the Mexican portion of the basin, partly because the federal and state governments have encouraged public participation in watershed governance, but have not provided the funding or technical support to do this on a regular basis.

7.1.4
Conclusion: Potential for Transboundary Collaboration

Bridging organizations such as the Binational Commission for Environmental Collaboration (CEC) have in the past provided the incentive and platform for addressing transboundary ecosystem problems [20]. Other organizations, such as the International boundary and Water Commission (IBWC/CILA) and the Sonora-Arizona Commission, more recently have offered a similar platform for building knowledge and understanding of institutional dynamics, but they do not actually have the capacity by themselves to study ecosystem dynamics and feed them into basin management practices. Together with the Transboundary Aquifer Protection Program (TAPP) they could bridge the institutional design deficits and link together the complex issues at multiple governmental levels, and provide the leadership needed to bring the Partnership and the San Pedro Binational Commission together to re-establish ecological and social system resilience [21–23].

References

1 Varady, R.G. and Morehouse, B.J. (2004) Cuánto cuesta? Development and water in Ambos Nogales and the Upper San Pedro Basin, in *The Social Costs of Industrial Growth in Northern Mexico* (ed. K. Kopinak), Center for U.S.-Mexican Studies, CA, pp. 205–248.

2 Priscoli, J.D. and Wolf, A.T. (2009) *Managing and Transforming Water Conflicts*, Cambridge University Press, New York.

3 Ostrom, Elinor (2005) *Understanding Institutional Diversity*, Princeton University Press.

4 Alley, R.B., Marotzke, J., Nordhaus, W.D., Overpeck, J.T., Peteet, D.M., Pielke, R.A., Pierrehumbert, R.T., Rhines, P.B., Stocker, T.F., Talley, L.D. and Wallace, J.M. (2003) Abrupt climate change. *Science*, **299** (5615), 2005–2010.

5 Young, O.R., King, L.A. and Schroeder, H. (2008) Summary for policy makers, in *Institutions and Environmental Change: Principal Findings, Applications, and Research Frontiers* (eds O.R. Young, L.A. King and H. Schroeder), MIT Press, Cambridge, pp. xiii–xix.

6 Adger, W.N. (2006) Vulnerability. *Global Environmental Change*, **16**, 268–281.

7 Adger, W.N. and Kelly, P.M. (1999) Social vulnerability to climate change and the architecture of entitlements. *Mitigation and Adaptation Strategies for Global Change*, **4**, 253–266.

8 Mendoza, J.E. (2009) Developing U.S.-Mexico border region: employment evolution & prospects on the Northern Mexico border, Baker Institute Research Project, http://bakerinstitute.org/publications/LAI-pub-BorderSecMendoza-041509.pdf (accessed 24 September 2010).

9 BBC (June 24, 2009). Mexico's drug-fuelled violence, http://www.bbc.co.uk/news/world-latin-america-10681249 (accessed 24 September 2010).

10 Wilder, M. and Romero-Lankao, P. (2006) Paradoxes of decentralization: water reform & social implications in Mexico. *World Development*, **34** (11), 1977–1995.

11 Scott, C. and Banister, J. (2008) The dilemma of water management 'regionalization' in Mexico under centralized resource allocation. *International Journal of Water Resource Development*, **24** (1), 61–74.

12 Glennon, R. (2002) *Water Follies: Groundwater Pumping and the Fate of America's Fresh Waters*, Island Press, Washington, D.C.

13 U.S. Fish and Wildlife Service (2011) Digest of Federal Resource Laws of Interest to the U.S. Fish and Wildlife Service. Endangered Species Act of 1973.

Available at http://www.fws.gov/laws/lawsdigest.html (accessed 24 September 2010).

14 Upper San Pedro Partnership August (1998) US Public Law 108–136, Section 321 http://thomas.loc.gov/cgi-bin/query/F?c108:1:./temp/~c108eLXrg1:e160121 and http://www.usppartnership.com/docs/AppendicesBCDEF.pdf (accessed 24 September 2010).

15 Tàbara, J.D. and Pahl-Wostl, C. (2007) Sustainability learning in natural resource use and management. *Ecology & Society*, **12** (2), 1–15.

16 Serrat-Capdevila, A., Browning-Aiken, A., Lansey, K. Finan, T. and Valdes, J.B. (2009) Increasing social-ecological resilience by placing science at the decision table: the role of the San Pedro basin decision support system model (Arizona). *Ecology and Society*, **14** (1), 37. Available at http://www.ecologyandsociety.org/vol14/iss1/art37/ (accessed 24 September 2010).

17 Mostert, E., Pahl-Wostl, C., Rees, Y., Searle, B., Tabara, D. and Tippett, J. (2007) Social learning in European river-basin management: barriers and fostering meachanisms from 10 river basins. *Ecology and Society*, **12** (1), 19–35.

18 Pineda Pablos, N., Browning-Aiken, A. and Wilder, M. (2007) Equilibrio de bajo nivel y manejo urbano del agua en Cananea, Sonora. *Frontera Norte*, **19** (37), 143–172.

19 Pliego, E. (2008) Tarifas autosuficientes para la prestación del servicio de agua potable. Caso: Comisión Estatal del Agua Unidad Operativa Cananea, Sonora. Master's thesis. Hermosillo: Colegio de Sonora.

20 Commission on Environmental Cooperation (CEC) (1998) Advisory panel report on the Upper San Pedro initiative: recommendations and findings presented to the CEC.

21 Berkes, F. (2009) Evolution of co-management: role of knowledge generation, bridging organizations and social learning. *Journal of Environmental Management*, **90** (5), 1692–1702.

22 Guston, D. (2001) Boundary organizations in environmental policy and science: an introduction. *Science, Technology & Human Values*, **26** (4), 399–499.

23 Cash, D.W., Adger, W.N., Berkes, F., Garden, P., Lebel, L., Olsson, P., Pritchard, L. and Young, O. (2006) Scale and cross-scale dynamics: governance and information in a multilevel world. *Ecology & Society*, **11** (2), 181–192.

Further Reading

Pahl-Wostl, C., Craps, M., Dewulf, A., Mostert, E., Tabara, D. and Taillieu, T. (2007) Social learning & water resources management. *Ecology & Society*, **12** (2), 1–19.

7.2
How Stakeholder Participation and Partnerships Could Reduce Water Insecurities in Shared River Basins

Elena Nikitina, Louis Lebel, Vladimir Kotov and Bach Tan Sinh

7.2.1
Introduction

Effective water governance in transboundary river basins appears to depend on a complex combination of domestic government policies and measures, sound multilateral and basin–specific agreements, and broader cooperation and dialogue involving multiple stakeholders. Stakeholder participation, in particular, is increasingly seen as an essential component of water governance in transboundary river basins [1]. Stakeholder inputs to deliberations are valued because they can help to identify new opportunities for cooperation as well as new burdens and risks that need to be taken into account in policies and agreements [2]. Coordination of stakeholder activities is often crucial for planning, implementation and evaluation. Coordination may be achieved through diverse mechanisms, including formal regulations between agencies or in river basin councils, through to much less formal networks and partnerships [3].

In this chapter we critically explore the opportunities and challenges involved in reducing water-related insecurities through expanding stakeholder participation and partnerships. We draw on evidence from selected shared river basins in Asia. Here we focus on floods and water shortages, but recognize that pollution issues are also important in many basins.

7.2.2
Stakeholder Engagement

During recent decades there has been worldwide debate on how to enhance the governance of shared rivers [1, 4]. According to a recent survey, there are 57 international rivers in Asia, and 39% of the land area lies in international river basins [4]. The debate involves both researchers and practitioners, with significant attention being paid to enhancing the design and implementation of transborder intergovernmental agreements and soft law arrangements. One of the messages is that many water related problems–conflicts between water users, controversies between upstream and downstream states, or regions of a shared river basin within the same country–are rooted in failure to establish good water basin governance [5]. Poor coordination within and across state borders is frequently identified as an underlying problem. Another is the lack of stakeholder involvement.

Governments have not conceived of water management in transboundary contexts as requiring the participation of actors apart from the state. Thus many international legal arrangements for shared river basins do not contain provisions for stakeholder participation.

For example, the Mekong River Basin Agreement, 1995, does not contain provisions for enhancing stakeholder involvement. However, subsequent development of this regime has encouraged broader stakeholder participation. Under the first phase of the Basin Development Plan (BDP), national governments were identified as the primary stakeholders by being referred to as 'internal'; non-state actors were referred to as 'external'. Non-state actors could engage only in national and basin or sub-area forums [6]. The early drafts of the Phase Two plan for the BDP were severely criticized for lack of attention to alternative views, but were eventually revised with commitments to 'institutionalize the participatory planning process established during BDP Phase I' [7]. In practice, however, expanding stakeholder participation has had modest impact, as the BDP process itself has failed to keep up with the plans and actions of individual country governments in the river basin [8].

The water border along the Amur River and its tributes is shared by Russia and China for 3544 km. Recent intergovernmental bilateral agreement on the protection of transborder watercourses between Russia and China, 2008, contains provisions on stakeholder involvement. It supports cooperation between non-government and research organizations on water use and protection of the Amur, ensures public access to information on water, and takes into account the interests of the indigenous populations. However, in many cases, such international legal provisions still remain a kind of window-dressing [9].

7.2.3
Stakeholder Roles and Participation

One of the messages from comparative studies is that it is helpful to disaggregate the policies, measures and actions undertaken by various stakeholder groups. The level of interest and actual capacity to be able to participate in decision-making, and the ability of government agencies, private firms, non-government organizations and individual households to take real action, often vary widely for any specific water insecurity issue.

Local public participation appears to be particularly important in reducing water-related risks to livelihoods from floods and seasonal water scarcity. Levels of participation vary significantly across countries and depend on culture and socio-economic conditions. For example, in the Amur Basin in Russia, there is a low level of public involvement, coupled with high reliance on the state and its 'paternalism' for issues concerning flood defence and environmental protection. Amongst the reasons for this is the traditionally low level of public awareness and participation in decision-making inherited from the Soviet regime. Another factor is the marginalized socio-economic situation in the Amur provinces: most of the Far East territories have always been developed as peripheral raw materials regions with vast sparsely populated land areas, poor infrastructure and low income groups. Jobs, health and security issues inevitably top the current public agenda. On the other hand, the representation of women in regional parliaments is the highest in the country, and appears to be an important driver for strengthening women's participation in water-related decision-making.

In the low-lying areas of the Mekong, a long history of experiences with seasonal flooding as part of the monsoon has often been embedded in local institutions and practices, providing a useful source of knowledge and norms to be applied more widely. In the Mekong delta in Vietnam, temporary 'emergency kindergartens', where parents can leave their children under organized supervision, operate in some areas during the peak flood season. In rice-growing areas social institutions associated with sharing water and the diversion of seasonal flood waters are also important. A high level of local participation in local management and decision-making, however, is not necessarily translated into a voice when it comes to large water infrastructure projects, such as hydropower dams or major irrigation schemes. Here, the only ways for poor farmers and fisherman to influence decision-making may be through protest, mass-media and networking [10].

7.2.4
Stakeholder Coordination and Partnerships

Coordination of the activities of stakeholders is usually regarded as an important function of effective water governance in shared river basins. Coordination does not imply consensus; disagreement and deliberation are often essential to improving governance arrangements. Coordination may be relatively formal, for example, between various government bodies and authorities within states, or between representatives in international negotiations, or may be more informal. Informal coordination has diverse forms, including networks and partnerships.

The extent of coordination varies significantly across countries and across river basins. Overall, there is significant experience in the management of small and intermediate sized watersheds [11]. Successful informal partnerships or coalitions for larger watersheds that cross national borders appear to pose more difficulties.

In many countries significant institutional reforms were undertaken to implement integrated water resources management through the formation of river basin or watershed organizations or councils [12]. Varying greatly in detail they are supposed to enhance stakeholder participation and improve coordination. Ideally, stakeholder representation is broad, covering different water users, businesses and non-government organizations in addition to representatives of government line agencies and local authorities. Basin organizations may be nested.

In Russia, for example, the recently established Amur Basin Council intends to ensure a broad representation of various water users, local civil society and indigenous people. However, in practice, it appears that it is highly dominated by government officials from various levels of authority. Similar observations have been made for committees of river basins in Thailand and Vietnam [11, 13].

Most river basin organizations face substantial difficulties in achieving their objectives. Several common problems are apparent. Firstly, most are dominated by bureaucrats and officials. This suppresses active deliberation with and by non-state actors. Instead, agendas are dominated by bureaucratic competition and rivalry for control over funds. Secondly, members on committees or councils often have multiple roles and competencies resulting in confusion about specific responsibilities within the river basin organization. A common consequence is a lack of clear responsibilities for implementation and practical results. Thirdly, coordination problems persist despite initial intentions. Different agencies represented in a council continue to pursue their own plans for the watershed or sector when not immediately engaged in council discussions. If overlooked, pre-existing water management institutions can greatly affect the implementation of new plans [14]. Negotiating of shared compromised visions has been proven difficult. How to improve stakeholder coordination in shared rivers often remains a serious challenge.

Collaboration and partnerships between local authorities, communities and other stakeholders is important for reducing water insecurities. Several lessons can be learnt from experiences in the Mekong delta of Vietnam. Following reunification in 1975, government resettlement policies brought people from the North to the least densely populated and regularly flooded areas of the delta. New economic zones were established for intensive rice cultivation: typically three crops per year, including one during the riskier peak flood season. Resettlement schemes were accompanied by technical solutions, including the construction of dams and embankments around new residential clusters. These structural measures resulted in prolonging floods, causing loss of earnings for farmers and social tensions. In response to public pressures, local authorities in An Giang province launched innovative projects with substantial local stakeholder inputs in design and implementation to diversify incomes during the flood season [15]. Partnerships between provincial departments of agriculture and rural development, private firms and farmers made the project highly successful.

7.2.5
International–Domestic Linkages

Stakeholder participation is a key element when addressing water-related insecurities in shared river basins, both within and across national borders. Stakeholders are the major drivers in routine domestic implementation of interstate accords once they are signed. The role of targeted stakeholders in complying with, or violating, the norms and behavioural prescriptions of intergovernmental agreements is crucial, but their commitment is often not properly secured, so implementation is poor. Of course, in some instances, stakeholders may not wish to be involved and resort to non-compliance or protest, where agreements do not adequately respect their interests. In other cases, the problem stems more from the lack of opportunities or incentives for stakeholder participation.

Finding mechanisms and tools to consolidate participatory capacities is one of the keys for institutional successes within and across state borders. Forging interactions in the triangle 'state–business–society' has been touted as an important instrument for domestic implementation of interstate accords. Partnerships amongst diverse stakeholders can help to reduce conflict and lead to pro-active approaches to water-related risk reduction.

Multi-stakeholder dialogues, by creating and supporting spaces for meaningful conversations, have the potential to play a significant role in improving the governance of regional and transboundary waters [2, 16]. Dialogues may, for instance, contribute to reducing water conflicts, ensuring equitable and fair allocation, and promoting ecological sustainable use and management. Dialogues may also be informative and help to shape more formal negotiations and decision-making processes, by bringing in a wider range of perspectives on needs, impacts and options, and having them deliberated openly [2]. Over the past decade, dialogues around water resource infrastructure programmes and management policies have sprouted around the world at many different levels of governance.

In Vientiane, Lao PDR, in July 2006, for example, an alliance of actors in the Mekong region came together to publicly discuss The World Bank's Mekong Water Resources Assistance Strategy, the Asian Development Bank's Greater Mekong Sub-Region Programme and the Mekong River Commission's Draft Strategic Plan [17]. The organizers hoped that the dialogues would lead to constructive analysis and adaptation of these strategies [18]. Non-state actors from local communities, academia, non-governmental organizations and the private sector participated in discussions along with government officials and representatives from multilateral agencies [19]. In a region with authoritarian states and partial democracies, regional forums sometimes allow a safer and more open discussion of projects and issues than would be the case in the individual countries.

Stakeholder participation is equally important in domestic cases so that agreements can be implemented and enacted upon, also in everyday practice, soft law and less formal arrangements such as joint declarations or memoranda of intent, cooperative action plans and programmes established within specific river basins. Stakeholder participation appeared to be particularly useful, for example, during the formation of the Amur basin regime and the long and controversial negotiation process between Russia and China.

The detailed pathways and mechanisms through which stakeholder participation and coordination reduce water insecurities in shared basins deserve further investigation [1, 19]. Potentially important mechanisms include: (i) making the interests, capacities and risks of the most vulnerable groups, otherwise marginalized from assessment and planning procedures, more visible within and across national boundaries; (ii) wider sharing and better understanding of knowledge and practices, which is critical for the reduction of disaster risks; (iii) social learning around risks and vulnerabilities leading to new management goals and more opportunities for collective responses, linking, where appropriate, domestic and international efforts; and (iv) higher public acceptance of policies and measures proposed by governments or under international agreements.

7.2.6
Conclusion

Meaningful participation by all relevant stakeholders is critical for the effective design, negotiation and implementation of policies and measures to reduce water-related insecurities in shared river basins. Careful attention to the processes by which stakeholder activities are coordinated implies that efforts must often go beyond formal institutional arrangements between state agencies and countries. Coordination through broader networks or specific partnerships may be particularly salient where more formal procedures are cumbersome or grid-locked, for example, in bureaucratic turf wars. Nevertheless, interventions should pay careful attention to pre-existing water management institutions, both formal and informal. Overall, expanded stakeholder participation and coordination appear to be important foundations for improved decision-making and effective implementation and verification of water governance arrangements in shared river basins.

Acknowledgements

The research leading to these results was funded by the European Community's Seventh Framework Programme (FP7/2007-2013) under grant agreement No 226571; the Asia-Pacific Network for Global Environmental Change, APN under international projects ARCP2009-03CMY-Nikitina and 2005-01-CMY-Nikitina; and Challenge Program for Water and Food under M-POWER Project, PN50.

References

1 Pahl-Wostl, C. (2007) Transitions towards adaptive management of water facing climate and global change. *Water Resource Management*, **21**, 49–62.
2 Dore, J. (2007) Multi-stakeholder platforms (MSPS): unfulfilled potential, in *Democratizing Water Governance in the Mekong Region* (eds L. Lebel, J. Dore, R. Daniel and Y. Koma), Mekong Press, Chiang Mai, pp. 197–226.
3 Nikitina, E., Ostrovskaya, E. and Fomenko, M. (2009) Towards better water governance in river basins: some lessons learned from the Volga, *Regional Environmental Change*, **9** (2) 1–13.
4 Conca, K. (2006) *Governing Water*, The MIT Press, Cambridge, p. 94.
5 Kotov, V. (2009) Russia: changes in water management and the water law, in *The Evolution of the Law and Politics of Water* (eds J. Dellapenna and J. Gupta), Springer Science–Business Media BV.
6 MRC (2005) The MRC Basin development plan. Stakeholder participation, BDP Library vol. 5. Mekong River Commission.
7 MRC (2006) Basin development plan. Programme phase 2, 2006-2010. 15 final version, August 2006, Mekong River Commission.
8 Dore, J. and Lazarus, K. (2009) Demarginalising the Mekong River

Commission, in *Contested Waterscapes in the Mekong Region: Hydropower, Livelihoods and Governance* (eds F. Molle, T. Foran and M. Käkönen), Earthscan, London, pp. 357–382.

9 Razzaque, J. (2009) Public participation in water governance, in *The Evolution of the Law and Politics of Water* (eds J. Dellapenna and J. Gupta), Springer Science–Business Media BV.

10 Molle, F., Foran, T. and Käkönen, M. (eds) (2009) *Contested Waterscapes in the Mekong Region: Hydropower, Livelihoods and Governance*, Earthscan, London.

11 Thomas, D.E. (2006) Participatory watershed management in Ping watershed: final report, Office of Natural Resources and Environmental Policy and Planning, Ministry of Natural Resources and Environment, Thailand, Bangkok.

12 Molle, F. (2008) Nirvana concepts, narratives and policy models: insights from the water sector. *Water Alternatives*, 1, 23–40.

13 Molle, F. and Hoanh, C.T. (2007) Implementing integrated river basin management: lessons from the Red River Basin, Vietnam, working paper, Institut de recherche pour le developpement, Mekong Program on Water Environment and Resilience, International Water Management Institute.

14 Mollinga, P., Meinzen-Dick, R. and Merrey, D. (2007) Politics, plurality and problemsheds: a strategic approach for reform of agricultural water resources

management. *Development Policy Review*, 25, 699–719.

15 Sinh, B.T., Lebel, L. and Tung, N.T. (2009) Indigenous knowledge and decision making in Vietnam: living with floods in An Giang Province, Mekong Delta, Vietnam, in *Indigenous Knowledge and Disaster Risk Reduction: From Practice to Policy* (ed. R. Shaw), NOVA.

16 Warner, J.F. (2006) More sustainable participation? Multi-stakeholder platforms for integrated resource management. *International Journal of Water Resources Development*, 22, 15–35.

17 World Bank, and Asian Development Bank (2006) WB/ADB joint working paper on future directions for water resources management in the Mekong River Basin: Mekong Water Resources Assistance Strategy (MWRAS), June 2006, The World Bank and Asian Development Bank.

18 IUCN, TEI, IWMI, M-POWER (2007) Exploring water futures together: Mekong Region waters dialogue, Report from regional dialogue, Vientiane, Lao PDR World Conservation Union, Thailand Environment Institute, International Water Management Institute, Mekong Program on Water, Environment & Resilience.

19 Sneddon, C. and Fox, C. (2007) Power, development, and institutional change: participatory governance in the lower Mekong Basin. *World Development*, 35, 2161–2181.

Further Reading

Lebel, L., Nikitina, E. and Manuta, J. (2006) Flood disaster risk management in Asia: an institutional and political perspective. *Science and Culture Journal*, 72, 1–2; 2–9.

Lebel, L., Nikitina, E., Kotov, V. and Manuta, J., (2006) Assessing institutionalised capacities and practices to reduce the risk of flood disasters, in *Measuring Vulnerability to Natural Hazards. Towards Disaster Resilient Societies* (ed. J. Birkmann),

United University Press, Tokyo, ISBN 92-808-1135-5.

Lebel, L., Dore, J., Daniel, R. and Koma, Y.S. (eds) (2007) *Democratizing Water Governance in the Mekong Region*, Mekong Press, Chiang Mai.

Nikitina, E. (ed.) (2006) Environmental risk management in large river basins: overview of current practices in the European Union and Russia. Executive summary, CABRI-Volga, EC.

7.3

Transboundary Stakeholder Analysis to Develop the Navigational Sector of the Parana River

André Hernandes

7.3.1
Introduction

Brazil is rich in natural resources, and mainly water resources (volume, quality and availability). This wealth has been subject to intense economic exploitation in recent decades, which has become an issue of great concern in some regions of the country. In some Brazilian regions, and mainly in the transboundary regions with large water bodies, this phenomenon is less noticeable. Signs of exhaustion and degradation are already perceived, but are not yet readily visible. Although there is an extensive river network in Brazil, nowadays inland navigation plays only a secondary role, particularly in regions where this form of transportation competes with road and railroad transportation infrastructure. However, in the Brazilian Amazon its use is massive due to particular local conditions. Overall, this sector has been losing importance since the 1960s when through highways and a predominantly land-based transportation infrastructure was established. The Regional Trade Agreement Mercosur, also known as the Common Market of the South, could play a role in this sector as a natural way of linking the countries that share the Parana transboundary river basin.

However, some difficulties emerge, as there is a lack of information about the national and international stakeholders' role in the shared use of this transboundary water resource. Studies show that, when properly performed, navigation can be a valuable instrument for environmental conservation, since navigation implies a constant presence along the whole water body, enabling monitoring programmes to be carried out, and possible threats to the environment to be quickly identified. In this way, the initial task of identifying the involved stakeholders starts with the conception of joint international policies for water resource management. This is what happened with the Guarani Aquifer System, a South-American transboundary groundwater resource. In this case there was an initial phase of data-collection to gain knowledge of the aquifer's characteristics, even though it was already managed by Brazil, Paraguay, Argentina and Uruguay.

7.3.2
Objectives

The proposal of this chapter is to establish an adjusted quantitative methodology to assess the relations between the diverse stakeholders involved with this water resource, in order to learn the importance of each one and understand how they interact. In this way it is expected that the forces acting in this sector will be determined, and that opportunities to improve the activities of the Federal Government organizations will be identified.

7.3.2.1 Home Organization

After the establishment of the Company of Ports of Brazil S/A (Portobrás), constituted in 1975 by law, the regional offices of the former National Department of Ports
and Navigable Ways (DNPVN) became known as Waterways Administrations and
were the responsibility of the ship owners Portobrás until 1990.

In 1990, with the dissolution of Portobrás, the company's obligations and rights
belonged to the Federal Government. The decentralization of the ports administration resulted in the Agreements of Decentralization of Waterways Services, between
the Federal Government, intermediaries from the former National Department of
Waterways Transport of the Ministry of Infrastructure, and the Dock Company (port
authority), with participation from Portobrás. In this way, the Dock Company
assumed responsibility for the activities and implementation of the agreements
and took over the functional activities of the former Portobrás. The National
Department of Waterways Transportation (Ministry of Infrastructure) was in charge
of coordinating, controlling and checking the execution of the agreement, especially
issues related to accounting, as foreseen in the ninth article clause. When Law #
10,233/01 was passed, the National Department of Transport Infrastructure (DNIT)
was created and amongst its other responsibilities it is in charge of managing the
waterways programmes.

7.3.2.2 The Transboundary Context

The Parana River Basin Region (PRBR) has certain particular characteristics. It is one
of the twelve Brazilian river basins regions defined by the Brazilian National Water
Resources Council (CNRH). The PRBR embraces the Brazilian stretch of one of the
river basin units of La Plata River basin, the High Parana, which corresponds to
the Parana River basin drainage area as far as the east estuary of the Iguaçu River, at
the triple border area between Brazil, Argentina and Paraguay. It corresponds to
approximately 59% of the transboundary basin of the Parana River and 29% of
La Plata River basin as a whole (Figure 7.3.1).

This large transboundary basin, however, plays a very small role in issues of
cooperation amongst the countries that share it, on account of the Itaipu hydropower
plant. This artificial barrier, constructed to supply Brazil and Paraguay with electrical
energy, does not have canal locks that would allow freedom of movement. This barrier
has resulted in the riparian countries being unfamiliar with the characteristics of the
basin as a whole, each country developing actions only within the scope of its own
borders, and not taking into account the potential of joint measures. Furthermore,
the rivers that form the Parana River basin tend to be seen in terms of their potential
for the generation of electricity. This creates a series of problems for navigation, such
as the necessity to maintain appropriate water levels in the dams of the hydropower
plants. In general these compromise the depth requirements of the shipping canals,
thus hindering navigation. This problem compromises confidence in the waterways
transportation system, hindering its development.

From the point of view of international navigation, very little is known by DNIT in
relation to other stakeholders who have interests in the river. This makes it difficult to
develop strategic actions since there are many entities, organizations and institutions

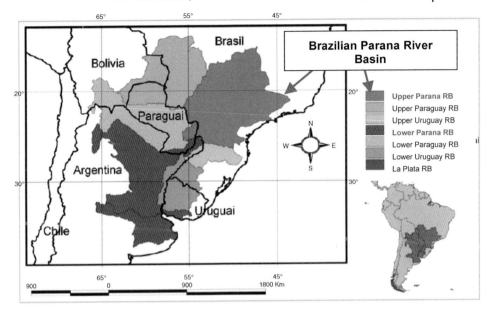

Figure 7.3.1 Parana River basin and sub-basins of the La Plata River [1].

that at first glance do not seem to be of relevance to the Ministry of Transport. To develop navigation in this context, and recognizing that intense load movement already occurs mainly for export purposes to the very countries that share the basin, it is important to have a better understanding of their role in the decision making process.

7.3.3
Key Stakeholder Analysis

The methodology used firstly analysed the stakeholders interested in using this water resource. This resulted in the realization that the international stakeholders were largely unknown, a fact that is not surprising given that cooperation with other countries had never been considered by the DNIT in the past. For more than 40 years the organization had been focusing on national road transportation, and the water-ways transportation sector only started in 2001.

7.3.3.1 Methodology
For this analysis, five stages were considered:

1) identification of the stakeholders that are direct users and those with interest in the water resource;
2) items to be evaluated;
3) establishment of the weights and values of each item;
4) graphical representation of stakeholders distribution;
5) analysis of relations between stakeholders.

7.3.3.2 **Results**

After the completed questionnaires were received from the study group, the results were tabulated and graphs were drawn to show the stakeholders' relative positions. The results were also evaluated to show general interest versus influence and specific interest versus influence and comparisons were made between the individual evaluations of each component.

The general interest evaluation tended to show values for neighbouring countries' interests (e.g. Argentina and Paraguay) that were not the same as when looked at from a specific interest point of view. It gives the impression that electrical energy generation is the most important issue, since there is a high degree of general interest shown in this by various stakeholders, such as the Itaipu Dam Administration, regulatory agencies and other system operators. The survey has shown that international stakeholders were considered generically as '*the government*', although they could be private or public institutions in their respective countries, demonstrating a clear lack of knowledge of the role of international stakeholders.

Stakeholders were definitely also interested in the issue of transportation infrastructure, on account of the fact that the waterways are an important alternative means of transport in the region. Environmental concerns are also present; evaluated organizations in the environmental sector have succeeded in increasing interest in water resources. This has resulted in environmental protection agencies, whether at a federal or regional level, having an increased influence on the decision making process for issues concerning water resources. As expected, the regulatory agencies also had an increased level of influence in this context. The general interest of neighbouring countries does not have much influence on the decision making process as this considered as being less important than the interests of Brazilian internal stakeholders. It is recognized that local stakeholders, such as agricultural producers, indigenous groups and groups with only minor economical influence have a weak role in this scenario. From the point of view of the research participants, the Brazilian Ministry of Foreign Affairs does not have much influence or general and specific interests, and is graded as being of low concern.

This is probably due to the stable international relations in the region, as far as electrical energy generation is concerned. In the past, this issue caused a great deal of conflict, and resulted in the establishment of the above-mentioned international agreements. At the present time, this issue does not cause any great disputes over the use of water resources. Nowadays, when facing issues concerning the demand for transportation in this international transboundary river basin, which is an important issue that concerns all countries of the region, stakeholders are approached and involved. Environmental organizations, regulatory agencies in diverse areas and electrical energy generation companies are evaluated as having a similarly ranked vested interest in the resource. Environmental organizations are concerned about harmful goods being carried by ships, which could put the environment at risk. This waterway is very dependent on the level of the dams, and the needs of the electrical power generating plants take precedence over transportation issues. Those stakeholders who interact directly with the organization are the ones that present higher values.

Comparing general interest and specific interest charts, there is a concentration trend of stakeholders in the latter, since the specific interests of local stakeholders are emphasized. Strengthening relations with regulatory stakeholders (such as the Brazilian Waterways Agency, the Brazilian Electrical Energy Agency and the Brazilian Water Agency) and environmental protection organizations could be a positive strategy, because these stakeholders share the same common interests. As these agencies are important policy makers, they have a high level of influence in this process. Waterways transportation companies and agricultural industry representatives should be encouraged to become stronger, since although they have a high level of interest, they have a low level of influence.

The DNIT's activities would be enhanced by closer cooperation with stakeholders. Acting at a local level could also be a useful strategy to strengthen the organization, for example, by getting better acquainted with those stakeholders considered as having low influence and developing actions aimed at improving their relative position, thus gaining allies in the search for the best conditions for the development of the activities. It was noticed that whilst the evaluations of the governments of Argentina and Paraguay show a high level of general interest, this is not reflected in specific interests. This analysis may be due to the fact that these two neighbouring countries have direct communication with maritime ports on the Parana River in the south, and the transport of their merchandise to Brazil is of little or no interest.

7.3.4
The Way Forward: Suggested Actions for Improvements

Based on the results of this research some recommendations were established that can be divided into two groups:

National stakeholders:
1) improve interaction with other stakeholders to better understand their performance and policies;
2) send a questionnaire to stakeholders to gauge their perception of particular issues.
3) improve relations with the Brazilian Ministry of Foreign Affairs to facilitate relations with international stakeholders;
4) increase the participation of the DNIT in all international and national forums related to this water resource, not only in forums related strictly to its main objectives but also to those that can provide supplementary information about all kinds of stakeholders;

International stakeholders:
5) gain a better understanding of the legal framework in Argentina and Paraguay;
6) introduce the organization to these countries;
7) initiate dialogues with these stakeholders;
8) participate, even as an observer, in similar committees and commissions in which Brazil already participates, such as the Paraguay Waterway Commission, which already has experience in international negotiations and agreements;
9) expand this research to international stakeholders if possible.

References

1 Tucci, C.E.M. (2004) La Plata River Basin water resources review–a regional point of view–Vol. I, CIC–La Plata River Basin Intergovernmental Coordination Committee.

7.4
Cooperation in the Navigable Course of the Sava River

Dragan Dolinaj and Milana Pantelić

7.4.1
Introduction

The spring of the Sava is located in the northeast part of Slovenia, upstream from the town of Radovljice. The Sava is created by two headwaters, the Sava Dolinka and the Sava Bohinjka. In its course through Slovenia the river Sava has made a composite valley (Figure 7.4.1).

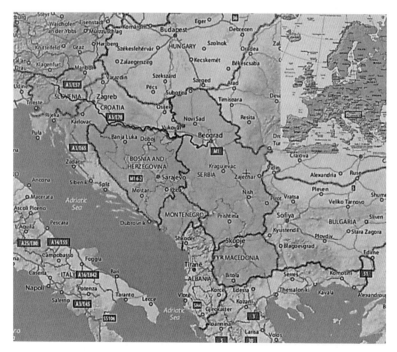

Figure 7.4.1　Position of the Sava River.

In its further course from the confluence of Krapina to Belgrade the river Sava is a typical plane river. Its course from the spring of Sava Dolinka to the confluence into the Danube is 945,5 km long and its length from the point where the Sava Dolinka and the Sava Bohinjka come together is 895 km [1]. The direction of its course is west-easterly and the river basin stretches in the same direction. Its course is wider nearer to the confluence of the river Drina, where it reaches its greatest width, whereas further towards Belgrade it becomes narrower. Asymmetry is one of the characteristics of the river Sava. The right-hand side of the river basin, which covers 77.8% of the entire basin area, is wider than the left-hand side, which covers only 22.2%. Its course is 446 km long in Croatia [2]. The river Sava also forms the natural border between The Republic of Croatia and Bosnia and Herzegovina. The length of this natural border is 300 km. (from the 507th to the 207th km of the river course) [2]. It enters the territory of the Republic of Serbia at Jamena village. Its course is 207 km long in Serbia. From the 207th km to a point 174.3 km further downstream the river Sava forms the natural border between the Republic of Serbia and Bosnia and Herzegovina. A confluence of the Sava's biggest tributary, the river Drina, occurs at the point of 174.5 km. Around 28% of the river Sava's basin is that of the Drina River basin. The river Sava gets around 21% of its water from the river Drina [1]. The Drina brings approximately $370 \, \text{m}^3 \, \text{s}^{-1}$ of water to the river Sava. The water regime of the river Sava is further influenced by the high water level regime of the river Drina, especially if this occurs at the same time as a high level of water in the Sava. Thus, in April, when the river Sava has a huge amount of water, it gets more than $640 \, \text{m}^3 \, \text{s}^{-1}$ from the Drina, whereas in May the flow of the river Drina is $620 \, \text{m}^3 \, \text{s}^{-1}$. The minimum flow of the river Drina is in September ($117 \, \text{m}^3 \, \text{s}^{-1}$). The Drina has fluctuations in both water level and flow, and on average its flow is more than $2500 \, \text{m}^3 \, \text{s}^{-1}$ for two or three years, and every 15th year its flow is more than $4000 \, \text{m}^3 \, \text{s}^{-1}$. A high level of water of the river Drina occurs during all seasons, but is more frequent in summer. The unstable regime influences the regime of the river Sava at its lower course in Serbia.

7.4.2
Navigable Course and Cooperation

The further development of nautical tourism in the river Sava is conditioned by the existence of a defined navigable course, piers and wharfs, as well as tourist capacities. During the 1990s the navigable course of the river Sava was neglected in Serbia, Croatia and Bosnia and Herzegovina. It is possible to sail down the river now, but the regulations of the captain's offices in Belgrade (Serbia), Sremska Mitrovica (Serbia), Slavonski Brod (Croatia) and Sisak (Croatia) have to be obeyed. The river Sava flows through the territory of the Republic of Croatia from the 635th to the 583th km and this is a category II navigable course [3, 4]. Thus, the maximum tonnage is 650 t [3]. From the 583rd to the 507th km the river Sava also flows through the territory of the Republic of Croatia. This is a category III navigable course, that is, the maximum tonnage is 1000 t [3, 4]. The minimum dimension below the bridges is 5 m if the water regime is high on the whole course of the river Sava through Croatia and in the border area between the Republic of Croatia and Bosnia and Herzegovina [2, 5]. At the 207th km the river Sava enters the territory of the Republic of Serbia and from the 207th to the 174.5th km it forms the natural border between the Republic of Serbia and Bosnia and Herzegovina. From the 207th km to the town of Šabac (103rd km) it is a category III navigable course. From the 103rd km to the confluence of the Sava into the Danube in Belgrade (Serbia) it is a category IV navigable course, that is, the maximum tonnage is 1450 t. Some problems can also occur when sailing through Serbia. From Sremska Rača (175th km) to Šabac (103rd km) shallows can occur if the water level regime is low. This makes sailing less safe and endangers the allowed sailing level of 2.5 m. There are more curves with a radius of less than 650 m upstream from Sremska Mitrovica, as well as two curves with a radius of less than 350 m (minimum value according to the EEC criteria). The navigable course of the river Sava through the territory of the Republic of Serbia has an unfavourable curve that makes navigation more difficult and sailing less safe. This also influences the sailing economy because in some parts it is possible to sail only in one direction. Two bridges near Šabac and one train bridge in Belgrade do not allow for sailing if the water level regime is high. The most favourable water level for sailing is when water gauges in Sremska Mitrovica and Šabac show a water level of + 25 cm and 150 cm. Waves can reach a maximum height of 0.30 to 0.50 m on the river Sava, and can cause lesser problems. According to the present conditions the river Sava navigable course is category III, that is category IV according to the dimension (it is a category IV navigable course to Šabac and a category III navigable course from Šabac). The riverbed of the river Sava is not suitable for sailing if the water level regime is low.

Since the navigable course has not yet been completed, only the temporary state can be analysed. From a navigational and sailing point of view, and according to the special characteristics of navigable courses and the dimensions of a navigable system, the river Sava can be divided into three sectors, and these sectors can be further divided into subsectors.

1) Upper Sava sector, Sisak (583 km)–Bosanska Gradiška (459 km);
2) Middle Sava sector, Bosanska Gradiška (459 km)–Sremska Mitrovica (136 km) with three subsectors, Bosanska Gradiška (459 km)–Slavonski Brod (364 km), Slavonski Brod (364 km)–Brčko (220 km), Brčko (220 km)–Sremska Mitrovica (136 km);
3) Lower Sava sector, Sremska Mitrovica (136 km)–Belgrade (0 km).

The upper Sava sector is characterized by numerous curves, the narrow width of its navigable course, numerous shallows and an insufficiently marked navigable course. All these factors influence the safety of sailing, the size of vessels and their drafts. Gauges in this sector are Galdovo (elevation '0' is at a height of 91.47 m above sea level), Sisak (elevation '0' is at a height of 90.95 m above sea level), Sisak (elevation '0' is at a height of 91.47 m above sea level) and Jasenovac (elevation '0' is at a height of 86.82 m above sea level). Shallows represent the biggest problem in this sector. The biggest shallows are from 582 to 580 km, 576 to 574 km, 546 to 544 km, 534 to 532 km, 509 to 506 km and at 493 km. There are 15 shallows that endanger sailing in this sector during all seasons (Table 7.4.1). Curves also endanger sailing in this

Table 7.4.1 Shallows in the Sava River.

Upper Sava sector		Middle Sava sector		Lower Sava sector	
km of river	Width of the shallow in metres	km of river	Width of the shallow in metres	km of river	Width of the shallow in metres
582–581	35	458–456	30	133–132	50
576–574	30	445–440	30	116–114	40
571–570	35	423–422	40	110–104	50
565–562	35	419–417	40	103–101	40
556–554	40	408–407	30	100–98	40
553–550	30	394–393	50	95–93	35
548–547	25	390–389	30	90–86	40
546–544	25	387–386	50	85–80	40
534–532	25	382–379	40	74–70	40
525–522	30	377–372	30	61–56	60
517	40	370–365	30	54–51	40
509–506	30	359	50	28–27	35
502–498	45	329–313	30	17–14	40
496–495	30	309–306	30		
493	45	305–303	30		
		295–292	40		
		294–293	25		
		289–288	50		
		287–284	30		
		278–276	40		
		274–270	50		
		269–264	30		
		255–254	30		
		238–236	40		
		221–220	25		
		215–213	30		
		212–209	30		
		207–206	30		
		201–198	50		
		188–187	35		
		182–181	35		
		177–176	55		
		168–166	50		
		165–163	30		
		158–155	40		
		153	50		

region. The most difficult curves are at 553, 539, 535, 530 and 478 km. There are 15 registered curves in this sector that endanger sailing. The middle Sava sector is the longest, stretching from 459 to 136 km. It has three subsectors. The position of the navigable course in the subsector Bosanska Gradiška–Slavonski Brod is much better than in the previous sector. The width of the navigable course is between 30 and 35 m in this subsector. The most dangerous shallows are from 458 to 456 km, 445 to 440 km, 419 to 417 km, 382 to 379 km, from 377 to 372 km and 370 to 365 km (Table 7.4.1). There are 13 shallows in this subsector that endanger sailing in the summer months. There are eight curves that can endanger sailing in this subsector. The curve at 412 km is the biggest problem in this subsector, since it can endanger vessels because of the whirlpools near the river banks. The subsector Slavonski Brod–Brčko is very difficult for sailing. There are 20 shallows in this subsector, and seven shallows are wider than 40 m or more. Shallows at 359 km and from 309 to 306 km, 305 to 303 km, 289 to 288 km, at 278 km, from 274 to 270 km and 238 to 236 km represent the biggest problem. There are ten curves that make sailing difficult in this subsector. The most dangerous curves with a small radius are at 299, 292 and 287 km. This subsector is characterized by an area that is extremely difficult to sail in from the 308th to the 289th km. The confluence of the river Bosnia, right tributary of the river Sava is at 306 km. The river Bosnia has a lot of alluvial material, which forms numerous shallows in the river basin. The position, size and the depth of the shallows varies. The subsector Brčko–Sremska Mitrovica is characterized by different sailing conditions. There are bigger shallows from 201 to 198 km, 177 to176 km, 168 to 166 km and at 153 km (Table 7.4.1). There are eleven shallows that can endanger sailing. There are three curves that influence sailing, at 173, 160 and 15 km. The lower Sava sector, from Sremska Mitrovica to Belgrade, has characteristics of low course. The course of the river is more stable, its width is bigger and curves are mild. There are gauges in Sremska Mitrovica (elevation '0' is at a height of 72.22 m above sea level), Šabac (elevation '0' is at a height of 72.61 m above sea level) and Belgrade elevation '0' is at a height of 68.28 m above sea level) in this subsector. There are 13 shallows that can endanger sailing in this part of the course of the river. The most dangerous shallows are near Šabac, at 110–104, 103–101, 100–98 and 95–93 km. There are four curves that can endanger sailing in this subsector. Insufficient depth and shallows cause the biggest difficulties to sailing on the river, and they constitute 11% of the river Sava's navigable course [2], from the border with Croatia to its confluence with the Danube. The greatest number of shallows occurs in the Rača and Šabac sector, which pose the biggest problems to sailing when the water level is low.

The insufficient height of two bridges on the navigable course of the river Sava in Serbia creates additional problems. These are the old train–road bridge near Šabac, where the height of navigable space according to navigable level is only 5.44 m (Figure 7.4.2) and the old train bridge in Belgrade, where the height of free navigable space according to high navigable level is 6.05 m (Figure 7.4.3). Bad marking and low maintenance are characteristics of the entire course of the river Sava, the navigable course is in a bad condition and the safety level is very low. This can be seen from the data on sailing restrictions during an average hydrological year that can last for 11 days. Not enough is invested into the maintenance and marking of the river Sava's navigable course, and the criteria for maintenance, marking and financing have not yet been set out. Thus, the optimal usage of sailing and pier capacities on the river Sava is not possible.

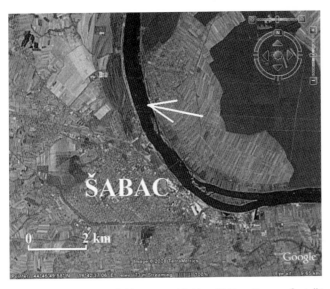

Figure 7.4.2 Position of old train–road bridge, 104 km. Source of satellite image: Google Earth.

7.4.3
Proposal for Further Actions

To solve the problem of navigation on the river Sava, it is necessary to commence activities that will ensure safe sailing. To achieve this, the whole navigable course should be inspected, and the critical parts of the navigable course with priority status

Figure 7.4.3 Position of old train bridge, 2.6 km. Source of satellite image: Google Earth.

should be immediately identified. In the sector of the upper Sava, with its numerous shallows and curves, it is necessary to redefine the navigable course, remove material from the river basin and cut some of the curves. In the middle sector of the Sava, material from the river basin from the 308th to the 289th km should be removed. In the sector of the lower Sava, activities should be related to the problems of shallows in the zone of Šabac, from the 110th to the 93rd km, which means removing material from the river bed and redefining the navigable course. There is the additional problem of the insufficient height of two bridges at Šabac and Belgrade. Later activities should be directed to marking the navigable course and maintaining it.

7.4.4
Conclusion

During the 1990s sailing on the river Sava stopped almost entirely. However, 13 years later in February 2003, the 'Raska', a ship belonging to a Yugoslav river ship company in Belgrade, transported its load from the Romanian pier of Konstanca to the Brčko district. This event reopened the river Sava as an international navigable river. Thus, the Sava became an international river traffic course, which was confirmed with the agreement signed in 2002 between the ministers for foreign affairs of Serbia, Croatia, Bosnia and Herzegovina, and Slovenia. The west Balkan countries still have numerous problems concerning their mutual economic and political affairs. For the river Sava to truly connect the countries it flows through and to become an attractive destination for navigators it is necessary for these countries to reach an agreement on mutual projects connected to the revitalization of the navigable course and to launch nautical and tourism projects. To realize these plans the International Committee for the river basin of the river Sava was founded on 3 December 2002. Its members are Slovenia, Croatia, Bosnia and Herzegovina, Serbia and Montenegro [6]. This agreement took effect on 29 December 2004. The International Committee was founded in order to try to achieve mutual goals: an international sailing regime on the river Sava and its navigable tributaries; the maintainance of its navigable course; taking measures for preventing danger; and removing damage caused by floods, drought and other accidents that could bring dangerous materials into the water [1]. This agreement also implies cooperation and information exchange between the participants, information on the water level regime, sailing regime, regulations, organizational structures, administration and technical support. Cooperation with international organizations is also included (International Committee for the Danube protection ICPDR, Danube Committee, European Economic Committee UN/ECE and European Union institutions). At a meeting of the Committee in July 2007 in Zagreb, the Committee decided to create a category IV international navigable course with the length of 600 km from Belgrade to Slovenia. If this ambitious project is put into practice, all members of the International Committee for the river basin of the river Sava would be connected by the navigable course of the Danube (European corridor 7). Thus, favourable conditions for sailing, tourism and economical development would be created, which would also contribute to better cooperation between the west Balkan countries.

References

1 International Sava River Basin
Commission (2007) Pre-Feasibility Study
for Rehabilitation and Development of the
Sava River Waterway–Final Report, Zagreb,
Croatia. Available at http://www.
savacommission.org/dms/docs/
dokumenti/public/projects/
pfs_sava_river/pre-feasibility_study.pdf
(accessed 1 April 2011).

2 International Sava River Basin
Commission (2008) Decision on
Classification of the Sava River
Waterway, Zagreb, Croatia. Available at
http://www.savacommission.org/dms/
docs/dokumenti/odluke_savske_
komisije/2008_decision_19-
08_classification_of_the_sava_
river_waterway/decision_
19-08_classification_of_the_sava_
river_ waterway.pdf
(accessed 1 April 2011).

3 International Sava River Basin Commission
(2006) Protocol on the navigation regime,
Zagreb, Croatia. Available at http://www.
savacommission.org/dms/docs/
dokumenti/documents_publications/
basic_documents/

protocol_on_navigation_regime.pdf
(accessed 1 April 2011).

4 International Sava River Basin
Commission (2007) Navigation Rules on
the Sava River Basin, Zagreb, Croatia.
Available at http://www.savacommission.
org/dms/docs/dokumenti/odluke_
savske_komisije/2007_decision_30-07_
navigation_rules_on_the_
sava_river_basin/decision_30-07_
navigation_rules_ on_the_ sava_river_
basin.pdf (accessed 1 April 2011).

5 International Sava River Basin
Commission (2007) Rules for Waterway
Marking on the Sava River Basin, Zagreb,
Croatia. Available at http://www.
savacommission.org/dms/docs/
dokumenti/odluke_savske_komisije/
2007_decision_31-07_rules_for_
waterway_marking_on_ the_sava_
river_basin/decision_31-07_rules_for_
waterway_marking.pdf
(accessed 1 April 2011).

6 International Sava River Basin Commission
(2007) Draft Strategy on implementation of
the Framework Agreement on the Sava
River Basin, Zagreb, Croatia.

7.5
Transboundary Cooperation through the Management of Shared Natural Resources: The Case of the Shkoder/Skadar Lake

Djana Bejko and Brilanda Bushati

7.5.1
Introduction

7.5.1.1 Ecological Values of the Lake Shkodra/Skadar

The cross-border site of the Shkodra/Skadar Lake lies within the 5500 km^2 large transboundary basin shared by the Republic of Albania and the Republic of Montenegro (Figure 7.5.1). The area of the targeted ecosystem is home for 350 000 people, who live in five municipalities in both countries. The lake, which is the largest on the Balkan Peninsula, is located in the karstic terrain of the south-eastern Dinaric Alps in Albania and the Skadar-Zeta valley in Montenegro.

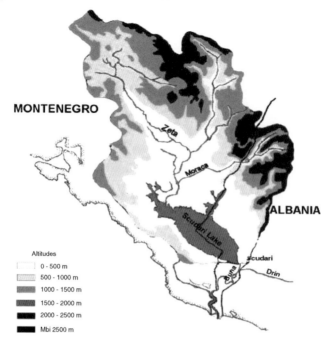

Figure 7.5.1 Map of Lake Shkodra ecosystem. Source: Reference [2].

Its surface varies between 540 km² in winter and 370 km² in the summer, and its shoreline is 168 km long. The Shkodra/Skadar Lake has an open drainage basin, receiving most of its waters through the Morača River and emptying into the Adriatic Sea through the 44 km long Buna/Bojana River. The large variety of habitats supports a rich number of species: 726 vascular plants, with more then 30 rare species, more then 50 species of fish, 15 of them being endemic to the water system, 15 amphibians, 30 reptiles and 271 species of birds, 90% of which are migratory species of international conservation concern [1].

Owing to its natural richness and ecological values, most of the lake and the adjacent wetland are designated as areas of official nature conservation and are also recognized as wetlands of international importance under the RAMSAR convention on both sides of the lake, consisting of 40 000 hectares in Montenegro and 23 027 hectares in Albania [3].

Agriculture, fishing and tourism are the most significant economic activities of the population around the lake. Fishing is the most traditional means of living, and even nowadays fish stocks are exploited at a considerable level. The catch is composed mostly of common carp (*Cyprinus carpio*), bleak (*Alburnus alburnus alborella*), perch (*Perca fluviatilis*) and eel (*Anguilla anguilla*) and shows a downward tendency due to overexploitation. According to past data [4] annual fish catches reached more then 700–1000 tonnes per year. In the lake basin both intensive and extensive forms of farming occur; however, small-scale subsistence production of vegetables and dairy products is more widespread. Products like the famous 'Vranac' grapes, Shkodra tobacco, 'Zeta' white corn or white goat cheese are very typical to the region and have both an economic and cultural value. Making baskets, boxes and other handicrafts out of reeds, willow and other plants is another traditional and widespread activity, especially on the Albanian side of the lake. As well as the possibility of visiting monuments of historical and cultural significance and witnessing traditional crafts, eco-tourism also has potential for the sustainable use of natural resources, as the area is very scenic [5].

After the 1990s, the anthropogenic pressures on the natural resources of the Shkodra/Skadar Lake increased and the range of human activities extended. As a result the natural values of the ecosystem, biological diversity and the water quality of the lake are threatened. These issues need timely addressing if they are not to become real obstacles for the sustainable development of the area. [4].

7.5.2
Working Method

This chapter was based on the work of Brown *et al.* [6] and the programme 'Transboundary cooperation through the management of shared natural resources' implemented in Albania and Montenegro from 2000 to 2008.

On the transboundary level, the first action was an extensive process of gathering and analysing available data and information, which directly involved relevant key players and academic institutions, to establish a database exchange network. A study on the local socio-economic conditions and joint surveys on different topics related to the protection of natural resources, biological diversity conservation and pollutants prevention were carried out on a transboundary level, resulting in an inventory of point and non-point pollution sources in the Shkodra/Skadar watershed, an assessment analysis of the polluters' impact on water quality, a database on biodiversity at a transboundary level and an assessment on changes in habitat.

The critical polluters and impacts on water quality are analysed, using the list and the inventory of the hot spots in the region [7]. The data analyses were compared with the Water Framework Directive [8] and the relevance of any differences were presented.

Based on the findings, the objectives for preventing point and non-point pollutants and for improving the water quality of the Shkodra Lake surface water are elaborated and set according to the priorities for reducing particular polluters. The concluding part of the study provides solutions for water quality protection and improvement. The study also indicates the costs involved based on the implementation of best management practices, according to the action plan developed.

As a direct follow-up to earlier efforts towards setting priorities, environmental planning initiatives were launched that reflect the local development context. Joint strategies were developed and environmental planning processes started in 2006. Different tools for the development of topic orientated action plans dedicated to building up local capacities on concrete environmental topics were developed. These initiatives formed the first actions in the programme, where local stakeholders addressed specific resource management topics, and took steps towards identifying measures [6].

Creating cooperation mechanisms for cross-border dialogue for the management of shared natural resources following the principles of democracy and transparency was amongst the key aims of the programme. To this end, the Transboundary Shkodra Lake Forum was established as a neutral platform where priorities could be set at a transboundary level. This functions as an informal cooperation body and neutral cross-border coordination platform, which gathers key local stakeholders together to set the priorities for intervention [9].

7.5.3
Conclusions

Based on the working results, the lake Shkodra/Skadar surface water is affected by point and non-point polluters, including a hot spot located in the north of Albania, only 500 m away from the lake shore. Chemical pollution dates back to 1991, when 480 tonnes of hazardous chemicals *en route* from Germany were left at the railway station of Bajza (North Albania). As of September 2009, 12 tonnes of hazardous chemicals and 6 tons of chemical compounds were still waiting to be shipped for destruction. The difference between the original 480 and the 18 tonnes collected 17 years later has been rinse-out during rainfall, which filtered into the lake basin, causing fish to die during 1997 on the shores of the lake [9].

The water basin is affected by organic and inorganic point polluters, like non-treated sewage waters from the main cities of Shkodra (120 000 inhabitants) and Kopliku (11 020 inhabitants) [4], and non-treated industrial waters, for example, from the aluminium plant located in Montenegro, that affect the water quality of the Shkodra/Skadar lake [10].

Most of the pollutants affecting the surface water of Lake Shkodra/Skadar, groundwater, soil and air originate from the city of Podgorica, situated on the Morača river terraces in the Zeta plain. On the Albanian side the main polluter is the city of Shkodra with its solid waste and wastewater. The main sources of pollution are the aluminium plant in Podgorica (KAP), the steel works in Nikšić, and wastewater and municipal waste from the cities and towns in the basin. [7]

According to the analysis conducted in 2001, the concentration of heavy metals in the water remains somehow within European Union (EU) norms (ISO 83-1) (Table 7.5.1).

Comparing the data presented in Table 7.5.1 with EU (ISO 83-1) standards it can be seen that the concentrations of aluminium (Al $= 45.8\,\mu g\,l^{-1}$) and cobalt (Co $= 1.09$ $\mu g\,l^{-1}$) are high. This finding permits priorities to be set for the reduction of industrial discharges in the watershed to protect water quality.

Even recent analyses of lake water and sediments show an increase in concentrations of heavy metals, for example, ammonium complexes in 2007. The concentrations are higher at the mouth of the Morača River, mainly due to the industrial wastes originating from the KAP. The highest Hg content in July 2005 was

Table 7.5.1 Metals content in the watershed of Lake Shkodra/Skadar. Source: Reference [11].

Metals	Concentration ($\mu g\,l^{-1}$)	EU standards ($\mu g\,l^{-1}$) (ISO 83-1)
Na*	1.85	<100
K*	0.36	<10
Zn	50.2	200
Mn	3.04	100
Cu	9.97	50
As	0.321	50
Al	45.8	0
Ni	<0.1	100
Pb	1.24	50
Cr	2.42	50
Co	1.09	0
Fe	0.12	1.0
Hg	6.29	1–10

$1.77\,\mathrm{mg\,kg^{-1}}$ in sediments ($0.40\,\mathrm{mg\,kg^{-1}}$ in fish), while this was undetectable from 1974 to 1977. Heavy metals are contaminant of the lake water. They are found in the sediments and the fish tissues too. [11]. The Hg in the sediment exceeded the EU standards in four of eight locations and Ni in two of eight locations in 2005. On the Albanian side of the Shkodra/Skadar Lake in 2003, at seven out of ten locations the concentration of Ni exceeded the EU standards.

In this context the guideline management plan for the Shkodra Lake watershed was prepared and a strategic goal was set in respect of protection and improvement of Shkodra Lake water quality. Meanwhile the vision for the future has been developed as presented below:

The Shkodra/Skadar Lake is transboundary with equal levels of protection on both sides. This level of protection is in accordance with high environmental standards, high water quality and rich biological diversity. The Shkodra/Skadar Lake is a suitable area for sustainable activities and it offers ecological, historical, cultural, rural and educational experiences with a lot of unique places to see and visit. The lake is used in a sustainable way, with cross-border cooperation and management and a high level of ecosystem protection.

Acknowledgements

Albanian and Montenegrin experts and academics from the universities of Albania and Montenegro contributed to the implementation of the programme 'Transboundary Cooperation through the Management of Shared Natural resources' for almost eight years. They have contributed with data, studies, presentations and the input they offered to the discussions both within and outside the workshops findings and priorities. Their contribution has been vital to my work and I would really like to express my gratitude to all of them.

Meanwhile, I would like to express my special gratitude to Professor Aleko Miho who supervised my degree from 2004 to date. His contribution, dedication and academic level are much appreciated. I also thank the authorities of both Albania and Montenegro for their commitment and support and say how much I appreciated their contribution and enjoyed our interactions. I would also like to thank all the actors that helped the work become a reality, and all others who, in various capacities, helped me reach a successful conclusion.

References

1 Dhora, D. (2006) *Liqeni i Shkodres*, University Luigj Gurakuqi, Shkoder, Albania.

2 Birdlife International (2001) *Important Bird Areas and Potential Ramsar Sites in Europe*, Birdlife International, Wageningen, The Netherlands.

3 Ziu, T. and Filipovic, S. (2002) A survey of Shkodra Lake. pp. 10–14, 21–33, 31–33.

4 Anonymous (2006) Information Sheet on Ramsar Wetlands (RIS) – Albania, Lake Shkodra.

5 Brown, E., Peterson, A., Kline-Robach, R., Smith, K. and Wolfson, L. (2000) *Developing a Watershed Management Plan for Water Quality: An Introductory Guide*, Millbrook Printing, Michigan, USA.

6 The Regional Environmental Center (2003) Risks and potentials assessment of Shkodra/Skadar Lake, Shkoder, Albania.

7 EC (2000) The Water Framework Directive – integrated river basin management for Europe, WFD, 60/EC. Directive 2000/60/EC of the European Parliament and of the Council establishing a framework for the Community action in the field of water policy, http://ec.europa.eu/environment/water/water-framework/index_en.html (accessed 04 October 2010).

8 Bejko, D. (2006) Veshtrim ekologjik i pellgut Liqeni i Shkodres, University of Tirana, Faculty of Natural Sciences, 196 pp.

9 APAWA, CETI, SNV, Montenegro, MoTE, MEFWA, GEF, & World Bank (2007) Lake Skadar/Shkodra integrated ecosystem management project. The strategic action plan (SAP) for Skadar/Shkodra Lake Albania & Montenegro, http://www.gov.me/files/1248091290.pdf (accessed 04 October 2010).

10 Bekteshi, A. and Beka, I. (2001) Integrated monitoring of Shkodra/Skadar Lake Project. Report on integrated monitoring and water quality, Heidelberg University, Germany.

11 World Bank and Royal Haskoning (2006) Lake Shkodra/Skadar Transboundary Diagnostic Analysis, Albania & Montenegro, http://www.gov.me/files/1248091671.pdf (04 October 2010).

7.6
How Far is the Current Status of the Transboundary Shkodra Lake from Requirements for Integrated River Basin Management?

Spase Shumka, Udaya Sekhar Nagothu, Eva Skarbøvik, Andrej Perovic and Sotir Mali

7.6.1
Introduction

Lake Shkodra (Figure 7.6.1) is the largest lake on the Balkan Peninsula in terms of water surface. The drainage area of the lake is about 5500 km^2 (4470 km^2 in

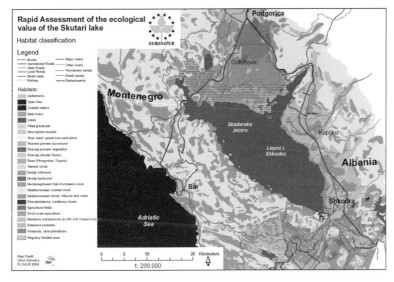

Figure 7.6.1 Transboundary Shkodra/Skadar Lake habitats. Source: Euronatur.

Montenegro and 1030 km^2 in Albania). The lake area varies between 353 km^2 in dry periods and 500 km^2 in wet periods (at the maximum level, 335 km^2 is in Montenegro and 165 km^2 in Albania). The basin of Lake Shkodra is a depression located south of the Dinaric Alps and orientated northwest-southeast, parallel to the current shore of the Adriatic coast. Sometimes the outflow from the lake in Buna-Bojana is impeded due to the increase in the flow in the Drin River [1]. This occurs mostly in the period from December to February, but may also occur during other months, depending on the water released from upstream reservoirs. The climate in the Shkodra basin is Mediterranean, but with higher rainfall amounts than in general in Mediterranean areas due to the mountains. Rainfall on the lake is between 2000 and 2800 mm per year, but within the basin some areas receive over 3000 mm annually. Humidity levels are low, sunshine hours and temperature in summer are high, giving high evaporation. The temperature in winter is low, due to the high elevations and predominant easterly and northerly winds.

From a zoogeographic perspective, the Shkodra Lake region is located in a zone where two major zoogeographic areas meet: the Palaearctic region (Europe, Asia, the Mediterranean and North Africa) and the Palaetropic region (Africa). Their linkage and influences can be seen amongst bird fauna, with incidences of African species (e. g. African cuckoo (*Cuculus gularis*), African black heron (*Egretta ardesiaca*) and winter migratory species of West Siberia (ducks and geese). During the last glacial period Lake Shkodra represented a refuge for several species existing at that time. This is evident from the relic and endemic animal and plant species found in the area. After the ice age, species such as the Dauric swallow (*Hirundo dauric*), Syrian woodpecker (*Dendrocopus syriacus*) and Spanish sparrow (*Passer hispaniolensis*) came to the region as they expanded their distribution area. Lake Shkodra is an important habitat for bird migration in the Mediterranean region. Every year large numbers of waterfowls use the lake for breeding and feeding. However, in recent years bird numbers have declined drastically due to uncontrolled hunting [2–4]. The Dalmatian Pelican (*Pelicanus crispus*) is highly endangered and is affected by flooding of nests and human disturbance [5, 6]. The avifauna provides a good potential to attract tourists and birdwatchers to the region.

Lake Shkodra's biodiversity has developed in a unique physical environment where geology, geomorphology, hydrology and climate provide a wide variety of habitats. Total biodiversity is high (species–area relationship = 0.875) and the region is considered to be a biogenetic reserve of European importance. In reality, it is currently faced with real threats and challenges [5]. In recent years, the lake has also started to attract a large number of visitors during the tourist season.

7.6.2
Survey Methods

A questionnaire survey was conducted in 2007–2008 on the Albanian and Montenegrin sides of the lake to collect data related to the socio-economic conditions, dependency of local people around the lake on tourism, the local perceptions and tourist trends. The study also looked at the secondary data related to institutional and policy measures in both countries. In total 200 respondents in Albania and 80 respondents in Montenegro were interviewed in 2007. The respondents from different villages and towns around the lake were selected based on random sampling. The questionnaire was separated into four categories. The first category consisted of questions on socio-economic characteristics of the household, followed by questions related to lake pollution, perceptions about tourism and environmental quality in the lake.

A series of water quality tests were conducted based on standard principles of chemical and physical analyses both *in situ* and in the laboratory.

7.6.3
Results and Discussion

Table 7.6.1 gives an overview of statistics on Lake Shkodra and the tourism industry, based on the questionnaires. The main attractions and priorities are also listed. The results of the questionnaires indicate that increases in agriculture, industry and tourism development are regarded as positive by locals, as these sectors provide employment and more income to the region. However, development in these sectors is also accompanied by various negative impacts on the catchment's environment, including:

- increased demand for irrigation water and water for the tourism industry, which may contribute to low water levels during summer in rivers and lake;
- inappropriate use of land resources, especially close to the lake shores, for construction of hotels and restaurants, and conversion of shore areas into beaches;

- pollution from untreated sewage and waste water, seepage from solid waste sites, which can lead to an increase of phosphorous and nitrogen content (Figures 7.6.2 and 7.6.3);
- cultivation of land close to rivers and lake shores with no vegetation zones to reduce particle, nutrient and pesticide runoff;
- pollution from farm machinery and pesticide spray equipment that are rinsed directly in river or lake waters;
- deforestation in the catchment's areas and so on.

The sampling strategy for chemical analyses of this project was designed to analyse the eutrophication challenges of the Shkodra Lake. The sampling point was based on the pelagic area of the lake.

The results of the water quality monitoring showed that the lake is as yet in a mesotrophic state, although some high concentrations of nutrients can be found. It is important to continue the monitoring of the lake due to the many anticipated changes in the catchment area of both countries. Increased focus on intensified agriculture, tourism and other activities will increase the pressures on the lake. Given the lake's unique biodiversity, preserving it by using sound and integrated management methods on a European basis will be of benefit to both countries. It is recommended that the principles of the EU Water Framework Directive are followed, in terms of

Table 7.6.1 Tourists and their preferences in Lake Shkodra.

	Albanian side	Montenegrin side	Lake Shkodra
Total number of tourists (in 2007)	45 000	95 000	140 000
Day visitors	45%	60%	52%
Tourists overnight	55%	40%	48%
Attracted by the lake	65%	71%	68%
Attracted by beaches	49%	3%	26%
Attracted by cultural/historical sites	62%	13%	38%
Attracted by food and hospitality	76%	22%	49%
Priority is good quality lake water	Highest priority 60% Medium priority 40%	Medium priority 30% Low priority 70%	Highest priority 30% Medium priority 35% Low Priority 35%
Priority is clean beaches	Highest priority 100%	Medium priority 24% Low priority 54% No priority 22%	Highest priority 50% Medium priority 12% Low priority 27% No priority 11%
Willingness to pay for clean environment	€26	€4	€15

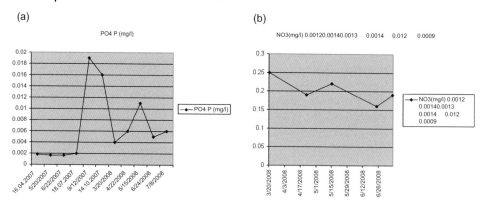

Figure 7.6.2 Concentrations of (a) phosphate and (b) nitrate in the analysed period.

monitoring, preparing river basin management plans and involving stakeholders [8]. In terms of monitoring, it has been shown that integration at a European level has still a way to go before it is harmonized [9], despite the fact that the EU Member States are presently implementing the requirements for monitoring according to the WFD Common Implementation Strategy [10]. As shown by Skarbøvik *et al.* [11], such harmonization across borders is difficult but not impossible to achieve, and in the case of the Shkodra Lake this should be a target issue along with EU integration of Albania and Montenegro.

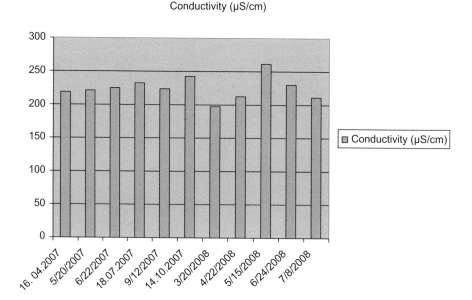

Figure 7.6.3 Conductivity unite μS/cm.

Tourism has been increasing in the region in recent years (Table 7.6.1), as evident from the surveys and secondary data. This might contribute to positive economic development in the region, but, at the same time, could affect the lake water quality and ecosystem. Regulations on building tourist infrastructure are weak, and differ because of the national regulatory frameworks within each country. At the same time tourists visiting Lake Shkodra are concerned about the water quality and clean beaches. This indicates that sustainable tourism development and ecosystem management are directly linked. To achieve sustainable management transboundary lakes like Shkodra need to have the cooperation of all parties sharing the waters. For Lake Shkodra, transboundary cooperation is very weak, and only a very few initiatives have been taken so far in this direction. From this study it is evident that although the intention to cooperate exists at different levels, in practice this is not adequate.

7.6.4
Conclusions

From the survey conducted (Table 7.6.1) there is evidently an increased interest in tourism development. This is expected to directly influence water quality, and the lack of integrated management is a challenging issue.

To achieve transboundary cooperation it is an advantage to have a set of commonly agreed principles. The EU Water Framework Directive should be useful in this context, and designating the Shkodra area as a pilot basin for implementation of the directive in the two countries is an idea that the authors wish to pursue [10].

Lake Shkodra is a unique lake with a rich fauna, including some endemic species. It is an important breeding ground for rare species and was therefore declared a Ramsar site in 2000 [13]. Being transboundary with the responsibility of management shared between two countries does not seem to have been an advantage in the past. It is hoped that the implementation of the WFD will cause renewed national responsibility for the relevant authorities in Albania and Montenegro, and as such, that harmonized monitoring can be implemented to ensure sound transboundary management of the lake.

Together with other initiatives, raising awareness and building capacity to face the challenges of transboundary cooperation is essential at this time.

Acknowledgements

The DRIMON Project (www.drimon.no) is funded by the Research Council of Norway. The grant is linked to a programme for research cooperation between Norway and a limited number of Balkan countries.

References

1 Kashta, L., Dhora, Dh. and Sokoli, F. (2001) Biological acknowledgment on the Albanian part of Shkodra Lake, in The Shkodra/Skadar Lake

Project, conference report 2001, pp. 41–46.

2 Van den Tempel, R. (1992) Verstoring van watervogels door jacht in wetlands. Technisch rapport Vogelbescherming 9, Natuurmonumenten, Zeist.

3 Schneider-Jacoby, M. (2000b) Freizeit und Entenschutz am Wasser – Sicherung der Brut- und Rastplätze von Kolbenenten und Moorenten in Deutschland, Schriftenreihe Landschaftspflege Naturschutz, 60, pp. 81–93, Bundesamt für Naturschutz, Bonn.

4 Schneider-Jacoby, M. (2001) Auswirkung der Jagd auf Wasservögel und die Bedeutung von Ruhezonen. ANL, Laufener Seminarbeiträge. *Störungsökologie*, **1/01**, 49–61.

5 Ruci, B. (1983) Data about the flora and vegetation of Shkodra Lake, Buletini i Shkencave Natyrore, Tiranë, nr. 3–4, pp. 109–113.

6 Saveljic, D., Rubinic, B. and Schneider-Jacoby, M. (2004) Breeding of Dalmatian pelican (*Pelecanus crispus*) on Skadar Lake. *Agrocephalus*, **25** (122), 111–118.

7 EC (2000) Directive 2000/60/EC of the European Parliament and of the Council of 23 October 2000 of establishing a framework for community action in the field of water policy.

8 Parr, T.W., Ferretti, M., Simpson, I.C., Forsius, M. and Kovács-Láng, E. (2002) Towards a long-term integrated monitoring programme in Europe: Network design in theory and practice. *Environmental Monitoring and Assessment*, **78**, 253–290.

9 EU (2003) Common Implementation Strategy for the Water Framework Directive (2000/60/EC) Guidance Document No 7. Monitoring under the Water Framework Directive Produced by Working Group 2.7–Monitoring, 160 pp.

10 Eva Skarbøvik, E., Shumka, S., Mukaetov, D., and Nagothu, S.U. (2010) Harmonised monitoring of Lake Macro Prespa as a basis for Integrated Water Resources Management. IRRIGATION AND DRAINAGE SYSTEMS, Volume 24, Numbers 3-4, 223–238, DOI: 10.1007/s10795-010-9099-1.

11 EU (2008) Commission Decision of 30 October 2008. Establishing, pursuant to Directive 2000/60/EC of the European Parliament and of the Council, the values of the Member State monitoring system classifications as a result of the inter-calibration exercise (notified under document number C (2008) 6016).

12 Schneider-Jacoby M., July (2005) Ramsar Site NP Scutari lake (Montenegro) and the potential Ramsar Sites (shadow list) Scutari Lake (Albania) and Bojana – Buna Delta (Albania – Montenegro). Euronatur Mission Report.

Further Reading

Regional Environmental Center for Central and Eastern Europe (2006) Bibliography on Skadar/Shkodra Lake. Promotion of networks and exchanges in the countries of the South and Eastern Europe. Available at http://albania.rec.org/Liqeni%20i%20Shkodre-Bibliografija.pdf (accessed 4 April 2011).

Regional Environmental Centre, (2001): Bibliography on Shodra/ Skadar Lake. Report from the Project: Promotion of networks and exchanges in the countries of the South and Eastern Europe. Podgorica. pp 147 (http://archive.rec.org)

Schneider-Jacoby, M. (July 2005) Ramsar Site NP Scutari lake (Montenegro) and the potential Ramsar Sites (shadow list) Scutari Lake (Albania) and Bojana – Buna Delta (Albania – Montenegro), Euronatur Mission Report.

7.7

Economic Governance and Common Pool Management of Transboundary Water Resources

Bo Appelgren

7.7.1
Introduction

Economic governance is focused on two complementary issues, namely, the problem of regulating transactions that are not covered by contracts or legal rules and the separate problem of rule enforcement. Institutions are made up of a set of rules that govern human interactions with the main purpose of supporting production and exchange as they influence human prosperity [1]. The rules protect property rights and enable the trading of property, as do the rules of the market. Other classes of institutions support production and exchange outside markets and in still another class are governments, which play a major role in funding pure public goods. This raises the following questions: (i) Which mode of economic governance is best suited for effective transactions for the enforcement of adopted measures in transboundary water management? (ii) To what extent are observed modes explained by efficiency and effectiveness? (iii) How should economic governance based on the common property concept be introduced and harmonized with the public regulation concept? (iv) How should economic governance for CPR management be reconciled and consolidated with integrated management processes frameworks including IWRM (Integrated Water Resources Management) and ICZM (Integrated Coastal Zone Management)?

Water resources management at an international level is usually implemented through treaties based on scientific evidence and engineering studies, together with institutions responsible for implementation that generally act as a social planner with allocation enforced at national and sub-national levels. However, this process is facing growing institutional challenges at the regional and national levels. With inefficient uses and demands exceeding available quantities there is constant pressure for costly supply projects to alleviate increasingly frequent national water shortages. Progress in institutional arrangements and international settlements for the regional reallocation in the most water scarce regions (such as the Middle East) is generally slow and is frequently being overrun by unilateral water transfers and infrastructure that reduce the scope and preclude future efficient reallocation in the region (e.g. The Disi Aquifer). Regional social planning allocation of transboundary water resources for maximum aggregate net benefit is less efficient and effective than reallocation through market mechanisms that, while also contested from the hydropolitical aspect and prevailing claims for rightful and equitable shares of the supplies, yield significant total and marginal economic benefits and water savings and respond to the demands in all countries irrespective of the initial ownership. Notwithstanding the opportunities and need for economic governance and while the scope of the physical problems in transboundary water management is increasing in scale and complexity, the range and opportunities of socio-economic and institutional responses for solutions are not immediately recognized [2].

Against this backdrop the critical issue and the opportunity for effective transboundary water resources management is the effectiveness of enforcement for sustainable usage at the national and local sub-national level. Transboundary water resources, such as large regional sedimentary aquifers, river and lake basins, inland seas and marine ecosystems, are common-pool properties and the usual situation is that they are poorly managed with the two primary solutions to the common-pool problem regulated by the central authorities (e.g. in socialism) that own the resource and levy a tax extraction. The solution entails coercion from disfranchising the original users. The other solution is privatization with the need for high initial investment and at very high transaction cost, especially for transboundary water resources. However, recently observed empirical research in economic governance has challenged these conventional understandings and demonstrated the efficiency and effectiveness of alternative common pools management of common resource systems including shared river, lakes and groundwater basins and large marine ecosystems. The final questions are (i) How to proceed and introduce common property based forms of governance as a supplement to national public regulation provided in transboundary water resources treaties and basin institutions? and (ii) How to coordinate and consolidate economic governance empirical research with integrated, multi-disciplinary mainstreaming management processes in sustainable transboundary water development and management, including IWRM and ICZM?

The motivation of the water sharing countries to participate and commit domestic financial and institutional resources in regional cooperation on transboundary water management is generally driven by an expectation for national economic benefits with sustainable national development for social and economic growth. The range and opportunities of socio-economic responses and economic rules for national social economic benefits in transboundary water resources management are becoming increasingly recognized and introduced in all regions. For example, the Organization of American States is well advanced in socio-economic management and supports its 35 members to improve the sustainable use of water resources for the purpose of sustainable social and economic growth [3]. The economic basis for transboundary and domestic water management is emphasized in the European Union (EU) Water Framework Directive (WFD), in Europe and the EU neighbourhood and enforced from the recovery of full financial and economic water costs including resource and ecological opportunity costs to river basin economic plans [4], and the water sharing Nile basin countries are currently introducing a benefit-sharing programme under the Nile Basin Initiative.

7.7.2
Economic Governance of Transboundary Water Management Systems

Regional development and transboundary water resources management depend to a large extent on functional national institutions with the capacity to establish and operate economic management instruments and central and local institutions, including flexible water allocation and use rights and mobilization of local users and communities sharing common property as common-pool resource groups under the public trust doctrine for the efficient allocation of transboundary waters resources at reduced transaction costs, built on a custom-based system of governance, with rules and enforcement mechanisms that benefit from knowledge about local conditions, and which therefore effectively produce acceptable outcomes and with design principles including rules with clearly defined entitlements, adequate conflict resolution mechanisms and a reasonable balance between duties and benefits, graduated sanctions and so on.

'Users of a commons, including unregulated transboundary resources, are caught in an inevitable process that leads to the destruction of the resources on which they depend' [5].

Under-valuation as the cause of depletion and pollution of a transboundary water resource refers to unclear property rights of common-pool resources with consequent problems of overexploitation. The empirical updated and compiled work of Elinor Ostrom [6], whose breakthrough is built on the handy insight of thousands of isolated case studies on the management of CPRs that allow for comparative reviews with the possibility of making strong interferences, has contributed an abundant variety of field evidence to the understanding of economic governance and common-pool natural resources management that includes the cases of transboundary lakes, rivers and ground waters. Orstrom's work shows that the established explanations of the statement 'users of a commons, including unregulated transboundary resources, are caught in an inevitable process that leads to the destruction of the resources on which they depend' [5] is over-simplistic and that in many cases, when common pool property is well managed, the commons users have created and enforced their own rules to mitigate overexploitation. The findings challenge the conventional understanding and solutions of privatization and regulation by central authorities and contribute new insights into conditions, with the focus on institutional diversity, to support resilient institutions and sustainable uses of common pool resources for the management of large-scale resources, including not only transboundary freshwater basins or groundwater aquifers and marine ecosystems but also pastures and fish stock.

7.7.3
Economic Governance Approaches to Transboundary Water Management

Socio-economic approaches are emerging as an increasingly necessary alternative to engineering projects based on scientific inputs and expert opinions for unilateral national water development projects and investments that become uncertain and conflicting hot spots in the transboundary water-sharing context, for example, the Disi Aquifer (Figure 7.7.1). If an international water resource in the territorial jurisdiction of two or more countries is unilaterally developed and

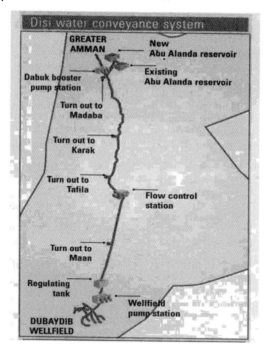

Figure 7.7.1 The Disi aquifer.

used within the national territory without the consent of the other countries, the resource does not form a regional parallel to the transformation of a common-pool resource into a private resource. It is therefore a combination of technological and institutional factors that determines whether resources are managed as common property.

The Disi aquifer development is a unilateral water transfer project in a transboundary groundwater management context that is likely benefit from an alternative joint economic governance approach. The Disi is a fossil, non-renewable aquifer resource shared by Jordan and Saudi Arabia. The aquifer is planned for the annual abstraction and transfer of $100\,000\,000\,m^3$ for a period of 50–100 years to support agricultural production and the supply of the capital city of Amman. The case involves socio-economic and environmental aspects, including rising water costs and values with increased water prices with consequent limitations in local social uses and conflicts between agricultural and domestic uses together with longer term issues in the perspective of critical sub-regional water scarcity and supply [7].

Economic water governance could provide alternative efficient and socially sustainable and environmentally friendly solutions and resolutions in response to the management of transboundary water sharing issue and conflicts. Efficient management of shared waters by local user groups, as common property under public trust and differentiated from public property resources, could provide the required flexibility and cooperative environment as a base for public water investment projects in water scarce transboundary basins and allow for local governance and control of growing withdrawals and pressures on transboundary water resources, and in the case of transboundary aquifers introduce economic institutions rather than economic incentives and water pricing to restrict dispersed proliferation of groundwater wells and other abstraction points.

Legitimate institutions comprise public regulation under government authority and the rules for common property based governance. Economic governance by water communities will reduce the transaction costs for the reallocation of water resources. The transboundary approach involves harmonization of the domestic systems for common property in public trust with agreed defined restrictions and easements for the use of transboundary resources distinguished from public property resources [8]. Under the public trust doctrine the resources in transboundary basins are subject to government regulation but without impediments for extractions for public purposes, including imposed easements for water and environmental conservation and therefore without involving the critical issue of compensations in water-sharing. As a consequence and depending on the State category (whether a guardian-State or a master-State [9]), property as public trust resources would be legitimate and accepted by water-sharing States for economic water governance at the regional transboundary level. At the centre are local stakeholders and water user communities in the socio-economic sectors, in agriculture, domestic supplies and other water-use sub-sectors with well-defined long-term social and economic interests in water resources management, use and conservation and with knowledge of the institutional and technical issues especially at local and operational level.

The possible gains from economic governance in transboundary water management are the opportunities for allocation with low transaction costs and limited coercion, the involvement of the national governments, reduced vulnerability to hydrological uncertainty and cost savings from limited data dependence and field monitoring. The mobilization of self-managed common-pool resources forms an attractive opportunity for institutional sustainability and resilience, to address resource under-valuation, handle hydrological uncertainty in transboundary water management and manage long-term increases in the water demands from population development, intensification of agricultural production and enhanced income and welfare. The basis is that common-pool groups depend on the productivity of the resources and are further motivated by customs and public trust property rights. The groups that occupy and live off the land overlaying the groundwater or adjacent to the surface water bodies represent actual human and institutional resources with the capacity to enforce measures and allocation schedules agreed at the transboundary level. The common-pool groups are legitimately enabled and mandated by national governments for sustainable enforcement of the international management allocation measures with very low transactional and operational costs.

7.7.4
Conclusions

Up-dated findings from empirical research on the economic governance of common-pool resources by local user groups have evidenced the viability of the rules for transactions and enforcement mechanisms as alternative informal institutions and as a solution for common resource allocation with low transaction costs. The proposed institution for the alternative solution for the management in transboundary water common pool resources under the public trust doctrine based on the concept of common property needs to be critically reviewed and further refined from a legal point of view.

The opportunity for efficient common pool governance supports the call for economic governance in response to human behaviour and interactions driven by productivity and exchange for enhanced prosperity under the national social dimension and priority goal of sustainable national development for social and economic growth and welfare. It can be expected to facilitate the progress in transboundary water resources management.

References

1 Williamson, O. (1985) *The Economic Institutions of Capitalism*, Free Press, New York.
2 Burke, J., Sauveplane, C. and Moench, M. (1999) Groundwater management and socio-economic responses. *Natural Resource Forum*, **23**, 303–313.
3 OAS, www.oas.org/dsd/waterresources. htm (accessed 08 October 2009).
4 Kraemer, A.R. (2007) A revolution in Europe; The Water Framework Directive. Presented at the first conference European Sustainable Water Goals (ESWG), Belluno, Italy.
5 Hardin, G. (1968) The tragedy of the commons. *Science*, **162**, 1243–1248.

6 Ostrom, E. (1999) Coping with the tragedies of the commons. *Annual Review of Political Science*, **2**, 493–535.
7 Ferragina, E. and Greco, F. (2008) The Disi project: an internal/external analysis. *Water International*, **33** (4), 451–463.
8 Barraqué, B. (2004) *Water and Ethics, Institutional Issues*, UNESCO/IHP, Paris.
9 Burchi, S. (1991) Current developments and trends in the law and administration of water resources: a comparative state-of-the-art appraisal. *Journal of Environmental Law*, **3** (1), 69–91.

Further Reading

Becker, N. *et al.* (1996) Reallocating water resources in the Middle East through market mechanisms. *Water Resources Development*, **12** (1), 17–32.
Royal Swedish Academy of Sciences (2009) Scientific background on the Nobel Prize in economic sciences, economic governance, http://nobelprize.org/nobel_prizes/ economics/laureates/2009/ecoadv09.pdf (accessed 29 October 2010).

7.8
Water Resources Management in the Rio Grande/Bravo Basin Using Cooperative Game Theory

Rebecca L. Teasley and Daene C. McKinney

7.8.1
Introduction

Water resources management is a complex issue that becomes more difficult when considering shared transboundary river basins. With over 200 transboundary river basins worldwide shared by two or more countries [1] it is important to develop tools to allow riparian countries to cooperatively manage these shared and often limited water resources. Cooperative game theory provides tools to help riparian countries in a transboundary river basin understand the benefits of cooperatively managing their shared and often limited water resources. Cooperative game theory is also useful for transboundary negotiations because it provides a range of solutions that will satisfy all players in the game and provides methods to fairly and equitably allocate the gains of that cooperation to all participating stakeholders, if that cooperation is shown to be possible. The objective of this chapter is to couple cooperative game theory with a comprehensive water management model for the Rio Grande/Bravo transboundary river basin to show the players what their increase in benefit may be under that cooperation and allows players to negotiate for a share of that increased benefit.

A large body of literature outlines cooperative game theory applications in water resources management. However, there is limited work on the application of cooperative game theory to transboundary river basins. Cooperative game theory has been applied to water sharing in the Ganges and Brahmaputra basin [2, 3], the Euphrates and Tigris [4–6] and the Nile Basin [7, 8] basins and for water trading from the Nile amongst the Middle East countries of Egypt and Israel, and the Gaza Strip and the West Bank [9]. In each of these cases, the individual countries were considered as the players in the game. For this research, cooperative game theory concepts are applied to the water scarce transboundary Rio Grande/Bravo basin in North America. Unlike the previous studies, this application specifies individual water users in the basin (i.e. irrigators, municipalities, etc.) as the players in the game that will obtain any increased benefits from cooperation. Additionally, the literature has outlined that the water planning models utilized in transboundary cooperative games tend to be simplified or lack extensive data.

The Rio Grande, or Río Bravo del Norte as it is known in Mexico, is home to over 10 million people and is considered to be one of the most water stressed basins in the world [10]. Rapid population growth, economic development and recent severe droughts have placed additional strain on already limited water resources of the basin. This transboundary basin has increased complexity due to its large size. The river is 3107 km from its headwaters in southern Colorado in the USA to the Gulf of Mexico. The Rio Grande/Bravo flows through the three US states of Colorado, New Mexico and Texas and the four Mexican states of Chihuahua, Coahuila, Nuevo Leon and Tamaulipas and forms over 2000 km of the international border [11].

A collaborative effort between technical and expert counterparts in Mexico and the USA is underway with the goal of improving management of the Rio Grande/Bravo's scarce water resources through the development and modelling of management scenarios. These management scenarios are evaluated in a hydrologic planning model that was developed in the software WEAP (Water Evaluation and Planning). The Rio Grande/Bravo hydrologic planning model is a demand driven model containing hydrologic and hydraulic data for 60 years with water rights and logic for legal institutions in the basin including international treaties and allocation rules. This model is used with the scenarios to demonstrate the effects of management changes on water availability in the basin. Additionally, this model is used for calculations in the cooperative game analysis. Details of the hydrologic planning model are contained in Danner *et al.* [12].

7.8.2
The Water Demand Reduction Cooperative Game

To illustrate the cooperative game theory application, a game utilizing water buy-backs is described. To reduce the water demand by a user, a portion of their water right is purchased and retired, or bought-back, hence the term, buy-back. The Water Demand Reduction game utilizes a scenario based on a water rights buy-back program named PADUA (Programa de Adecuación de Derechos de Uso del Agua y Redimensionamiento de Distritos de Riego) that was developed to reduce water allocations that most likely would not be met in drought conditions [13]. Details of the modelling of this scenario in the hydrologic planning model can be found in Sandoval-Solis *et al.* [14].

Cooperative game theory concepts are applied to the Water Demand Reduction game by identifying players who are able to form binding agreements, or coalitions, with other players. The players for this game are the three largest irrigation water users in the basin: Delicias Irrigation District 005 in the Rio Conchos, Mexican Irrigation District 025 and an aggregate of the largest irrigation districts in Texas below Falcon Reservoir (Watermaster Section 10), both located in the lower Rio Grande/Bravo basin (Figure 7.8.1). Although Districts 005 and 025 are from the same country, they are considered as independent economic units. The coalitions in this game range from non-cooperative coalitions where players act to maximize their individual benefits, to full cooperative coalition (Grand Coalition) where all players act collectively to maximize the coalitions' benefit beyond the non-cooperative solutions. Partial coalitions, or subsets of players, may also form. The players in the Water Demand Reduction game may form a total of seven coalitions. Each possible coalition and the presumed actions taken by those coalitions are described below:

Coalition {1}: This coalition represents Player 1, Irrigation District 005, acting alone with a total annual water demand of 1131 million cubic metres (MCM). Under this coalition, District 005 ensures their own delivery of water to satisfy their total demand.

Coalition {2}: This non-cooperative coalition represents Player 2, Irrigation District 025, acting alone with an annual water demand of 1127 MCM. In this coalition,

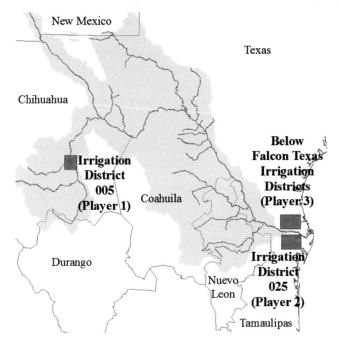

Figure 7.8.1 Approximate location of players in the Water Demand Reduction game [15].

Irrigation District 025 does not finance water buy-backs in the Rio Conchos and attempts to meet their water demand with the available water.

Coalition {3}: This coalition characterizes Player 3, Watermaster Section 10 (below Falcon reservoir), acting alone with an annual water demand of 647 MCM. Under this coalition, this player does not finance water buy-backs in the Rio Conchos and attempts to meet their water demands with the water available in the Rio Grande/Bravo.

Coalition {1, 2}: District 005 and District 025 work cooperatively under this partial coalition. District 025 provides the investment to buy-back 502 MCM of combined surface and groundwater water rights in District 005, reducing District 005s demand for water. Owing to physical losses in the system, as the water travels downstream, only about 20% of the water released from District 005 reaches the lower basin [14]. Additionally, according to the 1944 Treaty, any water from the Rio Conchos reaching the Rio Grande/Bravo is divided $^1/_3$ to the U.S. and $^2/_3$ to Mexico [16]. Some 418 MCM of surface water rights are purchased from District 005. District 025 is entitled to 56 MCM because of the system losses and treaty division. Only the surface water rights are available to the downstream players because there is no plan to pump the groundwater into the river.

Coalition {1, 3}: District 005 and Watermaster Section 10 work cooperatively, with Watermaster Section 10 providing investment to buy-back 502 MCM of surface and ground water rights, thus reducing District 005s overall demand for water. Owing to physical losses and treaty obligations described above, Watermaster Section 10 is entitled to an additional 28 MCM per year of the 418 MCM of purchased surface water rights after accounting for the treaty division and the system losses. The groundwater portion of the water buy-backs is not available to the Watermaster Section 10 because there is no plan to pump the groundwater into the river.

Coalition {2, 3}: District 025 and Watermaster Section 10 cannot increase their benefits without including District 005 because the water buy-backs occur strictly in District 005. Since they are not in this coalition, District 005 may continue to use water at their non-cooperative rate, which leaves District 025 and Watermaster Section 10 with access to the same amount of water they receive under the non-cooperative solution.

Coalition {1, 2, 3}: In the Grand Coalition, District 005, District 025 and Watermaster Section 10 all work cooperatively. Both District 025 and Watermaster Section 10 provide the investment to buy-back 502 MCM of water rights, thus reducing District 005s overall demand for water. District 025 and Watermaster Section 10 are entitled to share the 418 MCM of surface water rights bought-back, which is 84 MCM year after losses and treaty obligations. As in the previous coalitions, the downstream players do not have access to the groundwater rights bought-back.

7.8.3
Results

Each coalition's actions are modelled in the Rio Grande/Bravo hydrologic planning model to calculate their characteristic values, or coalition value. The minimum water delivery volumes to each player in the 60 year simulation are calculated and a monetary value is assigned to those water deliveries. For this game the monetary value of water deliveries is $86 000 per MCM (U.S. dollars) [15]. The characteristic values, expressed as v, for each coalition are shown in Table 7.8.1.

The Core is determined from the characteristic functions (Figure 7.8.2). The Core represents the feasible allocations to the individual players and also provides bounds for negotiation. The Core contains allocations (Ω_j) that improve each player's (player j) standing above their non-cooperative solution. To select a single allocation from the Core, the Shapley allocation method is used (Table 7.8.2). The Shapley

Table 7.8.1 Characteristic values of the Water Demand Reduction game.

Coalition	Characteristic value ($million per year)
$v(1)$	6.4
$v(2)$	34.4
$v(3)$	30.7
$v(1, 2)$	63.8
$v(1, 3)$	52.9
$v(2, 3)$	65.0
$v(1, 2, 3)$	97.6

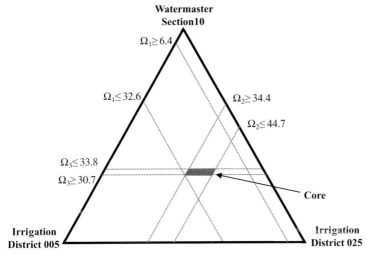

Figure 7.8.2 Core of the Water Reduction Cooperative game.

allocation is calculated from the marginal contribution of each player to the Grand Coalition. The Shapley allocation gives each player an increased allocation over their non-cooperative allocation (Table 7.8.2). Irrigation District 005 receives an additional $15.2 million, Irrigation District 025 an additional $7.3 million and Watermaster Section 10 an additional $3.7 million.

7.8.4
Conclusions

The cooperative game analysis of the Water Demand Reduction game demonstrates that under cooperation each player stands to receive increased allocations. Irrigation District 005 receives the largest increase over their non-cooperative solution largely because the water buy-backs occur in their district and the game does not exist without their participation. The increase in allocations to District 025 and Watermaster Section 10 is smaller due to the small increase in water volumes they may receive under the Water Demand Reduction game. Cooperative game theory provides information to water users to quantify their potential increased benefits through cooperation under water management scenarios.

Table 7.8.2 Non-cooperative and Shapley allocations in the Water Demand Reduction game.

Player	Non-cooperative allocation ($million per year)	Shapley allocation ($million per year)
District 005	6.40	21.6
District 025	34.4	41.7
Watermaster Section 10	30.7	34.4

Acknowledgements

The authors would like to thank Samuel Sandoval-Solis for his work on scenario modelling.

References

1 Wolf, A. (2002) International River Basins of the World, Oregon State University, http://www.transboundarywaters.orst. edu/publications/register/tables/ IRB_table_4.html (accessed 28 September 2010).

2 Rogers, P. (1969) A game theory approach to the problems of international river basins. *Water Resource Research*, **5** (4), 749–760.

3 Rogers, P. (1993) The value of cooperation in resolving international river basin disputes. *Natural Resources Forum*, **17** (2) 117–131.

4 Kucukmehmetoglu, M. (2002) Water resources allocation and conflicts: the case of the Euphrates and the Tigris. PhD dissertation, Ohio State University, Columbus, OH.

5 Kucukmehmetoglu, M. and Guldmen, J. (2004) International water resources allocation and conflicts: the case of the Euphrates and Tigris. *Environment and Planning A*, **36**, 783–801.

6 Kucukmehmetoglu, M. (2009) A game theoretic approach to assess the impacts of major investments on transboundary water resources: the case of the Euphrates and Tigris. *Water Resources Management*, **23** (25), 3069–3099.

7 Wu, X. (2000) Game-theoretical approaches to water conflicts in international river basin: a case study of the Nile Basin. PhD dissertation, University of North Carolina, Chapel Hill.

8 Wu, X. and Whittington, D. (2006) Incentive compatibility and conflict resolution in international river basins: a case study of the Nile Basin. *Water Resource Research*, **42**. W02417.

9 Dinar, A. and Wolf, A. (1994) Economic and political considerations in regional cooperation models. *Agricultural and Resource Economics Review*, **26** (1) 7–22.

10 World Wildlife Federation (WWF) (2007) World's top 10 rivers at risk, http://assets. panda.org/downloads/ worldstop10riversatriskfinalmarch13.pdf (accessed 28 September 2010).

11 Patiño-Gomez, C., McKinney, D.C. and Maidment, D.R. (2007) Sharing water resources data in the bi-national Rio Grande/Bravo basin. *Journal of Water Resources Planning and Management*, **133** (5), 416–426.

12 Danner, C.L., McKinney, D.C., Teasley, R.L. and Sandoval-Solis, S. (2006) Documentation and testing of the WEAP Model for the Rio Grande/Bravo Basin, CRWR online report, 06-08, revised February 2009 http://www.crwr.utexas. edu/reports/2006/rpt06-8.shtml.

13 SAGARPA–Secretaria de Agricultura Ganadería, Desarrollo Rural, Pesca y Alimentación (2005) Informe de beneficiarios del programa PADUA 2004.

14 Sandoval-Solis, S., McKinney, D.C. and Teasley, R.L. (2008) Water management scenarios for the Rio Grande/Bravo Basin, CRWR online report 08-1, http:// www.crwr.utexas.edu/reports/pdf/2008/ rpt08-01.pdf (accessed 28 September 2010).

15 Teasley, R.L. (2009) Evaluating water resource management in transboundary river basins using cooperative game theory: The Rio Grande/Bravo Basin. PhD dissertation, The University of Texas at Austin.

16 International Boundary Water Commission (IBWC) (1944) Treaties between the United States of America and Mexico: Treaty of February 3, 1944, IBWC online report, http://www.ibwc.state.gov/ Treaties_Minutes/treaties.html (accessed 28 September 2010).

7.9
Conflict Resolution in Transboundary Waters: Incorporating Water Quality in Negotiations

Eleni Eleftheriadou and Yannis Mylopoulos

7.9.1
Introduction

The urgent need for integrated water resources management has drawn attention to the issue of trans-national water bodies that comprise exploitable resources for two or more countries. It is estimated that 50% of the global surface lies in transboundary catchments and this percentage is continually increasing due to the creation of new nations, while around 40% of the earth's population resides in transboundary water catchments. According to historical facts, no record exists of hostility due to shared water resources, but it is anticipated that in the future water will replace oil as a primary cause of war.

It is a common phenomenon in transboundary water agreements for issues concerning water quality to be missing, while the majority of the agreements regard water quantity issues as the main priority [1]. The reasons behind this neglect of the water quality parameter is the shortage of appropriate tools for the estimation of the losses, the impediments in defining the pollution sources and the fact that it is more immediate to estimate water quantity deficiencies rather than water quality parameters. The omission of quality specifications has proved detrimental in many cases since water quality improvement brings numerous benefits to both sides of the borders.

7.9.2
Game Theory in Water Resources

In many cases negotiations can be assisted with the use of Game Theory, providing both sides with possible equilibrium states resulting from the players' specific options and preferences. An implementation example is described in Eleftheriadou and Mylopoulos [2] where the software tool Graph Model for Conflict Resolution II (GMCR II) [3] was applied as a support tool prior to the official negotiations. The aim of this study was to demonstrate that the use of such support tools can provide useful insights to the players regarding the possible outcome of the negotiations.

Game theory provides a systematic analysis of strategies and their interactions during any negotiation phase. One of the main characteristics of the theory is the cooperative approach that can be attributed even in competitive cases, proving the benefits of cooperation and converting the players' relations into 'co-opetitive' (a wordplay combining cooperation and competitive), a concept introduced by Brandenberger and Nalebuff [4], suggesting that players should compete and cooperate at the same time.

Game Theory aims at the study of four basic elements and their interactions [5]. These are (i) players: normally consisting of the decision-makers; (ii) interaction: the choices of one player affect those of the counter-player; (iii) strategy: each player has a strategy based on the interpretation of the players' interactions; and (iv) rationality: the players' choices are characterized by rationality.

All the above elements are often found in negotiations concerning the management of shared water resources. The interested countries act as 'players' with specific options and take decisions according to the payoffs that correspond to each option combination. In most cases the players have opposite interests as the benefit of one player entails loss for the counter-player.

7.9.3
Methodology

The Nestos/Mesta basin shared between Greece and Bulgaria was chosen as a reference area. This transboundary basin is of great interest as it is shared between two countries with many differences in economic and development level. The river under study is a very crucial asset for both countries as it supports important economic activities and has environmentally protected areas on both sides of the border. In this chapter a new methodology incorporating water quality will be connected to the water allocation model that was developed by Eleftheriadou and Mylopoulos [6] for the Nestos/Mesta River.

The proposed methodology (Figure 7.9.1) consists of the following steps:

1) The water quality simulation model: MONERIS (Modelling Nutrient Emissions in River Systems) simulation model was selected for the reference area and is applied in the whole catchment area.
2) Based on the outcomes of the simulation, that is, the concentrations of nutrients in the river (N-nitrogen and P-phosphorus), the losses due to water pollution are estimated for the productive sectors of each country.
3) Game Theory is applied to simulate possible negotiations between Greece and Bulgaria regarding water quality issues. The two players of the game have various options of actions and reactions, while each option corresponds to a specific payoff. The payoffs entered in the Game Theory matrix are the results of the second step of the methodology and correspond to the gain or loss of each player according to the combination of their options for a specific state. The possibility of inter-connected games is investigated in order to convert non-cooperative games into cooperative ones.
4) After several scenarios are analysed, compromising solutions are produced that will be acceptable to both countries, while ensuring environmental conservation.
5) The results of the above steps are then connected to the previous implementation that was focused on water allocation. The merging of the two components contributes towards the development of an integrated Negotiation Support System (NSS).

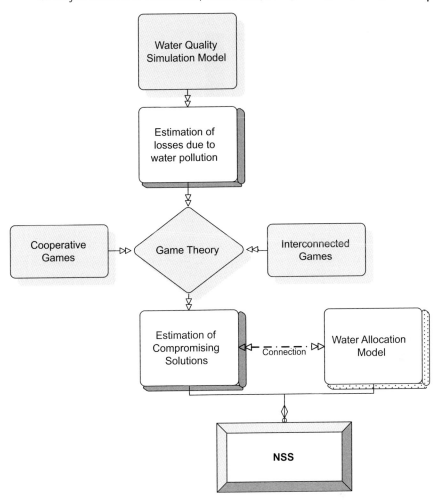

Figure 7.9.1 Graph of the proposed methodology.

7.9.4
Results

The results of MONERIS model can be used to provide input for the formulation of the players' options. As a demonstration example, the model predicted that future construction of water treatments plants in all the settlements of the Bulgarian basin would cause a high decrease (of approximately 82%) to the nitrogen concentration in the river (Figure 7.9.2). This fact can play a crucial role in the formulation of the games and the creation of additional options for each player, while it sets the basis for the incorporation of interconnected games.

The estimation of payoffs is a complicated procedure and has to be realized by using objective and realistic criteria. In cases where the water bodies receive heavy

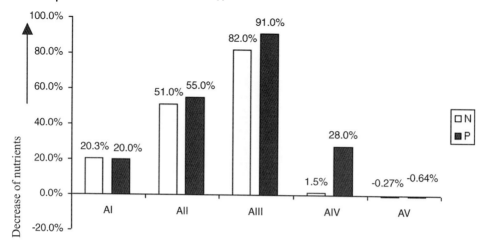

Figure 7.9.2 Variations of nutrients pollution in the river. Scenarios: AI: all Greek settlements connected to WWT; AII: half Bulgarian settlements connected to WWT; AIII: all Bulgarian settlements connected to WWT; AIV: change land use (agriculture → forests); AV: increase of population in both countries).

pollution with a resulting direct impact on the water uses, the losses of the productive sectors form the payoffs of the game. However, the Nestos/Mesta is considered a clean river and the observed pollution does not affect water uses (drinking, irrigation). For this reason, the payoffs for the Nestos/Mesta case are based on Environmental Protection Agency (EPA) recommendations that total phosphorus concentrations should be less than $0.1 \, \mathrm{mg \, l}^{-1}$ in rivers and nitrogen concentrations less than $2 \, \mathrm{mg \, l}^{-1}$. Payoffs depend on the difference between the predicted value of the nutrient in the river with the EPA recommendations; for example, for the case of phosphorus pollution, the payoff of each player will be equal to $P_P = [(0.1 - C_P)/0.1]\%$ and for nitrogen pollution $P_N = [(0.1 - C_N)/0.1]\%$, where C_P and C_N represent phosphorus and nitrogen concentration, respectively, predicted by MONERIS. The total payoff of each player is estimated as the average value of P_P and P_C.

Introducing as the basic game the case where Greece has the options of keeping the present situation or connecting all settlements of the basin to Wastewater Treatment Plants (WWTP) and Bulgaria has the options of keeping the present situation (no treatment at all), connecting half or connecting all the settlements to WWTPs, the matrix of this game is formed as shown in Figure 7.9.3.

The Nash solution for this game is the point where both countries agree to proceed to the connection of all settlements to WWTPs, a situation that is anticipated as it expresses a logical outcome.

The water quality issue should be approached in connection with water quantity in the building of any interstate water agreement. This can be realized through the use of interconnected games where the two issues can be linked and provide new equilibrium states that incorporate both issues.

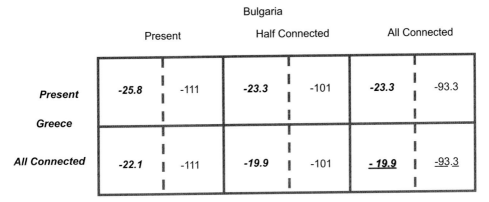

Figure 7.9.3 Matrix of the basic game.

For example, to simulate reality, the water quantity issue (where Bulgaria has the options of keeping 10% of the water flow to Greece or keeping the current levels of water consumption) should be 'brought to the table' and be linked with the basic game. Linking different games can create common motives for the opponents for cooperation and maintenance of agreements, but also can often explain 'irrational' behaviour that may be observed during negotiations. Moreover, each player is equipped with more options than in the isolated game, making feasible the creation of new states that were otherwise excluded from the game, while the new equilibrium state attained can differ greatly from the initial one.

In the interconnected game the Nash solution (*1.33, 1.22*) corresponds to the state where Bulgaria keeps 10% of the water flow and proceeds to the construction of WWTPs and is not Pareto optimal (Figure 7.9.4). There are other states in the matrix where both players' payoffs are higher than the Nash solution and could form possible new equilibrium states. To estimate a compromising solution acceptable to both players, tested methods can be applied such as the axioms of Nash and Raiffa's compromising solution.

7.9.5
Conclusions

Water quality degradation has direct negative consequences for all water uses. The quantification of this qualitative degradation and the calculation of economic figures will allow an in-depth incorporation of water quality issues in transboundary water management negotiations. It is anticipated that the present chapter will contribute to integrated transboundary water management where quantity and quality issues are confronted in parallel.

The outcomes of this chapter can be used in the building of agreements that incorporate water quality issues and can provide a proposed framework for the contents of such agreements. The achievement of cooperative solutions is not always feasible due to the absence of motives and mutual trust between the opponents [7];

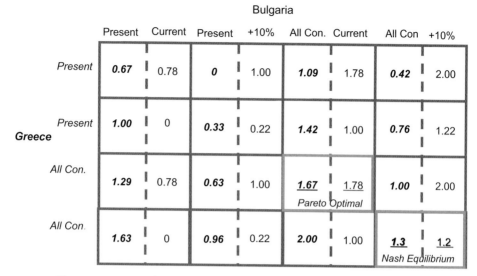

Figure 7.9.4 Matrix of the interconnected games.

thus interconnected games are applied to develop common motives and ensure the viability of any agreement signed by the two countries. Concerning the area under study, it is anticipated that the inclusion of water quality as a consideration in the current Greek–Bulgarian agreement would strengthen Bulgaria's position as it is estimated that the 29% of flow attributed to Greece is a quite favourable percentage [8]. Hence, even if Bulgaria causes transboundary pollution, the equilibrium states of the game will show that Greece should offer a kind of contribution to Bulgaria to reduce pollution in exchange for a higher agreed percentage of released flow at the borders.

The proposed methodology is expected to contribute to the development of sustainable transboundary water agreements where the principle of equity is implemented. The implementation of game theoretical methods is particularly useful in the analysis of various scenarios as it quantifies the consequences of each player's options and provides information that is indicative of the outcome of the negotiations. Moreover, the equilibrium states of the games, which were formed during the negotiation simulation, can form the basis for the commencement of the real negotiations between the two parties. It is of great importance that the options discussed at the negotiation table are realistic, non-detrimental to the environment and their impacts have already been assessed and are known to the concerned parties.

References

1 Bennett, L.L. (2000) The integration of water quality into transboundary allocation agreements: lessons from the southwestern United States. *Agricultural Economics*, **24**, 113–125.

2 Eleftheriadou, E. and Mylopoulos, Y. (2005) Conflict resolution in the management of

transboundary catchments: application in
the Nestos River. Presented at the 5th
International Conference on
Environmental Technology, HELECO 2005,
Athens, Greece 3–6 February.

3 Hipel, K.W., Kilgour, D.M., Fang, L. and
Peng, X. (1997) The decision support
system GMCR in environmental conflict
management. *Applied Mathematics and
Computation*, **83**, 117–152.

4 Brandenberger, A. and Nalebuff, M. (1999)
Co-Opetition, Kastaniotis, Athens.

5 Rasmussen, E. (1994) *Games and
Information: An Introduction to Game Theory*,
Blackwell, Cambridge.

6 Eleftheriadou, E. and Mylopoulos, Y. (2008)
Game theoretical approach to conflict
resolution in transboundary water
resources management. *Journal of Water
Resources Planning and Management*,
134 (5), 466–473.

7 Dinar, A., Ratner, A. and Yaron, D. (1992)
Evaluating cooperative game theory in water
resources. *Theory and Decision*, **32**, 1–20.

8 Kampragou, E., Eleftheriadou, E. and
Mylopoulos, Y. (2006) Implementing
equitable water allocation in
transboundary catchments: the case of
river Nestos/Mesta. *Water Resources
Management*, **21**, 909–918.

7.10
The Johnston Plan in a Negotiated Solution for the Jordan Basin

Majed Atwi Saab and Julio Sánchez Chóliz

7.10.1
Introduction

The equitable allocation of water resources is at the heart of conflict in river basins all over the world [1]. Moreover, the lack of any internationally accepted criteria for sharing water and the benefits thereof have fostered additional conflict. In its beginnings, the Arab–Israeli conflict may have been an argument about land, but little by little the scarcity of water and its capacity to boost the value added to farming have turned water from a relevant, if scarce, good into a key geopolitical resource and a factor in the evolution of the dispute. Thus, the control of water has been a high strategic priority in the policies of both Israel and the Arabs to the point where it was selected as one of the major issues of the multilateral talks that grew out of the Madrid Peace Conference in 1991. Theoreticians of *water wars* have frequently pointed to control of the river Jordan's water as one of the motives for the Arab–Israeli Six Days War in 1967. According to Haddadin, the June 1967 war between Israel and neighbouring Arab States was not a water war. Water contributed, however, to the tensions that prevailed in the region between 1964 and 1966 [2]. Nevertheless, in line with Wolf's arguments [3, 4] and those of other authors, we do not believe that water has to remain a permanent source of tension between people living in the same territory. Rather, we consider it can and should be a catalyst for cooperation, not conflict.

Taking an optimistic stance, we propose in this chapter a tentative model for water sharing in the Jordan basin using a negotiation game with two players, namely, Arabs and Israelis. We shall consider some possible and theoretical solutions in a simple framework: regular and general Nash [5], Raiffa–Kalai–Smorodinsky [6, 7] and the Johnston Plan (JP). These proposals should take other factors, such as environmental needs and political issues, into account before their application, but even by themselves may point the way to finding a future solution for the inhabitants of the Jordan basin. Section 7.10.2 sets out the ground rules for the game, reviews some of the admissible conditions for a fair allocation of water resources and gives an overview of the Johnston Plan, an unratified water agreement reached in 1955 in the Jordan basin between the riparian Arab countries and Israel. Section 7.10.3 proposes three highly significant solutions. These are then blended with certain other solutions. We end with our main conclusions (Section 7.10.4).

7.10.2
Key Elements of the Negotiation Game and Fairness Criteria

To obtain and interpret the different solutions to the game, we need the utility functions for each player, possible negotiating alternatives and a set of criteria allowing the evaluation of results. These matters are dealt with in this section.

7.10.2.1 Utility or Payment Functions for Arabs and Israelis
In this chapter, we shall use irrigated water returns as the basis for the utility functions. The aim of these calculations is to be able to make an economic assessment of possible transfers from one country to another. Nevertheless, such transfers affect the lowest areas on these curves, which undoubtedly correspond to the least profitable irrigation uses, since the urban–industrial uses are more profitable. For this reason, we have centred our efforts on the economic assessment of water use for agriculture.

By Arabs we mean Lebanon, Syria, Jordan and Palestine represented by the West Bank and the Gaza Strip. When analysing the agricultural uses of the Jordan River valley in the different countries we have not included Lebanon and Syria, given the difficulty of accessing significant data on use by these two countries. The omission of this analysis of water use in Lebanon is of little significance, due to this country's low current level of extraction from the upper Jordan. However, the case of Syria is more significant, due to its considerable involvement in the Yarmouk basin, the main tributary of the Jordan. Nevertheless, the lack of data on Syrian river valley agriculture, segregated from overall data on its agriculture, made it difficult to analyse the irrigated land fed by the waters of the Yarmouk. For this reason, we have centred on Israel, Jordan and Palestine for available analyses, uses and productivity.

The farm data for the three countries, Israel, Palestine and Jordan, have basically been taken from the websites of their respective statistical bureaus [8–12]. Based on an analysis of both the crop and water-use structures, and the revenues and costs generated by the different types of crops grown, we have estimated the standard gross margin generated by each crop type per cubic metre of water used as revenues, less direct costs (seed, fertilizer, pesticides, water, machinery, energy, etc.). In 2001, the total volume of water used by Israel in irrigation was $994.663\,hm^3$, compared to

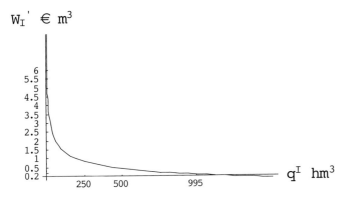

Figure 7.10.1 Standard gross margins on irrigation and adjusted function for Israel.

533.359 hm³ for the Arabs. These amounts of water are the status quo, which means what each country is using in irrigation, given Israel's control of the Golan Heights and most of the West Bank aquifers, and considering the frontiers before 1967 as a reference. The empirical results for Israel are given in Figure 7.10.1, and those for the Arabs in Figure 7.10.2. Both reflect a clear downward and roughly hyperbolic trend. For ease of mathematical operation we have therefore opted to adjust the data using hyperbolic curves, which are also shown in Figures 7.10.1 and 7.10.2.

The adjustments for Israel and the Arabs are respectively expressed as follows:

$$W_I'(q) = \left(-1.73262 + \frac{10.744}{q^{0.258446}}\right); \quad W_A'(q) = \left(-1.11024 + \frac{6.68398}{q^{0.261728}}\right)$$

$$(7.10.1)$$

Given $W_I'(q)$ and $W_A'(q)$, the utility functions for both players can easily be calculated by integration, giving:

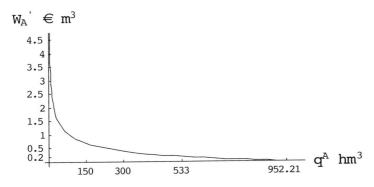

Figure 7.10.2 Standard gross margins on irrigation and adjusted function for Arabs.

$$W_I(q) = \int_0^q W_I' ds = \int_0^q \left(-1.73262 + \frac{10.744}{s^{0.258446}}\right) ds = 14.4885 q^{0.741554} - 1.73262q$$

$$(7.10.2)$$

$$W_A(q) = \int_0^q W_A' ds = \int_0^q \left(-1.11024 + \frac{6.68398}{s^{0.261728}}\right) ds = 9.05355 q^{0.738272} - 1.11024q$$

$$(7.10.3)$$

7.10.2.2 Negotiation Set

We have defined the negotiation set as the set of efficient points that do not result in utilities below the break-off point and no *player* will accept a payoff that is not at least equal to what it would receive at the break-off point. Because of the reality of the situation, we shall suppose that transfers are possible only in one direction–from Israel to the Arabs. We also assume that the Israelis will always transfer the water providing the lowest return; while the water received will be used by the Arabs to expand their agriculture without changing crop patterns and will be assessed at constant marginal return equals the average return of its current available water. Consequently, if x is the amount of water transferred, the water available to each player in an efficient solution will be: $q^I(x) = q^I_{act} - x$ and $q^A(x) = q^A_{act} + x$, where q^I_{act}, $q^I_{act} = 994.663 \text{ hm}^3$ and $q^A_{act} = 533.359 \text{ hm}^3$. And the utilities of each player, $W_I(x)$ and $W_A(x)$ will be as follows:

$$W_I(x) = 14.4885 \left(q^I_{act} - x\right)^{0.741554} - 1.73262 \left(q^I_{act} - x\right)$$

$$(7.10.4)$$

$$W_A(x) = \left[9.05355 \left(q^A_{act}\right)^{0.738272} - 1.11024 q^A_{act}\right] \left(1 + \frac{x}{533.359}\right)$$

$$(7.10.5)$$

The set of efficient solutions is given in Figure 7.10.3. The function is expressed as follows:

$$W_A(W_I) = \left[9.05355 (q^A_{act})^{0.738272} - 1.11024 q^A_{act}\right] \left(1 + \frac{f^{-1}(W_I(x))}{533.359}\right)$$

$$(7.10.6)$$

The following 7th order polynomial expression is an excellent approximation to the curve:

$$W_A(W_I) = 976.375 - 0.2484 W_I + 0.00568023 W_I^2 - 0.00006827693 W_I^3$$
$$+ 3.36779 \times 10^{-7} W_I^4 - 8.34019 \times 10^{-10} W_I^5 + 1.00573$$
$$\times 10^{-13} W_I^6 - 4.71692 \times 10^{-16} W_I^7$$

and for ease of use we shall apply the expression in the calculation and representation processes.

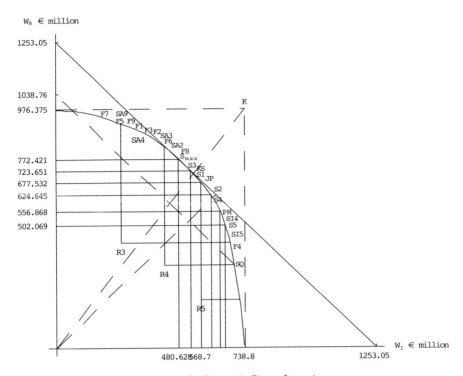

Figure 7.10.3 Efficient points curve and solutions (millions of euros).

7.10.2.3 Fairness Criteria
Table 7.10.1 sets out nine fairness factors inspired by the Helsinki Rules developed by the International Law Association in relation to the fair use of water. We have applied these factors to the present case in addition to the fairness criteria [13].

Mimi and Sawalhi [14] translate the principle of equitable utilization into the Jordan River basin by applying the nine equity factors that the International Law Association associated with equitable water use. The application of these factors yields nine alternative equity standards that served as benchmarks against which various possible allocation outcomes between all riparian parties were measured. Following the work of Mimi and Sawalhi we have obtained nine fair allocations, which are part of the negotiating set. These efficient points are given in Table 7.10.2 and represented graphically in Figure 7.10.3.

7.10.2.4 Johnston Plan (1953–1955)
Johnston, in the course of negotiations over the Plan, specified allocations for each riparian party. The basis for allocations in his proposals was agricultural use of water,

Table 7.10.1 Factors related to the fair use of water and their alternative equity standards. Source: own work based on Mimi and Sawalhi [14].

Factor	Definition	Criteria for the calculation of	Equity standards (%)	
			Israel	Arabs
F1	Geography	Proportional area of each country in the river basin	9	91
F2	Hydrology	Water contributed by each country in the river basin	12	88
F3	Climate	Rainfall of each country in the river basin	10	90
F4	Historic use of water	Current use of water[a]	60	40
F5	Economic and social needs of each riparian state in the river basin	Estimated demand for water from all sectors in 2025	8.4	91.6
F6	Population of each country	Projected population in 2015	16.4	83.6
F7	Ability to pay for alternatives	GDP of each country	6.7	93.3
F8	Availability of other resources	Water Stress Index (WSI): ratio of demand to availability of water	19	81
F9	Significant harm	Social, economic and environmental needs –similar to F5	8.4	91.6

a) Estimated by Mimi and Sawalhi [14].

calculated via the areas of irrigable land of each riparian party within the basin [15, 16]. The Plan estimated the flow of the Jordan River at around 1.28 hm^3, 31% of which was allocated to Israel and the remaining 69% to the Arabs. If these proportions are applied to the flows in our game (1528.02 hm^3), the result obtained is the point in the negotiating set represented by the Johnston Plan, or point JP = (575.79; 674.71).

7.10.3
Three Significant Game Solutions between Israel and the Arabs

In addition to the proposal of water allocation made in the Johnston Plan, we propose various solutions to the bargaining game between Israel and the Arabs as follows.

7.10.3.1 Regular Nash Solution without Lateral Payments and Break-off at (0;0)
In this case, regular solution without lateral payments, the negotiation set consists of the efficient points defined by $W_A(W_I)$ with non-negative components (see Figure 7.10.3).

Table 7.10.2 Fairness factors, Johnston Plan and utilities.

q_{act}^I: amount currently used by Israel (hm³) 994.663

q_{act}^A: amount currently used by the Arabs (hm³) 533.359

Total utilized by both Israelis and Arabs (hm³) 1528.02

Factor	F1	F2	F3	F4	F5 = F9	F6	F7	F8	JP
Israel (%)	9	12	10	60	8.4	16.4	6.7	19	31
Arabs (%)	91	88	90	40	91.6	83.6	93.3	81	69
Total (%)	100	100	100	100	100	100	100	100	100
Israel (hm³)	137.52	183.36	152.80	916.81	128.35	250.60	102.38	290.32	473.69
Arabs (hm³)	1390.50	1344.66	1375.22	611.21	1399.67	1277.43	1425.64	1237.70	1054.34
x (hm³)	857.14	811.30	841.86	77.85	866.31	744.07	892.28	704.34	520.98
$W_I(x)$	319.865	373.163	338.747	690.355	307.913	436.76	271.053	468.348	575.794
$W_A(x)$	889.829	860.494	880.051	391.133	895.696	817.469	912.319	792.046	674.705

The solution, therefore, is that of the problem:

Max $W_I W_A$

$$W_A(W_I) = \left[9.05355\left(q_{act}^A\right)^{0.738272} - 1.11024 q_{act}^A\right]\left(1 + \frac{f^{-1}(W_I(x))}{533.359}\right) \geq 0, \ W_I \geq 0$$

(7.10.7)

which is $S1 = (568.719; 677.532)$. These figures represent millions of euros, the first corresponding to Israel and the second to the Arabs. Comparing S1 to the status quo, $SQ = (697.444; 341.315)$, Israel would lose around €130 million, and the Arabs would increase their income by approximately €336 million, raising overall utility by 20%. Given these calculations, the transfer of water to the Arabs can easily be calculated using the utility functions $W_I(x)$ or $W_A(x)$, and would be in the region of 536.26 hm^3. As may be observed in Figure 7.10.3, point S1 is situated to the left of point SQ and is surprisingly close, indeed practically identical, to solution JP $= (575.79; 674.71)$, proving the validity of the proposal made in the Johnston Plan, which, after a half century, could still be rationally defended.

7.10.3.2 Nash Solution with Lateral Payments and Break-off at (0;0)
Assuming, in the context of negotiation games, that both Israel and the Arabs can make lateral payments, the Nash solution is then that of the following problem:

Max $W_I W_A$

$$W_I + W_A = \underset{W_I \geq 0}{\text{Max}}\left[W_I + \left[9.05355\left(q_{act}^A\right)^{0.738272} - 1.11024 q_{act}^A\right]\left(1 + \frac{f^{-1}(W_I(x))}{533.359}\right)\right];$$

$$W_I \geq 0; \ W_A \geq 0$$

(7.10.8)

As may be observed in Figure 7.10.3, the efficient solutions are situated along a line $W_A = 1253.05 - W_I$. The maximum level of joint utility is obtained at point $W_A(W_I)$, which is a tangential to the aforementioned parallel line. This is the point at which $d[W_A(W_I)]/dW_I = -1$. The point in question is $S_{max} = (480.628; 772.421)$. At this point, the joint utility obtained is €1253.05, which is greater than the €1246.251 obtained in the previous Nash solution. This point is not, however, the Nash solution. By symmetry alone, it can be seen that the Nash solution with lateral payments is $S2 = (626.525; 626.525)$.

7.10.3.3 Raiffa–Kalai–Smorodinsky Solution with Break-off at (0;0)
The Raiffa–Kalai–Smorodinsky solution is neither more nor less than the intersection between the negotiating set and the line joining the break-off point and an ideal point $K = (W_{Imax}; W_{A\,max})$, where W_{Imax} and $W_{A\,max}$ are the maximum utilities that can be achieved by the players within the negotiating set. In the present case, $K = (738.8; 976.375)$. Hence, we may affirm that the Raiffa–Kalai–Smorodinsky

solution is the intersection of:

$$W_A(W_I) = \left[9.05355\left(q_{act}^A\right)^{0.738272} - 1.11024q_{act}^A\right]\left(1 + \frac{f^{-1}(W_I(x))}{533.359}\right) \geq 0, \ W_I \geq 0$$

$$(7.10.9)$$

$$W_A = \frac{976.4}{738.8}W_I \tag{7.10.10}$$

which gives point $KS = (538.56, 711.75)$, as represented in Figure 7.10.3. The joint income produced is €1250.313 million, which is close to the amount obtained at S1 (€1246.25 million) and at S2 and S_{max} (€1253.05 million), and it results in a gain of approximately 20% in income for the Arabs compared to the status quo.

7.10.3.4 Other Solutions

In the above solutions we have assumed that the break-off point was the unavailability of any water, and that the players had no power to exert pressure. If these two factors change, so, too, do the solutions.

We have obtained three regular Nash solutions (S3, S4 and S5) with break-off points (R3, R4 and R5) $\neq$ (0; 0) for three cases. These break-off points are the utilities that each player would obtain if breaking off the game would ensure only half of the utility of fairness criterion F2 (contribution of each country to flows in the basin), the Johnston Plan and criterion F4 (current uses), all three of which could serve as the basis for an actual negotiation scenario (Figure 7.10.3).

We also have considered Nash solutions in which both players are treated asymmetrically with different levels of bargaining power. These solutions are called generalized Nash bargaining solutions. They can be seen in Figure 7.10.3 as solutions SA2, SA3, SA4, SA5, SA6 and SA9, in which the Arab power is greater, and in four other solutions, SI2, SI3, SI4 and SI5, in which the Israelis have greater pressure capacity. As might have been expected, the players increase their gains where they have greater negotiating power (Figure 7.10.3).

7.10.4
Conclusions

The most significant solutions obtained fall between S_{max} and $PM = (642.95; 556.87)$ where PM is the intermediate efficient point when the Arab utility is taken as the indicator. Meanwhile, the best solutions from a technical standpoint are those that are closest to S_{max}. We find the regular Nash solution with null break-off point, which is $S1 = (568.72; 677.53)$, practically the same as the solution $JP = (575.79; 674.71)$ for the Johnston Plan. Nevertheless, the overall utility of the four solutions is very similar, and all of them would result in an increase of approximately 20% compared to the status quo. All this suggests that the 1955 Johnston Plan could be revisited as a starting point for present-day negotiations. It would require some adjustment to make room for modern approaches to integrated, sustainable management, but the Plan's proposals appear to provide an acceptable combination of what is possible, technically feasible and socially desirable. In light of current circumstances and according to Elmusa, the Johnston Plan needs revision to include new elements and refinement of the allocation quotas. It should consider the question of water quality and protection and it should take into account the historic and ecological issues of the Jordan basin. The basin is central in the history and religions of Muslims, Christians and Jews.

As a last consideration, any good solution for water allocations amongst the riparian countries of the Jordan basin should consider the whole basin as a unit of management. Nevertheless, given the obtained results, Israel must scarify the irrigated low return crops to be able to transfer the water to the Arabs countries. However, for negotiations to be flexible and effective both players should take into account different options such as side payments, water markets, and searching for non-conventional water sources, including desalination and wastewater plants and, perhaps, water imports, an option that is only viable by reaching a full peace agreement in the region.

References

1 Wolf, A.T. (1999) Criteria for equitable allocations: the heart of international water conflict. *Natural Resources Forum*, **23**, 3–30.

2 Haddadin, M.J. (2003) The war that never was. *The Brown Journal of World Affairs*, **IX** (2), 321–332.

3 Wolf, A.T. (1995) Hydropolitics along the Jordan River: ScarceWater and its Impact on the Arab–Israeli Conflict, United Nations University Press, Tokyo.

4 Wolf, A.T. (1998) Conflict and cooperation along international waterways. *Water Policy*, **1** (2), 251–265.

5 Nash, J.F. (1953) Two person cooperative games. Econometrica, 21, 128–140.

6 Raiffa, H. (1953) Arbitration schemes for generalized two-person games, in *Contributions to the Theory of Games II* (eds H.W. Kuhn and A.W. Tucker), University Press, Princeton, pp. 361–387.

7 Kalai, E. and Smorodinsky, M. (1975) Other solutions to Nash.s bargaining problem. *Econometrica*, **43**, 513–518.

8 Central Bureau of Statistics (CBS) (2002) Agriculture. Statistical abstract of Israel No. 53, http://www.cbs.gov.il/archive/ shnaton53/shnatone53.htm (accessed 24 September 2010).

9 Central Bureau of Statistics (CBS) (2003) Statistical Abstract of Israel No. 54 Energy and Water, Israel Water Commission, Table no. 21.6, http://www1.cbs.gov.il/ reader/shnaton/templ_shnaton_e.html? num_tab=st21_06&CYear=2003 (accessed 24 September 2010).

10 Department of Statistics–Agriculture Surveys (DOS) (2002) The Hashemite Kingdom of Jordan, http://www.dos.gov. jo/dos_home/dos_home_e/main/index. htm (accessed 24 September 2010).

11 Palestinian Central Bureau of Statistics (PCBS) (2001–2002) Statistical Abstract of Palestine Nº 4. Agriculture statistics (12.3), http://www.pcbs.gov.ps/Portals/_PCBS/ Downloads/book997.pdf (24 September 2010).

12 Palestinian Central Bureau of Statistics (PCBS) (2002) Statistical Abstract of Palestine Nº 3 and 4. Natural resources statistics (22.3). Tables: available water quantity by source and region, http://www. pcbs.gov.ps/DesktopDefault.aspx? tabID=4059&lang=en (accessed 24 September 2010).

13 Eaton, J. and Eaton, D. (1994) Water utilization in the Yarmouk-Jordan 1192–1992, in *Water and Peace in the Middle East* (eds J. Isaac and H. Shuval), Elsevier Publishers, Amsterdam, pp. 93–106.

14 Mimi, Z. and Sawalhi, B. (2003) A decision tool for allocating the waters of the Jordan River basin between all riparian parties. *Water Resources Management*, **17**, 447–461.

15 Elmusa, S. (1998) Toward a unified management regime in the Jordan Basin: the Johnston plan revisited, in *Transformations of Middle Eastern Natural Environments: Legacies and Lesson*, Bulletin Series, no. 103, Council on Middle East Studies, Yale Center for International and Area Studies and the Yale School of Forestry and Environmental Studies (eds J.

Coppock and J.A. Miller), Yale University Press, New Haven, pp. 297–313. Available at http://environment.research.yale.edu/documents/downloads/0-9/103elmusa.pdf (accessed 4 April 2011).

16 Shamir, U. (1998) Water agreements between Israel and its neighbours, in *Transformations of Middle Eastern Natural Environments: Legacies and Lesson*, Bulletin Series, no. 103, Council on Middle East Studies, Yale Centre for International and Area Studies, and the Yale School of Forestry and Environmental Studies (eds J. Coppock and J.A. Miller), Yale University Press, New Haven, pp 274–296. Available at http://environment.research.yale.edu/documents/downloads/0-9/103shamir.pdf (accessed 4 April 2011).

Further Reading

Baer, G., Schattner, U., Wachs, D., Sandwell, D., Wdowinski, S. and Frydman, S. (2002) The lowest place on earth is subsiding-An InSAR (Interfermetric Synthetic Aperture Radar) perspective. *Geological Society of America Bulletin*, **114**, 12–23.

UN (1997) Convention on the law of the non-navigational uses of international

watercourses, http://untreaty.un.org/ilc/texts/instruments/english/conventions/8_3_1997.pdf (accessed 24 September 2010).

World Bank (2001) The Hashemite Kingdom of Jordan: water sector review update (main report), World Bank Report No. 21946.

Part Four
Bridging the Gaps

Transboundary Water Resources Management: A Multidisciplinary Approach, First Edition.
Edited by Jacques Ganoulis, Alice Aureli and Jean Fried.
© 2011 Wiley-VCH Verlag GmbH & Co. KGaA. Published 2011 by Wiley-VCH Verlag GmbH & Co. KGaA.

8
Capacity Building and Sharing the Risks/Benefits for Conflict Resolution

8.1
Capacity Building and Training for Transboundary Groundwater Management: The Contribution of UNESCO

Jean Fried

8.1.1
Field Experience or Specific Training?

The study and management of groundwater, already naturally complicated especially because it is a hidden resource and quite expensive to monitor, are further complicated when groundwater is transboundary. While the aquifer will retain its physicochemical and hydrogeological properties on both sides of a political boundary, the way in which it is perceived and used will often be different, both in legal and economic terms, on both sides of the boundary, increasing the difficulties of decision-making. When the boundary is international, differences in culture, language, education, institutions and political priorities further complicate management and decision-making. Therefore, the knowledge and the experience necessary to manage these aquifers in a sustainable manner requires many disciplines such as physical sciences, law and social sciences, which may be quite different from each other in terms of reasoning, scientific analysis and experimental methodology amongst others.

Two questions arise:

1) Is it sufficient for managers to have had field experience of groundwater resources management in order to be able to manage transboundary groundwater resources as well?

2) Alternatively, would it be more efficient to provide training for groundwater practitioners specifically on transboundary issues, by designing a transboundary groundwater management curriculum, which would be part of their continuous professional training?

Based on the experience gained on transboundary waters and more specifically transboundary aquifers from UNESCO programmes such as ISARM (Internationally

Transboundary Water Resources Management: A Multidisciplinary Approach, First Edition.
Edited by Jacques Ganoulis, Alice Aureli and Jean Fried.
© 2011 Wiley-VCH Verlag GmbH & Co. KGaA. Published 2011 by Wiley-VCH Verlag GmbH & Co. KGaA.

Shared Aquifer Resources Management) and centres such as IGRAC (International Groundwater Resources Assessment Centre), UNESCO decided that specific training was preferable and established a programme to identify the needs of such training and its scientific and methodological characteristics.

The UNESCO Training Programme in Transboundary Groundwater Resources Management aims at assisting Member States to build the knowledge and abilities necessary for the best integration of shared water resources into their national water budget. From a UN point of view, this activity should also contribute to the achievement of the Millennium Development Goals, essentially MDG 7 'ensure environmental sustainability' and MDG 8 'develop a global partnership for development'.

To achieve these objectives, UNESCO did set up a Think Tank, chaired by the author, and convened a workshop in November 2006 to answer the following questions: Why do transboundary groundwater issues necessitate specific training? Which target groups should be addressed? What should be taught? Answers to these questions were proposed at the conclusion of the Workshop and are presented hereafter.

Another conclusion of the Workshop was to organize and hold an experimental course to test the workshop results, both in terms of content and pedagogical method used, and to develop a curriculum that could be promoted by UNESCO in universities and higher education institutions. Called 'pilot', the experimental course was held as a side event of the IV International Symposium on Transboundary Waters Management, held in Thessaloniki, Greece, in October 2008.

As the results of the first pilot course were not fully satisfactory, a second pilot course was designed, and held as a side event of the ISARM Transboundary Aquifers Conference: Challenges and New Directions, held in Paris, France in December 2010.

Both these pilot courses had a simplified content, which did not cover all the disciplines. They were like laboratory experiments or models based on a limited number of assumptions and, therefore, more easily controlled, analysed and interpreted.

8.1.1.1 Training Objectives

There are many reasons why training is necessary. The usual water management activities, like the allocation of quantities or the prevention of pollution, become more complex when watershed or aquifers are transboundary in nature and therefore are affected by two or more different decision-making processes, and follow different institutional, administrative and legal systems. Those working in the management of these watersheds or aquifers do not speak the same language or share the same culture, and often have different political priorities. Up to now, practitioners and policy makers have not had the chance to benefit from education and training specifically in the area of transboundary groundwater and its many complex factors, but have just made the best use of their classical education in water problems and learned on the job, acquiring experience in the field. A systematic approach involving training and capacity building would be more efficient, and would provide the added benefit of identifying best practices, while also focusing on methodology and essential management tools. A comprehensive curriculum that extends from the scientific to the policy and legal aspects of managing transboundary groundwater is essential not only for those who manage these resources, whether currently or in the future, but also for those who study them, teach them and provide leadership, whether on a local, national or international level.

Why a specific groundwater course and not a more general course concerning the hydrological cycle of which groundwater forms a part? The logical integrated water management approach would be to address all aspects of the cycle, but the very specific characteristics of groundwater, such as its vulnerability to pollution and especially diffuse pollution on large time-scales, combined with its almost impossible rehabilitation due to geological conditions, the uncertainties over its physical, biochemical and flow dynamics properties due to the difficulties of its monitoring, lead us to recommend specific groundwater courses. Yet it was considered that a general introduction to the hydrological cycle should still be part of the syllabus, emphasizing the quantitative and qualitative interactions between surface and groundwater, and mentioning wastewater and agricultural return flow. Following this introduction, the courses should then focus on groundwater and specific issues related to transboundary conditions. In particular, the issues related to aquifer recharge areas located in different administrative or political regions were considered as being at the root of management difficulties, requiring that both the political establishment and the general public acquire a sound understanding of the hydrological cycle.

8.1.2
Training Target Groups

Two levels of water management activities have been identified: the policy level and the practitioners' level.

Activities at policy level consist in having a very general vision of transboundary groundwater problems, including language problems, in taking part in negotiations, in making political and managerial decisions, and in creating the institutions and instruments necessary to transboundary management, especially joint institutions and partnerships. The corresponding target group consists of *future decision-makers and planners, intergovernmental negotiators*: graduate students of political science, public policy and human science in general aiming at a political career, and students of business and management with an interest in resources economics and policies, amongst others.

Activities at practitioners' level consist of implementing the political, institutional and managerial decisions, from the scientific, legal and economical points of view. The corresponding target group consists of *confirmed practitioners*, for example, civil engineers, environmental scientists, geographers, political scientists, human scientists in general (e.g. historians, anthropologists, sociologists, linguists, translators and interpreters, amongst others), lawyers and economists.

Other groups also currently play some role in the management of water and should not be neglected in a general water education programme. Specific transboundary groundwater education may not be necessary in these cases but these groups should be included in an overall educational programme on water issues. The Think Tank has identified three such target groups:

1) Considering the fact that the general public should participate in decision-making and implementation processes, in accordance with UN sustainable development principles, and bearing in mind the role of elementary and high school teachers in education, *educational colleges* and similar institutions are a first target group.
2) Considering the significance of the role of the media in providing information to both the general public and to decision-makers, *schools of journalism* are a second target group.
3) Considering that the United Nations has given the armed forces new responsibilities in rebuilding regions devastated by conflicts, as part of their worldwide intervention system, *military academies and schools* are a third target group.

8.1.3
Communication as a Basis of a Transboundary Groundwater Curriculum

Good communication is a major issue in water management in general, but becomes an even more critical factor in the case of transboundary water management, where one object has to be managed by many very different people at the same time. Even within a single country, the relationships between those who conceive and plan the policies, those who have to implement these policies scientifically and technically, those who operate and maintain the systems and those who legally ensure their good operation and protection, amongst other stakeholders, are already difficult and often not efficient. When the situation is complicated by language differences and historical and cultural heritages, water management can be a real challenge unless a big effort is made on communication. Since designing a training programme that would be able to face these issues is an extremely complex task, it was decided to design and organize two pilot courses, as mentioned in the introduction of this chapter.

The pilot courses would deal with two levels of communication, communication between practitioners coming from different disciplines and communication between practitioners with different languages and cultural heritages. They would gather participants, both instructors and students, from different professional backgrounds and from different countries. The pilot courses would also experiment with two possible formats corresponding to the two main training methods that had been discussed during the November 2006 Workshop: (i) intervention of specialists of each concerned discipline, explaining their discipline, then integration based on the presentation of a general multidisciplinary example corresponding to a real case, or (ii) the presentation of a general case, emphasizing the multidisciplinary dimensions and completed by a discussion of specialists highlighting specific features from individual disciplines, if and when this was deemed necessary for understanding the general presentation.

8.1.4
Experimenting Transboundary Groundwater Curricula and Pedagogy: Two Pilot Courses

The first pilot course, organized in October 2008 in Thessaloniki, consisted of *a levelling stage*, ensuring that the participants become familiar with the basic concepts and terminology of each other's disciplines, and an *integration stage*, integrating different backgrounds and addressing all participants in the same way and dealing with specific transboundary groundwater issues (Annex I). As already mentioned, the pilot course was conceived as a model based on simplifications: in particular, only three disciplines were introduced: law, some aspects of political sciences and hydrogeology. Therefore, the levelling stage was divided into two parts: levelling of lawyers and policy-makers by scientists, namely, hydrogeologists, and levelling of scientists by lawyers and policy-makers.

As with every model, such a simplification has its weaknesses and it is the task of the modeller to compensate for these to enable a fruitful and correct analysis to be made. In this respect, the multidisciplinary Think Tank, established at the start of the Programme and before the Identification Workshop, is playing a constructive role but, of course, its work will be critically scrutinized.

The objectives of the pilot course were to teach hydrogeologists, lawyers and policy-makers how to work together as a team, provide them with common knowledge, develop a common language and make them more familiar with the methods of reasoning of their partners in order to facilitate dialogue and exchanges of ideas.

The pilot course was both informative and interactive. It began with a role-play to set the scene and to provide participants with a general view of both the transboundary groundwater issues and the necessary corresponding knowledge in the three disciplines covered on the course. The two half-day levelling sessions were followed by a half-day integration session. Each session consisted of two to four short lectures to introduce the basic concepts and terms, followed by a round-table discussion:

- The 'scientific levelling' session enabled the hydrogeologists to evaluate what the lawyers and policy-makers knew and understood of the hydrogeological aspects and to offer them the necessary complementary scientific knowledge.
- The 'legal and policy levelling' session enabled the lawyers and policy-makers to evaluate what the hydrogeologists knew and understood about the legal and policy aspects and to offer them the necessary legal and policy knowledge.
- The 'integration' session was essentially based on the presentation of case histories to view the theory from a practical angle. It aimed at offering practical training in multidisciplinary teamwork through round table discussions. The presented cases also mentioned other domains of knowledge, such as, for instance, financial mechanisms across boundaries (e.g.polluters pay, benefit sharing).

One of the conclusions drawn from the results of the first pilot course was that improvements should be made to the methodology used. This was because the 2integration session did not fulfil the expectations of the Think Tank, as the presented examples did not really make the best use of the levelling sessions. It was therefore decided to change the format for the second pilot course. This second course would be based on the presentation of several general transboundary groundwater cases, requiring the integration of the various disciplines. After the general presentation, specialists would then explain the role of these individual disciplines.

As such the second pilot course, organized in December 2010 at UNESCO Headquarters in Paris, consisted of four sessions based on the presentation by keynote speakers of case studies covering the various aspects of transboundary groundwater management and integrating the concerned disciplines, for example, hydrogeology, law, economy and policy (Annex II). A fifth session enabled the evaluation of the course.

For each session, the keynote presentations were followed by a discussion amongst experts (Annex III) who were specialists of a particular discipline, explaining the role of their discipline in the presented examples or other cases of their choice. This was followed by a general discussion, where the course participants were also brought in, to highlight the integration of disciplines.

The sessions dealt with the case of aquifers crossing regional administrative boundaries inside a given country (the USA), aquifers crossing boundaries of countries in a union of independent countries (the European Union), aquifers crossing international boundaries of countries without political conflict (Kenya/Tanzania and EU/Switzerland) and, finally, aquifers crossing international boundaries of countries with political conflict. This last case was presented as a role-play using a fictitious example.

The experimental phase of the development of a training course will be finished when participants' comments have been analysed and interpreted. Curricula for systematic courses for the various target groups identified above will then be designed by the Think Tank.

8.1.5
An Instrument for Training: a Manual Gathering the Contributions to the Pilot Courses

A practical result from both pilot courses will be the Manual for 'Transboundary Groundwater, Elements for a multidisciplinary approach Science/Law of Transboundary Aquifer Systems Sustainable Management'. This Manual will consist of the levelling parts of the first pilot course, respectively Part 1 'Levelling of lawyers by scientists' and Part 2 'Levelling of lawyers by scientists', include the examples treated during the courses and the corresponding comments of the experts in Part 3 'Integration of science and law: case histories'. The Manual is, at present, in a draft form and should be finalized in 2011. It is intended to complement the present book as a source of examples and case histories.

Annex I: First Pilot Course Thessaloniki 13–14 October 2008
The pilot course is conceived as a model to be tested, evaluated and adapted in the process of designing a systematic training program. Therefore, it does not cover all the domains of transboundary groundwater and does not address all stakeholders. Three domains of knowledge are covered in the pilot course: hydrogeology, law and policy-making.

Monday 13 October 2008
09h30–10h00: Introduction:
Rationale, objectives, methodology and expected output of the pilot course, and presentation of the moderator and the lecturers, by the director of the course
10h30–12h00: Session 1: Role-play between the lecturers and with the participants on a transboundary aquifer system
To establish a general image of the issues that can be expected when working on a TBGW system from the hydrogeological, legal and policy points of view, and have a general evaluation of the gaps in knowledge of the participants.
14h00–18h00: Session 2: Levelling of lawyers and policy-makers by scientists,
The scientists will explain what knowledge they need and use and how they approach TBGW issues from a hydrogeological point of view, introducing and explaining the current scientific and technical language in hydrogeology and its adaptation to transboundary groundwater:
- Introduction to groundwater (the hydrogeological cycle, vulnerability, saturated and non-saturated zones etc.)
- Characterization of transboundary aquifers (including identification and delineation)
- Management of transboundary aquifers, from the hydrogeologist's point of view: shared data collection and processing, information management, modelling, quality management
Pollution (dispersion, residence times, effect of the non-saturated zone on the classification of pollutants, etc.) will be interwoven in these three chapters
Uncertainty and precautionary principle, and their application to transboundary aquifer systems will be treated in Session 4, integration.

Tuesday 14 October 2008
08h00–12h00: Session 3: Levelling of scientists by lawyers and policy-makers
The lawyers and policy-makers will explain what international water law consists of and what specificities they experience when dealing with transboundary aquifers systems:
- terms and concepts: a short review of the terminology currently used by water lawyers
- sources and definition of international groundwater law
- evolution and emerging principles of international groundwater law (in particular, the work of the International Law Commission)
- compliance mechanisms
- procedural rules and dispute settlement mechanisms
13h30–17h30: Session 4: Integration of science, law and policy: case histories and round table discussions
Two case histories, related to the emergence of new nation-states and the creation of transboundary aquifers in that particular context, will be presented and discussed in a round-table discussion that will take most of the session: Central Asia within the creation of new republics from the former Soviet Union and the Balkans within the creation of new republics from the former Yugoslavia.

Annex II: Second Pilot Course UNESCO Paris 9–10 December 2010

Thursday 9 December 2010
09h00–09h15 Introduction
09h15–09h30 Presentation of the methodology of assessment of transboundary aquifers
09h30–12h30 Session 1: Administrative versus hydrological boundaries within a given country

- 09h30–10h00 Keynote presentation: 'Mississippi vs. Memphis; the curious case of the Memphis sand aquifer'

- 10h00–11h Expert Panel presentations after the keynote

- 11h00–11h15 Break 15min

- 11h15–12h30 General discussion

14h00–17h00 Session 2: Transboundary groundwater issues within an association of countries, evolving towards a federal organization, the case of the European Union

- 14h00–14h30 Keynote presentation 'The EU Groundwater regulatory framework and its application to transboundary groundwater bodies'

- 14h30–15h30 Expert Panel presentations after the keynote

- 15h45–17h00 General discussion

Friday 10 December 2010

09h00–12h30 Session3: International transboundary groundwater issues 'without' a political conflict

- 09h00–09h30 Keynote presentation: 'The Transboundary Aquifers of Kenya: their characteristics, status, management, and legislative, socio-economic and political contexts for their wise and sustainable use'

- 09h30–10h00 Keynote presentation: 'The transboundary aquifer of the Geneva region (Switzerland and France): a successful 30-year management between the State of Geneva and French border communities'

- 10h00–11h00 Expert Panel presentations after the keynotes

- 11h15–12h30 General discussion

14h00–16h00 International transboundary groundwater issues 'with' a political conflict situation: Role-Play on a fictitious example

16h30–18h00 Course Evaluation and Conclusion

Annex III: Summary of Instructions to the Experts

Guidelines for the expected contributions of the Expert Panel
Structure of the Expert Panel
The expert panel will consist of six experts, representing the following disciplines: hydrogeology (2), law (2), economy (1), policy (1). The Chairperson and the Moderator will lead the sessions. Each expert will emphasize significant aspects of his/her discipline concerning sustainable transboundary groundwater management that are included in or missing from the keynote presentation of each session, and during the general discussion guide the students to better understand the specificities of his/her discipline as applied to sustainable transboundary groundwater management
Tasks of the experts:
Before the course the experts will receive the keynote papers of each session and write comments, analyse the documents, highlight issues related to their speciality and illustrate with examples and solutions. If accepted by the Task Force, each contribution will later be adapted to be included in the Manual (first draft prepared after the first pilot course) and distributed to the students in the follow-up of the pilot course The experts will also decide which items concerning their discipline should be covered by a levelling course
During the course the experts will contribute to the presentations and discussions and help identify knowledge, tools and methods that enable practitioners to manage transboundary aquifer systems, and communicate, negotiate and work together across boundaries and disciplines. In particular, they will identify the specificities of transboundary versus classical groundwater management and illustrate the necessary multidisciplinarity of transboundary groundwater studies and management

8.2
A Risk-Based Integrated Framework for Conflict Resolution in Transboundary Water Resources Management

Jacques Ganoulis and Lena Salame

8.2.1
Managing Transboundary Water Resources: Quantity and Quality

In today's complex and globalized economy water resources play a strategic role. In addition to the fact that fresh water is essential to all kinds of life, it is also used in agriculture and industrial processes. Fresh water supply is utilized in human conurbations to meet household demands, in municipal and industrial wastewater treatment plants and also in agriculture, to dissolve and remove dirt and waste. A sufficient supply of fresh water has become a necessary condition to ensure economic growth and development. At the same time, preservation of good water quality in rivers, lakes, aquifers and coastal waters is necessary to protect public health and ecosystems.

As demands for water by different users increase and the quality of water deteriorates because of pollution, transboundary water resources have become increasingly important. The needs for economic development of riparian countries is put under stress and conflicts result not only between countries, but also between different 'direct' and 'indirect' water users, shown in Figure 8.2.1 for surface transboundary water resources. The problem is further exacerbated in transboundary regions when long-term droughts have decreased the available amount of water, while at the same time the needs for water have increased.

The importance of transboundary water resources and problems of transboundary water quantity and quality may be better perceived by analysing the economic importance and the opportunities in the *market of water*. In the European Union it was estimated [1] that in the 1990s the costs for operating the municipal water supply and waste water systems alone were about 14 billion Euros per year. For the implementation of the municipal waste water and drinking water directives over the whole of Europe it is estimated that a total investment of several billion Euros will be needed in the years to come. To face the problems of future water demand and to combat growing pollution it is expected that new technologies, new investments and new management tools will be introduced.

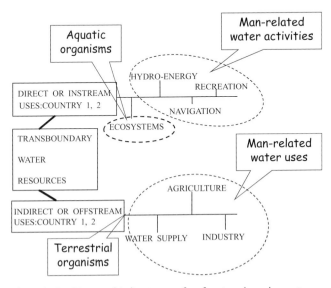

Figure 8.2.1 Direct and indirect uses of surface transboundary water resources.

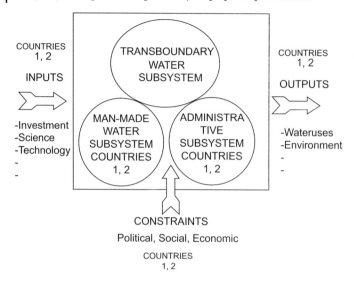

Figure 8.2.2 Description of a transboundary water resources system.

When examining the management issues of transboundary water related problems, one realises that besides the scientific and technical components there are also
social, economic and institutional considerations. If transboundary water resources
are defined as a system (Figure 8.2.2), apart from the natural water subsystem, the
man-made water subsystems (channels, distribution systems, artificial lakes, etc.)
should also be included, as well as the administrative systems of the riparian
countries. These subsystems are interconnected and are subject to different
constraints, such as social, political and economic, in the riparian countries
(Figure 8.2.2). At the next level up, the decision making system, which is distinct
from the administrative system, is subject to different prerogatives, incentives and
motivations. Inputs to the system are data, investment, science and technology and
outputs are decisions on water uses, environmental issues and new economic
activities for each riparian country.

In many cases of transboundary water resources management, water quality
plays an increasingly important role, equal to that of water quantity [2]. This is
especially true in developed areas, but less so in developing ones, where water-
quality issues are low on the list due to lack of finances. In fact, as pollution of
surface, coastal and groundwater worsens, it becomes increasingly important to
adopt an integrated approach encompassing both water quantity and quality. In this
sense, protection of water ecosystems requires the water resources system to be
considered as composed from both abiotic and biotic elements ([2], Figure 8.2.3). As
shown later on in this chapter, in recognition of the interconnection of the land-
water and surface-groundwater relationships, and the various economic sectors,
disciplines, actors, and policy instruments involved, an integrated approach should
always be taken.

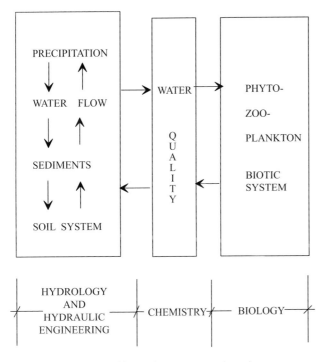

Figure 8.2.3 Abiotic and biotic elements in transboundary water systems.

8.2.2
The Risk Analysis Framework in TWRM

Uncertainties in TWRM (Transboundary Water Resources Management) may be distinguished between:

1) aleatory or natural uncertainties or randomness,
2) epistemic or man-induced or technological uncertainties.

8.2.2.1 Aleatory Uncertainties or Randomness
It is postulated that natural uncertainties are inherent to the specific process and they cannot be reduced by use of an improved method or more sophisticated models. Uncertainties due to natural randomness or aleatory uncertainties may be taken into account by using the probabilistic or fuzzy logic-based methodologies, which are able to quantify uncertainties [2].

8.2.2.2 Epistemic or Man-Induced Uncertainties
Man-induced uncertainties are of different kinds: (i) data uncertainties, due to sampling methods (statistical characteristics), measurement errors and methods of analysing the data, (ii) modelling uncertainties, due to the inadequate mathematical models in use and to errors in parameter estimation, (iii) operational uncertainties,

which are related generally with the construction, maintenance and operation of engineering works, (iv) unpredictability of human and societal behaviour – including individual, cultural, political, economic, and institutional variability, (v) influences of international relations, as they relate to TWRM, (vi) variations in vulnerability and modes of adaptation to forces of global change and (vii) degree of development and influence of civil society.

Contrary to natural randomness, man-induced uncertainties may be reduced by collecting more information or by use of improved conceptual and mathematical models. Alternatively, when data are scarce, the fuzzy logic-based theory may be useful in order to handle and quantify imprecision [2].

Different types of uncertainties may be distinguished as follows:

- **hydrologic uncertainty:** this refers to various hydrological events such as precipitation, river and groundwater flows;
- **environmental uncertainty:** these are uncertainties related to water quality, biodiversity and ecosystem distribution;
- **structural uncertainty:** this means all deviations due to material tolerances and other technical causes of structural failure;
- **economic uncertainty:** all fluctuations in prices, costs, investments, which may affect the design and management processes;
- **human related and institutional uncertainty:** these are all the above cited but mainly (iv)–(vii) in Section 8.2.2.2, and also those related to different institutional settings responsible for implementing TWRM in riparian countries.

Methods and tools for assessing and quantifying such uncertainties in TWRM could be incorporated in the risk analysis process.

8.2.2.3 Risk Assessment and Management

Risk and reliability have different meanings and are applied differently in various disciplines related to TWRM, such as engineering, environmental sciences, economics, public health and social sciences. The situation is sometimes confusing because terminologies and notions are transferred from one discipline to another without modification or adjustment. This confusion is further amplified as scientists may have different perceptions about risks and use different tools to analyse them.

Risk Definition   Risk has different connotations and interpretations depending on the socio-economic context and the historical developments of the riparian countries. Different societies have developed their own perceptions, beliefs and modalities to interact with uncertainties, to manage unforeseen incidents and to deal with potential losses. Generally speaking, risk is defined as 'the possibility of loss or the possible or the mean or average loss'.

Table 8.2.1 summarizes some more specific disciplinary definitions and mathematical estimations of the notion of risk [2]:

- In economic sciences risk is related to possible economic losses. The expected economic loss or the variance of loss may be used to estimate the economic risk.

Table 8.2.1 Risk definition and risk estimation in different scientific disciplines.

Discipline	Risk definition: possible loss of ...	Risk estimation
Economy	Money, capital, markets, investment, infrastructure	Expected loss: E [capital < 0] or deviation from target Ta: Ta-E [capital < Ta]
Social sciences	Revenues, jobs, security, property (incl. land), social cohesion, institutional capacity, coping and adaptation capability	Expected loss of revenues, possible number of jobs lost
Public health	Lives, human health, conditions for vector disease	Number of deaths or casualties or sick days (loss of work) per million of population
Ecology	Species (animal and plant), habitat, streamflow, landscape, air quality	Species mortality, bio-diversity index, eco-integrity characteristics
Environmental (abiotic)	Air, water or soil quality characteristics	Expected deviation from regulated quality standards, possible consequences
Engineering	Technical security, reliability	Probability of failure, possible damage

- In social sciences the risk is related to the possible loss of revenues and the expected number of jobs lost.
- In public health the risk is related to the number of people infected (not all disease is infectious – for example chemical pollution and radioactivity can cause disease) and expressed per thousand or per million of population.
- In environmental sciences risks are distinguished between abiotic and biotic/ecological environment (Table 8.2.1).
- In engineering sciences the probability of a failure and the related consequences are taken into account.

In a typical problem of a failure under conditions of uncertainty, there are three main questions, which may be addressed in three successive steps:

1) When could the system fail?
2) How often is failure expected?
3) What are the likely consequences?

The first two steps are part of the uncertainty analysis of the system. The answer to question 1 is given by the formulation of a critical condition, producing the failure of the system. To find an adequate answer to question 2 it is necessary to consider the frequency or the likelihood of failure. This can be done by use of the probability calculus. Consequences of failure (question 3) may be accounted for in terms of different losses or benefits.

As explained in Reference [2], we should define as *load* (ℓ) a variable reflecting the behaviour of the system under certain external conditions of stress or loading. There is a characteristic variable describing the capacity of the system to overcome this external load. We should call this system variable *resistance* (r). A *failure* or an *incident* occurs when the load exceeds this resistance, that is:

FAILURE or INCIDENT: $\ell \geq r$
SAFETY or RELIABILITY: $\ell < r$

In a probabilistic framework ℓ and r are taken as random or stochastic variables. In probabilistic terms, the chance of failure occurring is generally taken as a first definition of risk. In this case we have:

RISK = probability of failure = $P(\ell \geq r)$

A distinction should be made between hazard, risk, reliability and vulnerability:

Hazard: means the potential source of harm. The term evolved from the Arabic 'al zahr', meaning the dice. The negative connotation and the associated notion of danger arose in Western Europe from games first learned during the Crusades in the Middle East, where the use of loaded dice usually resulted in losses.
Risk: the possibility of adverse effects like loss and injury caused by exposure to a hazard.
Reliability: the possibility of the system to perform under a given hazard.
Vulnerability: denotes the susceptibility of the system to cope under a given hazard.

The vulnerability of a system (natural or social) may be estimated by measuring the possible degree of damage it incurs or the severity of the consequences if a given incident occurs, for example, a natural hazard (flood, earthquake, tsunami, etc.). Also very important is the vulnerability of social systems due to events other than natural hazards – for example war, political unrest or economic collapse.

If we consider the first definition of risk as being the possibility of losses (injuries, deaths, economic losses, environmental damages), we may see that two elements are essential for describing risks: (i) the severity of the hazard and (ii) the susceptibility of the system to sustain the hazard. This is why risk may be defined by the following product equation:

RISK = (Hazard) × (Vulnerability)

Risk is usually taken as the 'mean' or 'expected value' of consequences or damages expressed by the product of probability and its consequences, that is:

RISK = (Probability) × (Consequences) = (Mean or Expected Consequences)

As shown in Figure 8.2.4, application of Risk Analysis (RA) in TWRM consists of two main parts:

1) the assessment of risk,
2) the risk management.

The assessment of risk is mainly based on data collection and modelling of the physical and socio-economic system, including forecasting of its evolution under

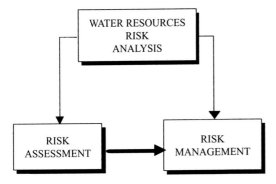

Figure 8.2.4 Risk assessment and risk management as elements of Risk Analysis (RA).

risk. Although the main objective of RA is the management of the system, this is possible when risks have been previously quantified.

The risk assessment phase involves the following steps:

Step 1: risk or hazard identification,
Step 2: assessment of loads and resistances,
Step 3: uncertainty analysis,
Step 4: risk quantification.

As shown in Figure 8.2.5, the actual approach in TWRM planning aims primarily to reduce technical and economic risks by achieving two main objectives [2]:

1) technical reliability or performance
2) economic effectiveness.

8.2.2.4 Institutional and Social Issues

In recent years special attention has been paid to institutional and social approaches in TWRM planning (Chapters 7.1 and 7.2). The institutional or administrative framework may be conceived as being the set of state owned agencies, private enterprises and also NGOs and community-based organizations dealing with

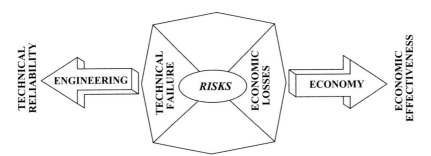

Figure 8.2.5 Technical and economic performances in TWRM.

production, distribution, treatment and also policymaking, planning, management, enforcement of regulations, revenue-generation and collection, juridical decision making and transnational negotiations.

Of particular importance is their scale of operation (local, regional or state), their degree of autonomy from the central administrative body, and the involvement of different water stakeholders in the decision making process. The administrative systems and water laws and regulations, together with social perception on the use of water and traditions involved, make the issue of water resources management very complex.

From the perspective of the engineering profession, the water crisis is also methodological. Reducing environmental and social risks together with technical and economic reliability in planning is a major challenge. To develop an integrated methodology it is necessary to define a new conceptual framework or, in other words, shift to a new scientific paradigm.

8.2.3
Towards an Integrated Risk-Based Sustainable TWRM Approach

As shown in Figure 8.2.6, four main *objectives* or *criteria* are to be taken into consideration:

1) **Technical reliability:** some measures for technical performance are technical effectiveness, service performance, technical security, availability and resilience.
2) **Environmental safety:** environmental indicators may be positive or negative environmental impacts such as increase or decrease in the number of species, public health issues, flora and fauna modifications, losses of wetlands, landscape modification.
3) **Economic effectiveness:** costs and benefits are accounted for, such as project cost, operation and maintenance costs, external costs, reduction of damages benefits, land enhancement and other indirect benefits. Unless contingent valuation is used carefully, cost–benefit ratios are likely to prove inaccurate because of the traditional undervaluing of environmental resources and goods (such as clean air, clean water, running streams, healthy habitat and wildlife, diverse forests).
4) **Social equity:** social impacts are, for example, related to risk of extremes, duration of construction, employment increase or decrease and impacts on transportation. Here, too, estimates are extremely difficult and notoriously unreliable.

After the definition of the objectives, the steps to be undertaken for the Integrated Multi-Risk Composite Method planning are the following [2–6]:

1) Define a set of *alternative actions* or *strategies,* which includes structural and non-structural engineering options.
2) Evaluate the outcome risks or *risk matrix* containing an estimation of the risks corresponding to each particular objective (technical, environmental, economic and social).

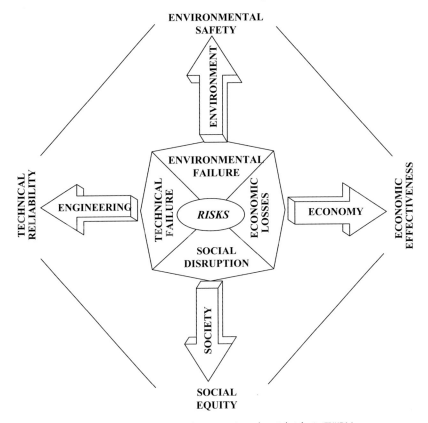

Figure 8.2.6 Engineering, environmental, economic and social risks in TWRM.

3) Find by use of an averaging algorithm the *composite risk index* for technical and ecological risks (eco-technical composite risk index) and the same for the social and economic risks (socio-economic composite risk index).
4) *Rank the alternative actions*, using as a criterion the distance of any option from the ideal point (zero risks).

As shown in Figure 8.2.7, in the two-dimensional plane with the composite eco-technical and socio-economic indices as coordinates, strategies 1, 2 and 3 are ranked 1-3-2 using as the criterion the distance of any strategy from the ideal point (0, 0).

8.2.4
Modelling Transboundary Water-Related Conflicts

Useful definitions of terms used are as follows [2]:

Goals: Broadly speaking, every state has social, economic and political goals linked to water resources development, conservation and control and protection of the river basin. Economic goals may be to obtain new water resources to

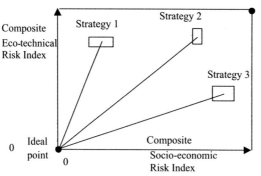

Figure 8.2.7 Ranking different strategies based on eco-technical and socio-economic risks.

increase food production, conservation goals may be to control water pollution, and control and protection goals may concern defence against floods or drought control. These goals may be achievable by jointly building water reservoirs. This would entail the states involved cooperating together and solving possible areas of conflict.

Purposes in accomplishing goals: Goals are accomplished by various water resources developments, transfers of water from the water-surplus adjacent river basins, water conservation, control and protection. Each particular goal means satisfying some particular purpose, which may have to do with irrigation, drainage, hydropower production, navigation, water supply, water pollution control, flood defence, drought control, or other.

Objectives and attributes in accomplishing purposes and goals: Finally, to satisfy the purposes of state goals in water resources development one must define and then maximize or minimize particular economic, social, monetary and political attributes. The particular purposes, attributes and interests in water resources development of the river basin should be strictly taken into consideration in any future cooperation on conflict resolution between the states.

Conflict situations in TWRM may occur on at least two levels:

1) conflict amongst specific attributes, in particular economic, environmental and social ones;
2) conflicts of goals or general interests between countries and amongst groups of actors involved.

Two different cases are shown in the following two figures: In the first one, each country proceeds separately and evaluates alternatives according to its own objectives (Figure 8.2.8). In the second, different attributes used by the two countries are first traded-off and then alternatives are ranked according to the composite objectives (Figure 8.2.9). More details of this methodology may be found in References [2, 7].

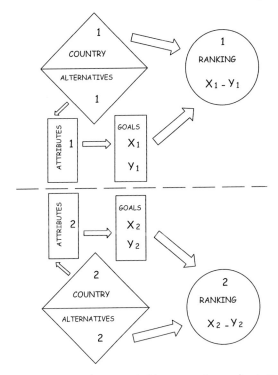

Figure 8.2.8 Each country decides separately according to its own objectives.

8.2.5
Hydro-Politics for Conflict Resolution: The UNESCO PC-CP Initiative

Given the diversity of needs and interests that surround water, disputes and conflicts over the resource are normal and not unexpected. We should therefore be prepared to anticipate, prevent and address water conflicts as and when they arise [8]. We should also be able to turn risks into opportunities and share them between riparian stakeholders.

The main obstacles to effective cooperation often originate from the lack of trust and perceived mutual interests at the top of the political sphere. Overcoming obstacles necessitates having a global understanding of the potential outcomes of cooperation at local, national and regional levels. Such a task can hardly be achieved without the involvement of many players with various visions and areas of expertise.

This is one reason why alternative solutions to traditional diplomacy have been implemented to foster cooperation on transboundary water management. Known as Track II initiatives, they consist of formal or informal interactions between public and/or private stakeholders from different groups and nations to facilitate discussions and exchanges in order to foster peace and cooperation [9].

UNESCOs Water Programmes IHP and WWAP (the International Hydrological Programme and the World Water Assessment Programme) have developed a conflict

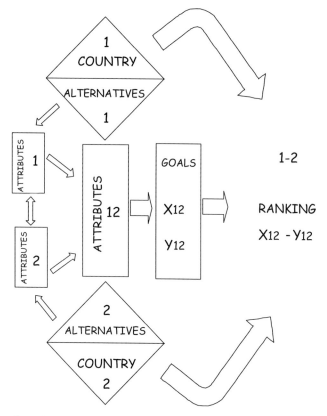

Figure 8.2.9 Compromising countries' different attributes.

resolution component that uses Track II initiatives as intervention mechanisms to help in tipping the balance away from risk and conflicts towards cooperation and mutual benefits. The UNESCO initiative PC-CP (From Potential Conflict to Cooperation Potential) uses research and training to facilitate multi-level and interdisciplinary dialogues to foster peace, cooperation and development related to the management of shared water resources. Examples of Track II initiatives using (i) research and (ii) education to manage risk and conflicts and enhance cooperation between parties sharing the same water resources are given below.

The PC-CP process of producing case studies, for example, follows an innovative approach. The research process supports the cooperative process in managing a transboundary basin or aquifer. The case studies therefore not only increase knowledge surrounding a water body, but promote cooperation amongst the riparian states. This is achieved by involving high-level players, governmental advisers, experts and stakeholders, who participate in preparing a consensus document reflecting the status of conflict and cooperation in the transboundary water body.

The case study process provides a forum where sensitive issues related to the transboundary water body can be discussed, in addition to supporting cooperation, exchange of data and information, and development of the shared resource. Lastly, the process offers stakeholders an opportunity to build a shared vision for the future management of their water resources.

8.2.5.1 Examples of Track II Initiatives

The Mono River Case Study   The Mono River basin, an area of $24\ 300\ \text{km}^2$, is shared between Benin and Togo. A bi-national case study committee was created to participate in two workshops in the framework of the PC-CP programme. The PC-CP workshops, held in Benin and Togo, gave research guidance to the experts.

The case study recommended putting in place an institutional framework for cooperation between the two riparian countries for sustainable and peaceful management of the transboundary water resources. The process that led to the completion of the case study under the aegis of PCCP was a first step in this direction.

Lake Titicaca Case Study   Located in the south of Peru and north-west of Bolivia, the Lake Titicaca basin is spread over an area of $56\ 270\ \text{km}^2$. Decades of cooperation led to the creation of the Bi-national Autonomous Authority of Lake Titicaca (ALT) (Autoridad Binacional del Lago Titicaca, ALT) in 1992. The Ministries of Foreign Affairs of Peru and Bolivia selected experts to write the Lake Titicaca case study and the lessons learned from the joint management of the basin resources were then disseminated through the PC-CP programme.

Although much can still be improved in the management of the basin, the PC-CP research on Lake Titicaca helped avert unilateral decision-making, which could have generated conflicts between the two riparian countries. Amongst the recommendations made through the cooperation process was the 'need to continue the integration, as well as to create broader and more efficient coordination between the ALT and the local institutions so that they can participate more actively in the management of the Lake Titicaca basin' [10].

The PC-CP training activities bring together parties concerned with the management of transboundary water bodies that usually work on different sides of the border. They aim at building or enhancing the skills of these parties in anticipating, preventing and resolving conflicts that may arise amongst them and assist them to understand each other's positions and interests as regards the water they share. They also help them to build a bridge between their 'apparent' diverging objectives as well as to create a real and genuine working relationship.

Training of Trainers from the Middle East   A four-day advanced training course for trainers in cooperation-building skills in the use of transboundary water resources in the Middle East took place on the premises of the WWAP Secretariat, in Perugia, Italy from 28 to 31 October 2008.

The course provided new insights into the art of cooperation and trust-building required for the use of transboundary water resources and aimed to enhance the teaching skills of the participants. It was also a forum where participants shared their respective skills and experiences with each other.

8.2.6
Conclusions

One of the main causes of inefficient TWRM is a lack of political commitment. Leaders need to be willing to use their power, influence and personal involvement to ensure that issues related to TWRM receive the visibility, leadership, resources and ongoing political support that is required to ensure effective action. This commitment needs to be underscored by a solid and accurate understanding of the dimensions of the problem. Political commitment in its broadest sense means leadership commitment that is political and governmental leaders and managers, but it also includes civil and community leaders at all levels of society, the private sector, NGOs and those involved in education. The more that these actors can be engaged in policy dialogue, planning and evaluation, the greater the chances for an effective response to TWRM issues, and the lesser the chances of potential conflicts arising.

The current lack of political commitment in TWRM issues has an immediate, direct and negative effect on the level of collaboration between riparian countries. In this sense we are referring to both official collaboration that is acknowledged by the governments concerned, and non-official collaboration that has no governmental involvement. Riparian country collaboration might include projects, programmes or partnerships [11], with a river basin or aquifer as a geographic focus involving organizations and representatives (acting in an official or non-official capacity) from two or more countries that share the international water body. When such collaborative efforts are missing, effective TWRM is extremely difficult to achieve.

Apart from the lack of political commitment, there are other reasons for the low level of collaboration between riparian countries on TWRM issues. Actors may view shared water bodies in a purely competitive light, and not be able to envisage the benefits of cooperation to achieve any kind of compromise solution to problems. There may also be a lack of motivation and mutual trust between involved parties, who see each other as opponents rather than parties with the same overall interests at heart.

The way in which people behave and their attitudes are influenced both by their cultural background and the socio-economic environment of the country from which they come. There may be distinct and wide differences in these areas between people from riparian countries, factors that can severely hinder the development of effective joint TWRM policies. Such differences are usually deep-rooted and result in neighbouring countries having not only different visions but different priorities, goals and aspirations. Such obstacles do not easily pave the way towards collaboration.

A two-pronged attack to any problem enhances the chances for success and in this sense both top-down and bottom-up approaches to TWRM are recommended. Of course both of these approaches have the same ultimate goal. Therefore, for effective TWRM practices to emerge, there is a need not only for movements orchestrated by traditional power structures but also for more spontaneous efforts originating by the politics of a small community. The active involvement of civil society, NGOs regional partnerships and stakeholders in all stages of these processes, from their inauguration to completion, will encourage a sense of ownership, which is vital for their sustainability and ultimate effectiveness [11].

Specific projects at local, regional and international scales involving the cooperation of scientists and experts from a variety of disciplines will encourage collaboration and help overcome the obstacles of limited political commitment and cultural and socio-economic differences. If such cooperative efforts are linked in a larger cooperative network, they may form a force for collaboration at the global level.

At this level tools of hydro-diplomacy such as enquiry, negotiation and mediation can help reach conciliation and stimulate cooperative processes between riparian countries, in order to achieve effective TWRM. As a first step a problem must first be defined, then negotiations can begin, followed by systematic management of inter-organizational relations and monitoring of agreements. Such strategies should lead to collaborative policies that are comprehensive, participatory and environmentally sound. Formal or informal interactions, known as Track II initiatives, between public and/or private stakeholders from different groups and nations to facilitate discussions and exchanges in order to foster peace and cooperation are also effective tools, especially when political commitment is lacking. Track II initiatives use research projects or educational courses to manage risk and conflicts and enhance cooperation between parties sharing the same water resources.

A risk-based integrated framework is undoubtedly a very useful tool enabling practitioners to think logically about problems of risk management. However, it cannot be considered as a flawless template for actually managing real risk, largely because it is very difficult to quantify social values with the same degree of accuracy as physical goods can be quantified, due to the vagaries and unpredictability of human, social and institutional behaviour. This means that this sort of modelling has limited efficiency because of the extremely high degree of error to which the predictability of the social elements is subject. However, such errors may be reduced by involving stakeholders in the process and by ensuring a maximum level of political commitment.

References

1 Williams, H. and Musco, D. (1992) Research and technological development for the supply and use of freshwater resources. Strategic dossier, SAST Project No. 6, Commission of European Communities, EUR-14723-EN.

2 Ganoulis, J. (2009) *Risk Analysis of Water Pollution*, 2nd edn, Wiley-VCH Verlag, Weinheim, 311 pp.

3 Goicoechea, A., Hansen, D.R. and Duckstein, L. (1982) *Multiobjective Decision Analysis with Engineering and Business Applications*, John Wiley & Sons, Inc., New York.

4 Roy, B. (1996) *Multi-criteria Methodology for Decision Aiding*, Kluwer, Dordrecht, The Netherlands.

5 Tecle, A. and Duckstein, L. (1994) Concepts of multicriterion decision making, in *Multicriterion Decision Analysis in Water Resources Management* (eds J. Bogardi and H.P. Nachtnebel), IHP, UNESCO, Paris, pp. 33–62.

6 Duckstein, L. and Szidarovszky, F. (1994) Distance based techniques in multicriterion decision making, in *Multicriterion Decision Analysis in Water Resources Management* (eds J. Bogardi and

H.P. Nachtnebel), IHP, UNESCO, Paris, pp. 86–112.

7 Ganoulis, J. (2006) Water resources management and environmental security in Mediterranean Transboundary River Basins, in *Environmental Security and Environmental Management: The Role of Risk Assessment* (eds B. Morel and I. Linkov), Springer, pp. 49–58.

8 Salamé, L. *et al* (2009) Developing capacity for conflict resolution applied to water issues, in *Capacity Development for Improved Water Management* (eds M.W Blokland, G.J Alaerts, J.M. Kaspersma and M. Hare), Taylor and Francis, London, Ch. 6.

9 Unver, O., Salamé, L. and Etitia, T. (2010) Best practices in transboundary water management. Ingeneria y territorio, Water management – the United Nations and Water, No. 9, p. 28.

10 UNESCO PC-CP (2011) PC-CP Brochure.

11 Ganoulis, J. *et al* (eds) (2000) *Transboundary Water Resources in the Balkans: Initiating a Sustainable Co-operative Network, NATO ASI SERIES, Partnership Sub-Series 2: Environmental Security*, vol. **47**, Kluwer Academic, Dordrecht, Boston, London, 254 pp.

United Nations
Educational, Scientific and
Cultural Organization

9

The Thessaloniki Statement

At the IV International Symposium on Transboundary Waters Management
held in Thessaloniki, Greece, from 15 to 18 October 2008

We the Participants from 42 Countries and International and Regional
Organizations, Having

- reviewed the current situation of different transboundary surface and ground-water bodies;
- realized the common obstacles that in many cases detract from the best uses of those resources;
- recognized that the quantity and quality of those resources are affected by various human activities;
- considered that water resources are subject to the increasing influence of global and climate changes;
- considered the scientific, technical, economic, financial, policy and legal aspects involved in the management of transboundary water resources;
- recognized that it would be beneficial to broaden the scope of cooperation amongst states sharing water resources by involving the various stakeholders;
- recognized that transboundary water resources should be regarded as a shared resource for satisfying the basic human as well as ecosystem needs and enhance sustainable socio-economic development of the basin populations;
- recognized that, in order to implement international legal obligations, it is necessary for states to take action within their domestic contexts

Are of the View that in Order to Face the Above Challenges and Maximize the Advantages from Cooperation Amongst Countries

1) The states sharing transboundary water resources should:
   a) enter into agreements and understandings for cooperation in the management of these resources, consistent with principles of international law on water resources;
   b) put in place sustainable institutional arrangements at the transboundary river basin or aquifer level;

Transboundary Water Resources Management: A Multidisciplinary Approach, First Edition.
Edited by Jacques Ganoulis, Alice Aureli and Jean Fried.
© 2011 Wiley-VCH Verlag GmbH & Co. KGaA. Published 2011 by Wiley-VCH Verlag GmbH & Co. KGaA.

    c) put in place monitoring and evaluation mechanisms to continuously assess the transboundary river basins and aquifers;

    d) enhance the knowledge base relating to these resources in order to develop effective water resources protection and management mechanisms;

    e) explore alternative institutional approaches for the governance of transboundary water resources at the local users' level.

2) In order to implement international and regional legal obligations within their national contexts, states should put in place adequate national water management policies, legislation and institutions.

3) Arrangements should be made in order to build the capacity of states' multilateral, bilateral and national administrations to implement the provisions of the agreements entered into, and also to promote water education and organize coordinated multidisciplinary training programmes.

4) Sustainable financing mechanisms should be established for transboundary water management.

Therefore

We the participants appeal to international institutions, as well as regional, national and local authorities, in partnership with the private sector, for assistance, technical support and financing in these endeavours.

Index

Transboundary Water Resources Management: A Multidisciplinary Approach, First Edition.
Edited by Jacques Ganoulis, Alice Aureli and Jean Fried.
© 2011 Wiley-VCH Verlag GmbH & Co. KGaA. Published 2011 by Wiley-VCH Verlag GmbH & Co. KGaA.